D1246921

The Network Nation
Human Communication via Computer

The Network Nation
Human Communication via Computer

Starr Roxanne Hiltz
Department of Sociology, Upsala College, East Orange, New Jersey

Murray Turoff
Department of Computer Science, New Jersey Institute of Technology, Newark

With Forewords by

Suzanne Keller
Department of Sociology, Princeton University

and

Herbert R. J. Grosch
President, Association for Computing Machinery, 1976–1978

1978

Addison-Wesley Publishing Company, Inc.
Advanced Book Program
Reading, Massachusetts

London·Amsterdam·Don Mills, Ontario·Sydney·Tokyo

Background cover illustration shows routes for television broadcast, microwave, coaxial, local multi-wire cable, and the underwater and radio channels. Dots represent local switching offices. Reproduced with permission of A.T. & T. Co.

First printing, 1978
Second printing, 1979
Third printing, 1981

Library of Congress Cataloging in Publication Data
Hiltz, Starr Roxanne.

The network nation.

Bibliography: p.
Includes index.
1. Information networks. 2. Computer networks.
3. Telecommunication. I. Turoff, Murray, joint author.
II. Title.
TK5105.5.H54   001.6'44'04   78-12365
ISBN 0-201-03140-X
ISBN 0-201-03141-8 pbk.

Manufactured in the United States of America

ABCDEFGHIJK – 8987654321

For the next generation

ANDREA DELPHI TUROFF
KATHERINE AMANDA HILTZ
JONATHAN DAVID HILTZ
MONIQUE OFRA TUROFF

May they live our dreams

Contents

Foreword

by Suzanne Keller

After having for so long heard about the coldness and impersonality of the computer, here is a book that proposes exactly the opposite thesis, namely, that computers may become the source of a special and new form of human community.

In their intriguing new book, Hiltz and Turoff explore the emergence of a new form of communication called computerized conferencing. Computerized conferencing refers to "any system that uses the computer to mediate communication among human beings."

The authors expect it to revolutionize not only communications but social and intellectual life as well. By increasing both the speed and the amount of communication, it will encourage networks of support and exchange for a great variety of groups ranging from shut-ins to research scholars. For scientists in particular such networks can decrease the isolation experienced by those who find themselves at out of the way institutions and increase connections among specialists who may have difficulty finding colleagues close at hand. The potential payoff for science is thus considerable both in terms of interdisciplinary research and by making possible instantaneous exchanges of ideas, insights, advice, and suggestions.

Changes in computer and communications technology have made computer conferencing the "cheapest, most convenient, and potentially most powerful option for geographically dispersed groups of people who must regularly exchange information and opinions." Cheaper and more convenient (though not necessarily more enjoyable) than travel, it is an extremely flexible mode of communication as to time, place, and pace. At eight dollars per hour per participant, it is already less costly than an ordinary long-distance phone call.

The authors expect computerized conferencing systems, as yet used by only a few thousand people, to be in wide use by the mid-eighties. They will become as ubiquitous as the telephone and have an even more revolutionary impact. For today, it is evident, there is a distinct cultural lag between our communications needs and modes. Who in the scholarly community, especially though not exclusively in the natural sciences, has not felt frustrated at some point about the huge time gap that elapses between the creation of new knowledge and its publication? Computer conferencing, by providing virtually instantaneous exchanges among far larger groups than has hitherto been possible, can accelerate this process.

Its potential extends far beyond science and research from creating networks of inner city and suburban isolates to providing new learning opportunities for the disadvantaged. It may even, at least initially, alter the power structure of science by decreasing the distance between those who have and those who do not have access to the inner circles through equalizing access to such networks. Given the resiliency of stratification, however, one suspects that ways will be found for curtailing such access eventually.

Except for face-to-face communication, groups have so far been extremely limited in their capacity to relate to one another. Spatial considerations keep the size of interacting groups fairly small or promote a hierarchical structure in which the few participate while the many watch from the sidelines. Computer conferencing is thus a major, perhaps *the* major, departure from traditional models of communication among groups. Moreover, as it becomes more expensive and time-consuming, hence less satisfying, to travel to conventions and conferences, it will increasingly be possible to participate in them via this remarkable innovation.

The network nation could thus re-unite individuals and groups dispersed over wide distances by the jet plane and car and recreate emotional bonds among family members, friends, and professional colleagues whom fate has separated.

Despite its unique potential, the authors warn that there are dangers to such a system unless wise decisions and policies are made today. Censorship and invasions of privacy are one concern and a widening gap between the better and the lesser educated is another. In thus alerting the reader to the pitfalls as well as the benefits of this new system the authors introduce a needed balance into their presentation by forcing us to consider the multiple impacts of this new technology.

I found the book imaginative and exciting. Indeed, I consider it the first in-depth study of a futurist technology that does not simply describe that technology but attempts, with considerable success, to explore its ramifications and wider impact for individuals and institutions.

I enjoyed reading the illustrative pages of the *Boswash Times* of February 14, 1981 and I found the system described in sufficient detail to give us a fairly comprehensive picture of the human meaning of this prospective addition to our culture.

There are those who may wonder whether eventually people will be as glued to their computers as they are to the television sets of today. While this is a possibility in lives bereft of meaning and joy, it is not inevitable. Just as the radio and telephone have been incorporated into an expanding storehouse of communications modes, so computer conferencing will be added to our other social linkages and become yet another—albeit powerful—source of community.

Suzanne Keller, Acting Chairperson
Department of Sociology
Princeton University

Foreword

by Herbert R. J. Grosch

I am writing this foreword while flying over Greenland, en route from an invitational workshop about what the computer profession calls "software engineering" — the technology of making very complex computer programs responsive to user needs, free of errors, and economical. That workshop was in Aarhus, Denmark, and now I go directly to Anaheim, California, to participate in a monster technical meeting and exhibit, the largest in history: an annual gathering of the clan held for nearly thirty years. There will be fifty thousand of us, members of societies like my own which make up the American Federation of Information Processing Societies, and members of similar societies around the world.

When I made my first trip to Europe, to look at British and continental progress in the scientific computing field, and to see the pioneer venture in commercial data handling which had just been set up by the Lyons chain in England, I flew in a propeller plane, a Douglas DC-6B — and had a berth! Primitive though that plane seems today, its airframe had been designed in Santa Monica on punched-card calculators which already used vacuum tubes to speed up designs and production paperwork.

That was 1954. Now the planes fly twice as high, hold three times as many passengers, go twice as far in fewer hours. And they carry inertial computers, digital instrumentation and controls, and are designed (along with their engines) by extraordinarily sophisticated computer techniques. And I have not even *mentioned* the Concorde!

Yet these improvements in air transport, and in the design and control processes involved, are dwarfed by the fantastic progress in computers themselves, and in the communications by which computers interact. In 1954 the early electronic machines I managed performed perhaps 600 operations in a second, and the key-driven machines of the thirties and the electromechanical punched-card equipment of the forties had been thousands of times slower still. Today the biggest and most powerful computers do tens of millions of operations per second, and further gains of at least a hundredfold in speed are clearly in sight. A large metal cabinet held 100,000 cards in the forties; in 1954, a reel of magnetic tape that comfortably fit under one's arm held ten times as much; in 1965, the first hints of microelectronics promised whole new universes of data storage capabilities; today those capabilities are for sale in computer hobby stores in every large city in the western world. By the end of the century, we are promised ten million characters of information — more than a million English words — on a single solid-state chip smaller than a postage stamp.

From the forties to the sixties the central computers grew from a few interconnected cabinets to whole rooms, even whole buildings, filled with electronics. Then they began to shrink, and to become very, very much cheaper. People who had been constrained to share the giant machines began to speculate about having their own minicomputers. The possibility of satellite and optical-fiber data communications meant that one-application or one-user machines could be interconnected easily, and tied to data banks of expensive or proprietary or very specialized information.

The prospects of new ways of using computers are entrancing. Much of those excitements, much of those hopes can be felt in this book. I share them, take pleasure in them. I have read science fiction for almost 50 years — enjoyed futures where far stranger computer triumphs and hazards have been commonplace. On the other hand, I have been responsible for the budgets and the personnel, and the organizational placement and prestige, of big shops in Washington and elsewhere. The kind of glee with which the "heroes" and "heroines" of this book abuse the university administrations, the government departments, the foundations, the taxpayers, is painful to such a person.

Take ARPANET. It was supposed to demonstrate how large and small computers, produced by a dozen manufacturers, operated all over the world, and linked by a variety of communication circuits, could be made to work together. But practical networks doing real workaday tasks, and larger and more diverse than the initial stages that the Advanced Research Projects Agency funded, were already in existence.

The real reason for many such ventures was resentment of control: a desire to access more computers, enjoy intellectual and financial freedom, "do your own thing," as the vernacular had it. Contrast that with the enormous discipline required to put man on the moon — computer discipline, as well as managerial and operational and training discipline. Contrast it with the concern for detail required to design, install, and maintain a giant airline reservation system.

So, I look at this book two ways. As a dream of new ways to use computer and communications power, to achieve completeness of recall for human interactions, to free people from the temporal urgencies of telephones and face-to-face conferencing on the one hand, and the delays and unresponsiveness of mail and messenger methods on the other — well, I am excited. Of course, I see problems, and will return to them briefly a little later. But it *is* a dream, a piece of science fiction come true.

Yet, I also see the lure of indiscipline, the chance to play intriguing games without a referee. I worry about youngsters who may never have to accept the older realities. Why learn to spell? The system will correct you. Why learn to speak forcefully and persuasively? The computer will democratize discourse. We already have people, and not only school children, who must use a pocket calculator to do the very simplest sums.

But I have been asked to look mostly at the technology, so let me return to that. The processors, the fantastic memory devices, the inexpensive but sophisticated terminals, are available right now, and will become even better year after year for a decade or more. The appropriate communications channels are somewhat more problematical, not because of technology — there will be satellites, and lasers, and quartz and glass fibers instead of copper wire — but because of policy and politics. Tariffs, data flows across international borders, privacy requirements, all have to be considered. Nevertheless, what is needed will come, and at attractive prices — in the end.

What is much, much harder is the software — that is, the programs that make the computer and its communications actually function. Each year the hardware improves; each year the software seems less satisfying. Nor do I see much improvement ahead. There is an almost impenetrable barrier between man and machine, or rather, between human language and computer language. The former is rich, inexact, constantly changing; the latter is sparse, precise, standardized (or should be!). The conferencing expounded in this book uses human/ human language. But to make it work, literally millions of computer instructions will need to be written, in human/computer language. That will be unbelievably difficult, and in the long run, will cost many fortunes, even by U.S. Defense Department or IBM standards. The Murray Turoffs of this new world welcome the challenge; I am very much aware of the difficulties.

Would I have them stop? Of course not; and they would pay no attention to me or to any other hardhead if we asked them to, even though there is a very high probability that these particular directions may not open into the new world of better human interaction and more flexible access to knowledge of which the authors and their friends dream. Remember the collapse of the videophone experiments. Remember the lack of acceptance of the bar codes on all those groceries. Doing something on a computer does not automatically guarantee success, or even a fair trial.

What it does usually guarantee, in my long and varied experience, is excitement. It is fair to say that after the dream comes drudgery and disappointment — reality; for much of the drama of the last 30 years, though, computer dreams have *changed* old realities. I hope it will be so with the dream embodied in this book.

Herbert R. J. Grosch
President, Association for
Computing Machinery, 1976-1978

Preface

As we write this Preface, two dozen programmers and their manager are discussing the documentation and design of a new computer system. About 30 people from around the United States and Canada are exploring the social and policy implications of human communication via computer. At another meeting, several administrators from the U.S. Geological Survey offices in Menlo Park, California, are evaluating a contract proposal. In a fourth conference, econometricians and experts in systems dynamics modeling are debating the theoretical and methodological issues involved in constructing national and world models of social and economic changes.

What makes these exchanges worthy of investigation is that though they have been under way for several months now, the individuals involved have never been in the same place at the same time. Some are participating from their homes as well as from their offices. These conferences are *computerized conferences*, a new form of human communication utilizing the computer. We believe that it will eventually be as omnipresent as the telephone and as revolutionary, in terms of facilitating the growth and emergence of vast networks of geographically dispersed persons who are nevertheless able to work and communicate with one another at no greater cost than if they were located a few blocks from one another.

Until now, postindustrial society, despite its heavy reliance on the generation and exchange of information, has been almost entirely dependent on forms of communication that have not changed in a generation and that are inadequate for many purposes:

The mails—for the reasonably rapid exchange of written letters or messages between individuals;

The telephone—for the immediate exchange of spoken information between two people; can be stretched by the conference call to three or four simultaneous participants;

Television and radio—for the one-way transmission of images and spoken information from a single source to a mass of undifferentiated persons;

Printed and published matter (books, periodicals, newspapers)—for the one-way transmission of printed information to a mass or a group.

Despite such improvements as an increase in the speed of mail transmission through the use of the airplane and the increasing sophistication of the mass telemedia, such as the introduction of color television and stereo radio, the technological communications developments of the last generation have left a big gap in available communication capabilities. There has been no means for a *group* of people to adequately exchange information among themselves and reach decisions, other than to meet frequently face to face and talk it out.

This represents no change since the cavemen gathered around campfires to discuss what to do about hungry saber-toothed tigers. As people have begun to play many roles (belong to many groups simultaneously) and as organizations have become decentralized in many geographically separated locations, this requirement to gather at the same time and in the same place has become expensive and inconvenient.

This book is concerned with the recent emergence of a new alternative for conducting communication among groups or networks of persons or organizations such as meetings, study groups, and teaching–learning exchanges. It uses computers and computer terminals to provide a written form of discussion or meeting among a group of people. In the United States, it has most often been called computerized conferencing, and in Canada, "computer-mediated interaction." Sometimes it is known as "teleconferencing," although this term is also applied to audio and television conferencing.

To participate in computerized conferencing, the members of a group type their written comments or contributions into a computer terminal attached to a telephone, which then transmits the material to the host computer. Instead of a face-to-face meeting in which only one person can talk at a time and everyone must be present at the same time and place, the insertion of the computer into the communications net means that all individuals may enter and receive the materials at a pace, time and place of their own choosing. The computer stores each entry and delivers it to those to whom it is addressed the next time they join the discussion and are not busy making an entry. Thus, the participants could conceivably all be making entries simultaneously; they could be spread out in locations all over the world; and the sending and receiving of the material could occur minutes, hours, days, or even weeks apart.

Moreover, in 1978 this could be accomplished for nationwide communication at a cost of under $8.00 per hour per participant, which is much less than the hourly rate for an ordinary long-distance telephone call. The advance expected in computer and communications technology can probably reduce this cost to a few dollars per hour in the early 1980s. Whether or not such efficiencies and the resulting potential benefits to society and the public will be realized is more likely to be a function of policy and regulatory decisions than of technological capability. The mere existence of a sizable potential savings over telephone, letters, and travel will bring about widespread usage in industry and government. If other opportunities are to be realized, there must be a greater awareness of the possibilities offered and their consequences.

Helping you to understand the nature and social implications of this new technology is the purpose of this book. The overall effect, we predict, will be to hasten the transformation of the social structure to what we term the "Network Nation."

What Is the Network Nation?

In "The Network City," Paul Craven and Barry Wellman point out that the network approach is characterized by its analytical emphasis on

> the primacy of structures of interpersonal linkages, rather than the classification of social units according to their individual characteristics . . . . [It] gives priority to the way social life is organized, through empirically observable systems of interaction and reliance, systems of resource allocation, and systems of integration and co-ordination [Craven and Wellman, 1973, pp. 1, 2].

While sharing with Craven and Wellman their insistence upon the functional importance of social networks in urban life, we propose that the widespread availability of human communication via computer will mean the ultimate replacement of urban networks as a basic form of social organization in postindustrial society by national and eventually international networks.

One way to view computerized conferencing systems is as an electronic communication network that obviates the necessity to be co-located in a dense urban area in order to have sufficient cheap communication ties. Melvin Webber foresaw this possibility when he wrote about "community without propinquity" in 1963:

> The unique commodity that the metropolitan settlement has to offer is lower communication costs. This is the paramount attraction for establishments and, hence, the dominant reason for high-density agglomeration.
> The validity of this proposition would be apparent if we were to imagine a mythical world in which people or goods or messages could almost instantaneously be transported between any two establishments . . . . One could then place his home on whichever mountaintop or lakeside he preferred and get to work, school, or shops anywhere in the world . . . . The spacial city, and its high-density concentrations of people and buildings and its clustering of activity places, appears, then, as the derivative of the communications patterns of the individuals and groups that inhabit it . . . . Here, a person is best able to afford the costs of maintaining the web of communications that lies at the heart of complex social systems [Webber, 1963, pp. 537, 540].

With the development of jet planes and the long-distance telephone, it has become possible to communicate or visit with persons in another part of a nation in the same time as it takes to reach a person on the other side of the same metropolitan area; but not at the same cost. In a society with increasingly specialized occupations and a great deal of geographic mobility, this has caused less than optimal communication patterns. Physicists still tend to talk most with colleagues at the same lab, even though there are other persons whose work more closely resembles their own in other parts of the nation or world. Family members decrease the frequency of communication when distance intervenes. The functional and emotional ties are there, building a complex network of mutual interests and acquaintances; but the flow of communications through this network is not very high because of the costs.

Just as the physically discrete nature of cities has developed into a meshed megalopolis of solid settlements, so too the functional independence of individual urban areas began to disappear with advances in transportation and communication. As Webber (1963) puts it, "The networks of interdependence among various groups are becoming functionally intricate and spatially widespread" [p. 534]. Or, as Jessie Bernard puts it in *The Sociology of Community*, the "community" has become the whole nation, rather than the town or city:

> Improvements in communication, especially the mass media, and most especially television, as well as in transportation, have profoundly changed the significance of space for human relationships . . . . Once individual mobility has reached a certain point, once speed and feasibility of communication have reached a certain level, and once economic and political integration have reached a certain level, we do not need the concept of community at all to understand how a society operates [Bernard, 1973, p. 181].

As functional communities have increased in size and spread beyond the geographic entities of cities to encompass individuals separated by considerable geographic distances, then, as Granovetter points out (1976, p. 1287),

> the fact that not all community members have (direct) social relations with one another has become a matter of prominent theoretical focus. The metaphor most consistently chosen to represent this situation is that of the "social network"—a device for representing social structure which depicts persons as points and relations as connecting lines.

Thus in place of thinking of a nation or society as a collection of communities, we need to think of it as a complex set of overlapping networks of actual or potential communication and exchange. Unlike a group, not all of the members of a network are directly in communication with, or even directly aware of, one another; but they are connected by communication and relationships through mutually known intermediaries, and thus, the *potential* for direct communication or exchange is there.

Computerized conferencing systems offer the possibility of conveniently and cheaply communicating with large numbers of people. It is our view that these systems allow a person meaningful, frequent, and regular communications with five to ten times more people than is possible with current common communication options.

It costs no more to communicate from Los Angeles to New York than it does from Boston to New York on such a system. (In fact, one irony of current regulatory policies is that it is currently often cheaper for two people from widely separated states to communicate on such systems than for two people from the same metropolitan area but in different suburbs.) It is just as easy to send a communication to 20 or 50 persons on such systems as to one.

When such systems become widespread, potentially intense communication networks among geographically dispersed persons will become actualized. We will become the Network Nation, exchanging vast amounts of both information and social-emotional communications with colleagues, friends, and "strangers" who share similar interests, who are spread out all over the nation. Ultimately, as communication satellites and international packet-switched networks reach out to other cities and villages around the world, these social networks facilitated by computer-mediated communications will become international; we will become a "global village" whose boundaries are demarcated only by the political decisions of those governments that choose not to become part of an international computer network. An individual will, literally, be able to work, shop, or be educated by or with persons anywhere in the nation or in the world. Because many more people can communicate well in written English than by speaking, the technology is likely to come to dominate international communication.

The first computerized conferencing system was created in 1970 and the use today is limited to tens of organizations and a few thousand people. The experience to date is of such a nature as to warrant the following general views on the parts of the authors:

• Computerized conferencing will be a prominent form of communications in most organizations by the mid-1980s.
• By the mid-1990s, it will be as widely used in society as the telephone today.
• It will offer a home recreational use that will make significant inroads into TV viewing patterns.
• It will have dramatic psychological and sociological impacts on various group communication objectives and processes.
• It will be cheaper than mails or long distance telephone voice communications.
• It will offer major opportunities to disadvantaged groups in the society to acquire the skills and social ties they need to become full members of the society.
• It will have dramatic impacts on the degree of centralization or decentralization possible in organizations.
• It will become a fundamental mechanism for individuals to form groups having common concerns, interests or purposes.
• It will facilitate working at home for a large percentage of the work force during at least half of their normal work week.
• It will have a dramatic impact upon the formation of political and special interest groups.
• It will open the doors to new and unique types of services.

- It will indirectly allow for sizable amounts of energy conservation through substitution of communication for travel.
- It will dramatically alter the nature of social science research concerned with the study of human systems and human communication processes.
- It will facilitate a richness and variability of human groupings and relationships almost impossible to comprehend.

However, if we are not careful about the decisions we are making today with respect to regulations, laws, policies, and the public interest, these systems may:

- Further widen the gap between small and larger institutions in terms of concentration of wealth and power.
- Further sharpen the distiction between the well-to-do, the educated, and the disadvantaged.
- Further the decay of the central city.
- Allow for comprehensive forms of censorship and regulation of information exchange, while destroying the press economically.
- Allow for the invasion of privacy in rather unique ways.
- Be regulated out of existence or emerge as a shadow of the potential inherent in the technology.

These are the assertions of the authors and one objective of this book is to convince the reader that these views are credible. To do this we shall describe in detail the capabilities of computerized conferencing systems; review the history of this technological innovation; discuss what is known about impacts on those who have had the opportunity to utilize these systems; look at the potential utility for a wide range of societal purposes; and review the present and future technical and policy issues related to their use. Finally, we will look at the current barriers to the introduction of this technology and the impacts on society that will result if and when these barriers are overcome.

For three years the authors have been carrying out the principal part of their professional research and communication over a computerized conferencing system. To a significant extent they have, in the language of anthropological fieldwork, "gone native." They have come to appreciate the unique kinds of social processes observed within the computer-mediated social system and wish to explain them to the outside world. It is quite clear from the tone of the book that there is a blurring between scientific objectivity and advocacy for this new form of communication. The authors admit to a degree of vested interest in their principal area of research, and the reader will have to weigh the content within the context of this acquired bias. The authors admit to conclusions and projections that are as yet unsupported by adequate experimentation.

The subject matter and data included in this book constitute a cross-disciplinary discourse that cuts across the boundaries of sociology, psychology, computer science, operations research, management science, information science, communications, and futures research. As a cross-disciplinary study, it may not be completely satisfying to those who prefer to remain within the demarcated subject area and methods of any one of these disciplines. However, we think it will prove enlightening and useful to those who have discovered the challenge and the problems of working in the grey area that characterizes the interfaces among disciplines.

Because of its interdisciplinary nature, the book may be used for courses in each of these areas. We recommend that all readers start with Chapters 1–3 and include Chapters 12 and 13. The ordering and most important chapters for various disciplines are as follows:

Computer science—9, 10, 11, 4, 7, 8, 12, 13;
Business—4, 11, 12, 9, 8, 13;
Sociology—5, 6, 7, 8, 9, 12, 13;
Technology assessment—5, 6, 11, 12, 13.

For courses in the areas of communications, futures studies, or man and technology the book should be read in its current order, with the instructor skipping over those chapters not felt to be pertinent to the specific course.

Acknowledgments

We are most of all grateful to the many members of EIES who answered our questions and questionnaires, let us monitor and use their communications, and gave us permission to include our observations in this book. Many people have also been most cooperative in reading drafts of chapters and making suggestions about how they might be improved. In particular we would like to thank Robert Bezilla, Carl Hammer, Suzanne Keller, and Harold Linstone.

We would also like to thank Fred Weingarten and Harold Bamford of the National Science Foundation for their encouragement of research efforts in this area. Among the other individuals to whom we are grateful for help in completing the book are Anita Rubino at NJIT and Elizabeth Rumics of the Upsala College Library.

The Scenarios

How can we place you, the reader, in the future and let you envision some of the possible applications and impacts of computer based communications systems? How can we give you some of the flavor and details of events that might occur, without seeming too concrete about outcomes that are projected from an inadequate data base? We have chosen the scenario device: Each chapter begins with a fictitious issue of the Boswash Times, the computerized news service of the Boston–Washington megalopolis. The "news" stories presented are meant to represent those that might appear during the next two decades, as the Network Nation emerges.

For now, imagine that it is breakfast time in 1994, and you have settled down with a cup of coffee-substitute heated on your solar stove, to read your computer-generated equivalent of the daily newspaper, including all the news that is fit to display on your home terminal.

Acronyms

ARN	Astronomical Resource Network
ARPANET	Advanced Research Projects Agency Network, U.S. Department of Defense
CASNET/Glaucoma	An artificial intelligence system to diagnose glaucoma
CLEP	College-level equivalency program
CONCLAVE	A computerized conferencing system developed in England
CONFER	Computerized conferencing system developed at the University of Michigan
EIES	Electronic Information Exchange System (for Scientific Research)
EMISARI	Emergency Management Information System and Reference Index developed at U.S. Office of Emergency Preparedness
ERDA	Environmental Research and Development Administration
EPRI	Electric Power Research Institute
FORUM	Computerized conferencing system developed at The Institute for the Future
HEW	U.S. Department of Health, Education and Welfare
INTERNIST	An artificial intelligence diagnostic system being developed at the University of Pittsburgh
IRIS	Incident Reporting Information System developed at OEP
IRMIS	Internal Revenue Management Information System
ISM	Interpretive Structural Modeling *or* Interpretive Structure Model
JOSS	One of the first time-sharing computer languages designed for people unfamiliar with computers, developed at RAND
MAC	Multiple Access Computer; an early time-sharing effort at Massachusetts Institute of Technology
PERT	Program Evaluation and Review Technique; a Critical Path Analysis System
ORACLE	A computer conferencing system developed at Northwestern University for use in education
PILOT	A computer language for writing educational lessons
PLANET	"Planning Network"; a simplified version of the Forum computerized conferencing system

PLATO	Computer-assisted instructional system developed at the University of Illinois
RIMS	Resource Interruption Monitor System
RIPS	Resource Interruption Projection System
SED	Systems Evaluation Division
SIMSCRIPT	A computer language for writing simulations
Telenet	A pocket switched telephone network
TIMS	The Institute of Management Science
Tymnet	A digital data network
Tymshare	A company which offers time sharing services, including Tymnet

Short Forms

CAI	computer-assisted instruction
EFT	electronic funds transfer
EPC	editorial processing center
CRT	cathode-ray tube
CSG	(the) Communications Studies Group
FPA	Federal Preparedness Agency
GNP	gross national product
GSA	General Services Administration
IEG	Information Exchange Group (of the National Institutes of Health)
IFTF	The Institute for the Future
IQ	intelligence quotient
LVR	latency of verbal response
NGT	nominal group technique
NLS	on-line system
NMUD	Non-medical Use of Drugs (Canadian government group)
OEP	(the U.S. President's) Office of Emergency Preparedness
RFP	Request for proposal

The Network Nation
Human Communication via Computer

PART I

The Nature of Computerized Conferencing

THE BOSWASH TIMES

The Computerized News Summary Service of the Megalopolis

ISBN 0-201-03140-X, 0-201-03141-8 (pbk.)

Supreme Court Hears FBI Conference-Screening Case

Hearings continued today on the landmark civil rights case now before the Supreme Court. The FBI argued that in order to prevent computer conferencing networks from becoming sources of organized dissidence and possible revolution, they must have the power to routinely impound tapes for any computer communications system, and screen them for the presence of suspect words, such as "assassination." The argument has been that this is no different than a policeman routinely patrolling a beat, looking and listening to see if any activity seems suspect. Since no human looks at the messages unless the computer provides cause for obtaining a search warrant, current laws are not violated.

Civil liberties lawyers are expected to argue that such power of surveillance is a fundamental intrusion into the constitutional guarantee of freedom of speech. They have also revealed that all computer-mediated messages to or from approximately 500,000 Americans who now are considered potential security risks are stored on tapes in the FBI library.

Postal Service Requests Additional Subsidy; Jones Urges Abolishment

With the first-class postal rate at $1.00 per ounce and deliveries cut to once a week, the Postmaster General argued again today that Congress must increase the postal subsidy by an additional $1 billion a year. "It is simply not possible," he stated, "to raise rates or cut services any further, and still maintain a U.S. Postal Service."

Senator Marilyn Wu of Hawaii urged that the Postal Service be abolished instead. "All really important mail now gets delivered electronically by computer systems," she stated. "There is no reason why U.S. taxpayers should continue to subsidize an outmoded communication system, merely for the benefit of those few fanatics who refuse to have a terminal in their home."

Subcommittee hearings will continue tomorrow, when the Postal Service will present a proposal for a nationwide effort to increase interest in stamp collecting. Under pressure from industry, the Postal Service will also unveil its new type of mail, in which the receiver pays the cost of the postage.

ISBN 0-201-03140-X, 0-201-03141-8 (pbk.)

4

Figure 1-1 Person-to-person networks: Human communication and information exchange via computer.

Illustration by Joe Dieter, Jr. Permission to reprint granted by A.S.I.S.

ISBN 0-201-03140-X, 0-201-03141-8 (pbk.)

CHAPTER 1

An Overview: Computerized Conferencing and Related Technologies

"See ever so far, there is limitless
space outside of that,
Count ever so much, there is limitless
time around that"
from Walt Whitman, "Song of Myself"

Computerized Conferencing: What It Is and How It Works

A computer conferencing system (CCS) uses the computer to structure, store, and process written communications among a group of persons. When something is entered through one terminal, it may be obtained on the recipients' terminals immediately or at any time in the future until it is purged from the computer's memory.

Just as it would be difficult to explain to someone who has never observed or participated in a face-to-face discussion or decision-making group the communications processes and social dynamics involved, so the best way to learn about computerized conferencing is to participate. As a second-best substitute, we will try to lead you through the experience in these pages.

Imagine that you are seated in front of a computer terminal, which is like an electric typewriter, with either a long scroll of typed output ("hard copy") or a TV-like screen (a cathode-ray tube, or CRT) for display, or both. The terminal is connected to an ordinary telephone.

You dial the local number of your packet-switched telephone network service, which provides a low-cost link to the computer-host of the conferencing system. (For example, in 1978 TELENET, one such service, enabled U.S. customers to dial into a single CCS with local calls in any of 90 major cities for an average of $3.50 per hour.) You type in a few code words to identify

Starr Roxanne Hiltz and Murray Turoff, The Network Nation: Human Communication via Computer

yourself and are then given the following sort of information:

WELCOME

JOHN DOE ON AT 1/25/79 11:02 A.M.

PEOPLE YOU KNOW NOW ON TERMINALS

System Monitor (100)
Robert Johansen (708)
Linton Freeman (745)
Elaine Kerr (114)
Robin Crickman (727)

LAST ON 1/24/79 7:25 P.M.

WAITING: 14 private messages
                16 comments in conference 172: Therapy Group
                3 comments in conference 253: Chinese Recipes

INITIAL CHOICE?

This sample printout brings you up to date on communications directed to you since you last accessed the system (in our example, the night before). It tells you that there have been 14 messages entered for you. In addition, you see the names and numbers of those now "on line" with you, who could receive materials immediately. It reminds you that you belong to two "conferences," which are like written group discussions on a specific topic. In one of these discussions, there are 16 new entries that you have never read, and there are three in the other. Now, you are asked what you would like to do first. Among the options you have are to

- Accept the full text of some or all of the private messages;
- Scan just the title line of the messages, showing author and subject, before deciding which to read first;
- Choose to go directly to a conference and receive the new discussion entries there;
- Send a message or enter a conference comment first;
- Search and review earlier materials by such criteria as author and/or date, and/or subject. (Perhaps, for instance, you remember that there was a message sent to you by Elaine two days ago that you never answered. Rather than search through your file drawer, you may retrieve it by author and date, and review it before answering.)

Some of the unique capabilities provided to the participant in this remote, written communication form are the following.

1. Time and distance barriers are removed. Participants can send and receive communications whenever it is convenient for them, with the material enduring through Whitman's "limitless time" to be received or

ISBN 0-201-03140-X, 0-201-03141-8 (pbk.)

revised not only once, but again and again. As for "limitless space," messages can be sent and received wherever the user has a telephone and can carry a portable terminal. Geographically dispersed persons in different time zones may send and receive at different times, since the transactions are separate and the message is stored, or they may communicate in "real time" if they are at their terminals simultaneously.

2. Group sizes can be expanded without decreasing actual participation by either member. A single computer can accommodate from hundreds to thousands of users, whereas the mechanisms of finding a comfortable room and getting everyone together for a face-to-face meeting of such a group are expensive and discouraging. Specific conferences can accommodate from 2 to 100 participants, depending on its purpose and the communication structure provided by the computer software. This is because each person can "talk," or input, whenever she or he wishes, rather than having to "take turns" as in face-to-face verbal communications. Rather than only one person "having the floor," all participants could be typing in messages simultaneously. No one can be interrupted or "shouted down." In addition, since it is possible to read much faster than to listen, much more total information can be exchanged in a given amount of time.

3. A person participates at a time and rate of his or her own choosing. You need not leap out of the bathtub to answer a ringing telephone, or drag yourself out of a hotel bed at 7 A.M. to make a meeting that begins at 9 A.M.

4. The computer can offer you services and options not available in other forms of communication:

- Editing routines make it easy to make corrections and line up entries to make them appear neat.
- You can sign entries anonymously or with a pen name. This enables you to say controversial or critical things that might otherwise be embarrassing.
- Data or analyses stored on other computers can be brought into the discussion.
- Votes on issues can be administered, tabulated, and reported back to the group automatically.
- You can "whisper" to other conferees (via private messaging) without others knowing that the side conversation is going on.
- You can delay messages and set fixed delivery times.
- You can engage in a bridge game—with the computer dealing and keeping score for you and the other players (a very specific communication structure)—or in other "social" or "recreational" activities, such as a game of chess or a jokes conference.

ISBN 0-201-03140-X, 0-201-03141-8 (pbk.)

• The indexing and searching capabilities provided by the computer mean that the "subject" may serve as the "address." That is, the computer can link you to others with common interests, without your having to specify those individuals at the time you enter the communication.

One of the biggest disadvantages that most people see in this form of communication is that the entries must be typed in. It is possible to get around this; you can handwrite or dictate to a secretary, for instance, who might then enter the material directly through the keyboard, or type it into a cassette for high-speed entry through a special kind of terminal. For those who do not have easy access to a secretary, there remains the barrier of fumbling fingers on the keyboard, with entry rates much slower than talking. To compensate for this, however, there is the possibility of reading or skimming through the communications received at a much faster rate than is possible with the spoken word. Most important, the computer's text-editing capabilities are utilized to make corrections to text as flexible and simple as possible. You work in a "scratchpad," or composition space, in which all lines and pages are numbered. Whenever you want to make a correction, you indicate the line number and what you want to do, such as insert a line or a paragraph, or even a whole page from some previously composed material; delete a word or a phrase; correct the spelling or insert some words. The ability to edit and reuse old material means that even fairly poor typists are able to reach an input rate that is respectable. For example, the 17 scientists who were extensive users of EIES (Electronic Information Exchange System, pronounced "eyes") during its test period in 1977–1978 achieved an effective input rate of about 28 words per minute.

To further ease the transition to this new mode of communication, EIES and most other current systems are designed on the following principles:

1. It is not necessary to understand anything about computers to learn and utilize the system.
2. The user can make productive use of the system after a short period of practice (about half an hour).
3. The design is "segmented," so that the user can learn those particular advanced features he or she wishes to use when a need for them is felt. These are contained in on-line explanation files as well as in printed documents.
4. There is nothing the user can do in the way of mistakes that will either hurt the system or cost the user more than the little time needed to re-answer questions. The system is therefore designed to be "forgiving" in terms of a user error.

Before looking at the features of a particular CCS in more detail, let us review the technological and sociological developments that made possible

ISBN 0-201-03140-X, 0-201-03141-8 (pbk.)

the economical and flexible systems for communication via computers that are emerging today.

Technological and Historical Factors

The implementation of computerized conferencing and related technologies such as electronic mail is the culmination or product of a host of technological, sociological, and economic factors and their evolution over the past three decades. In a very real sense, the hindsight we can now apply makes the occurrence of these systems not only possible but inevitable. The factors that need to be considered include developments related to:

- Computer and communications technology, and associated cost trends;
- Trends in operations research and systems analysis;
- Interactive computer systems, such as CAI (computer-assisted instruction);
- Industry structure and government regulation;
- The increasing economic and social importance of information.

With these considerations in mind we would like to weave for you a rationale for the emergence of computerized conferencing at this time and thereby provide a basis for the rest of the book and our projections into the future. No doubt many of you have reflections on the past few decades that differ from ours, but history is no easier to interpret than the future is to forecast.

Computer Technology

When computers first emerged, a fairly simple computer by todays standards cost millions of dollars. The physical hardware and its associated maintenance represented 80–90% of the costs of operating such a computer system. The other 10–20% were associated with the people cost, the development and use of software, and the other care and feeding functions for satisfying the "number-crunching" needs of these "entities." As a result, the early days of computers led to a conditioning of the people associated with these systems in either managerial or technical positions. The emphasis was on "efficient" use of the hardware: "Let us not waste the time of the machine; people are cheaper, we can waste their time instead." A more subtle but direct consequence of such economic considerations was the centralization of computer services into an organizational entity so that economies of scale in the use of hardware could be obtained and cost could be better controlled and credited. Indirectly this centralization usually led to a further machine-oriented conditioning of "computer types," since they became in

ISBN 0-201-03140-X, 0-201-03141-8 (pbk.)

some sense isolated from the people who had to utilize the output of a computer—the "user" as she/he is "fondly" called in the trade.

Another phenomenon or subcultural trait that was felt in these early days was the tendency to set up the computer almost as an object of worship within a cult that could be compared to a form of religion. The image the industry initially established was one of a very complex device enshrined in its own special room, understandable only by a priesthood trained to speak its mysterious language, and able to deal with any and all problems brought to it. In the early days the marvelous capabilities of the machine were easily demonstrated by means of quite simple problems that would have required huge amounts of repetitive human effort but that could be handled effortlessly by the computer and those who knew how to program it. To the uninitiated what was accomplished in terms of the saving of money and time was awe inspiring, which of course, was somewhat ego gratifying as well as profitable to those working with or selling computer systems.

Current Changes

The primary technology shift has been the declining cost of computer hardware relative to other costs. As of the early seventies the hardware costs of major computer installations had dropped to below 50%, and should be below 25% by the early eighties. The cost of computers is such that there is a growing hobby market and the line between a sophisticated calculator and a full-scale computer, both costing in the neighborhood of hundreds of dollars, has become very thin. With computers comparable to those that once cost millions now available for only tens or hundreds of thousands of dollars, it is not valid to assume that computers should be centralized in an organization, or that a single computer should be all things for all people. In fact, such practices tend to drive up, in a nonlinear manner, the software and people costs, which are now the majority of total costs of computer systems. In 1955, one dollar bought the execution of about 100,000 single operations on a computer (like a single addition); in 1960 this rose to 1,000,000 and in 1970 to about 100 million. Many similar trends can be defined for other parameters that measure computer performance, such as cost of memory storage. These trends are likely to continue over the next decade. Contrary to those trends, the cost of software on a per-unit basis increases nonlinearly as one tries to maximize utilization of the hardware for a single machine.

The result has been a growing awareness that the cost-effective principle is not the highest possible efficient use of the machine, but efficient use of the people who work the machine. This means more software to translate the capabilities of the machine to functions and tasks that are directly relevant to the end user, who through a computer terminal can directly use the machine for his or her own purposes. The lower absolute costs of the machines also

ISBN 0-201-03140-X, 0-201-03141-8 (pbk.)

mean that a single machine can be dedicated to a single class of users and tailored to their job even though it will not operate at full capacity 24 hours a day. That this can limit the size of someone's computer empire in an organization, or spread out or decentralize control over data processing functions; that it violates rules learned by managers who did not understand the reasons for the rule; that it becomes ambiguous when a dedicated computer such as a word-processing machine is still a computer; and that it generally confuses all the established guidelines, has led to a fascinating transition state with many interesting and amusing situations and developments of which computerized conferencing is one example. We ultimately expect the outcome of all this to be a "computer in every home," which may become a future politician's equivalent of "a chicken in every pot."

Communications

In the past few decades computer technology has significantly penetrated the communications industry. With the need for computers serving different but related functions, computers have been interconnected with communication lines. With the need for many terminals to be able to access the same computer, it was self evident that use of telephone lines alone would prove to be far more costly than the actual computer costs. What has emerged is "packet switching," a technology for the low-cost transmission of digital information. Small computers in different cities collect data from many different terminals in that city via a local telephone connection, and package that information to send over very high-capacity transmission lines to other cities, where another small computer unpacks the data and sorts out what is to be sent on or dropped off at a local computer also attached to the network. Because these high-capacity lines are shared by many different users and computers are fully utilized, the cost reduction over other alternatives has already driven the charges for long-distance digital transmission far below that of voice or regular telephone transmission lines.

There are a growing number of such digital networks emerging, both private and under FCC regulation as "public" networks. Unfortunately, the conception of "public" currently used by most regulators and legislators is business and government use, and it is not yet conceived that the "general public" or ordinary citizens will be a market for such systems and that regulations or laws in the public interest should facilitate this. This is a significant concern for computerized conferencing, and the potential for citizen use will be indirectly discussed throughout this book. The costs of computerized conferencing via the computer that supports it is lower with today's technology than that of the communications network to which it must be tied. As might be expected, innovation in this area has been from new enterprises rather than the regulated phone companies. Now that big market

ISBN 0-201-03140-X, 0-201-03141-8 (pbk.)

potentials can be seen by even the most conservative managers and organizations, the next decade promises to be a regulatory, legislative, and legal battleground with billions in yearly revenue at stake.

Operations Research and Related Analytic Specialties

Associated with the developments just described has been the emergence of operations research, management sciences, systems analysis, and techniques of modeling, simulation, cost-effectiveness, benefit-cost analyses, and so on. Once there existed a technological device that could perform large numbers of very simple mathematical operations at great speed and with great precision, analytical techniques were sought for any problem that humans had to deal with, reducing the problem to a set of quantities whose interactions are governed by set logical procedures. In other words, how do you take a complex human or human systems problem and reduce it to a well-specified series of simple trivial problems with which computers can be taught to deal? During the development of operations research and "management science" (and still today, to a large extent), there was an honest conviction among many professionals that this was possible to do. It was, in fact, doable for certain classes of problems faced by organizations and it is very easy to feel that if it works in some cases, it can ultimately work in all cases.

As a result there were a considerable number of attempts to develop sophisticated models that would generate the most "efficient" decision or pick the "best" option from a set of alternatives. Particularly in the government this attitude reached its peak in the McNamara era and the general policies of cost-effectiveness analyses. A significant number of people quickly realized that these decision-oriented computer models could never account for the complexity in high-technology situations.

One of the most dramatic incidents involved the F-111. The computer-estimated savings for the Navy and Air Force to utilize the same plane rather than two different designs were later washed out by such detailed technical problems as the differences in specifications (i.e., a high-salt environment for the Navy), and separate logistic support chains for the two services. However, even the inclusion of the technical details at a fine level would still not account for the necessity to consider social-psychological factors related to the organizations involved in the decision process and in the implementation of the resulting decision. The human environment, including bureaucratic politics, has to be a part of the analysis in decisions governing human systems and their function. As a result there came about a growing concern for how to factor in detailed and qualitative considerations in a manner that could compete with the impressive outputs of computer models. We shall return shortly to these new problem-solving technologies, the development of which laid another foundation for the emergence of computerized conferencing

ISBN 0-201-03140-X, 0-201-03141-8 (pbk.)

because they were concerned with how to deal with qualitative statements and subjective judgments.

At an organizational and sociological level there was a distinct match between the organization of human groups and the way the computer organized problems. Having humans required to deal with large numbers of repetitive tasks leads to hierarchically structured organizations with fairly simple, well-specified unambiguous tasks to be performed by the occupants of each position and very specific rules to follow at all the lower levels in the organization chart. Bureaucratic and authoritative or feudal organizations* therefore can easily adapt to the use of computers because the organizational processes (the authority relationship and information transfers) between organizational nodes are well defined. Once again this led to an image of major success without much consideration of why such success could be achieved and its possible limitations. Today, most repetitive tasks have already been computerized, and the problems that remain usually involve lateral cooperation and coordination among organizational units.

Given this perspective on the early years (fifties and sixties), we are today in a very mixed situation. The seventies and probably the eighties are likely to be viewed as a period of transition in which ideas, concepts, approaches, and presumptions that emerged from the early computer days are being tempered and (most important) better understood, in terms of the conditions and circumstances that make analytical computer-based approaches appropriate or timely for a particular situation. In part this represents a growing maturity of a young profession that has learned from its mistakes, but also it is significantly influenced by advancement in the technology and our growing understanding of the nature of the problems confronting groups, organizations, and society.

The Information-Based Society

Increasingly, as we have entered a postindustrial society, we have moved from an economy in which goods were produced by physical labor to an economy that is not only dominated by the "service" or "white collar" sector, but within this, by the production, transfer, processing, and use of information. For example, Drucker (1968, p. 263) says that by 1965 the "knowledge sector" (consisting of the production and distribution of ideas and information) accounted for one third of the gross national product, and predicted that it would account for one half by the late 1970s. Daniel Bell states (1973) that whereas energy production drove industrial society, information production drives the postindustrial society. Thus, its class structure is based on access to information and control of decision-making processes, rather than

ISBN 0-201-03140-X, 0-201-03141-8 (pbk.)

---

*See Chapter 4 for an explanation of these terms.

on ownership or control of property. This has been a tragedy for the educationally disadvantaged; few jobs require physical skills anymore, and people with limited educations have not been given the opportunity to acquire the training and skills needed to obtain the new white collar jobs. (Some remedies for this situation will be discussed in Chapter 5.)

Most medium and large employers have computerized almost all their information-related operations. For example, a large insurance company that once employed acres of clerks at desks to post premium payments and other transactions has a computerized system that does a constant update of all policies. The people who once made multiple records and filed by hand have been largely replaced by keypunchers and programmers who feed the data to the machines and keep the system running smoothly. Keypunches are now being displaced with terminals, word processing, and data entry.

During the last few years, computer technology has changed so that most companies have data and programs entered on dozens or hundreds of remote terminals, for execution on a centralized "time-sharing" computer. Now emerging are major information transfer and transaction networks, EFT (electronic funds transfer) being one example. When it is generally realized that we can integrate communications into these systems, we begin to visualize specific communication structures to replace what we do now via such things as stock exchanges, classified advertisements, etc. This inherent merger of communications with what we now conceive of as information systems is a technology several orders of magnitude more powerful than either the telephone or TV. Information is a resource with very real economic value that is dependent on its organization, timeliness, and relationship to what we normally perceive as human communications. It will be a potential source of economic dominance as we move even further toward an "information economy." Who shall control, regulate, and have access to a technology merging information and communications will emerge as a crucial economic and political issue. Both the benefits and the dangers of various alternatives are very important in terms of preserving our democratic institutions in this country and abroad. We have gone through the evolution of property rights and civil rights; the current battle for equality involves information rights. The new professional class is going to include data brokers, information peddlers, commercial scholars, "experts," knowledge promoters and communication agents, though of course they will not use such delightfully accurate names for themselves. Investors may well end up buying and selling information futures.

Computer-Assisted Instruction

This teaching resource is an example of the interactive, user-oriented computer systems that have developed. For over a decade, CAI has been used in the classification and education of college students, medical students, and

ISBN 0-201-03140-X, 0-201-03141-8 (pbk.)

both normal and physically handicapped children. The advantages of these "programmed learning" packages over group instruction in a classroom are (1) pacing determined by the student (who asks for the next frame or question when he or she is ready), (2) more direct and unambiguous stimulus–response cueing, (3) provisions for instantaneous feedback of knowledge or results (dear to the hearts of everyone who believes in the reinforcement theory of learning); and (4) allowance for infinite repetition without taxing the patience of a teacher or other students listening to the process.

In computer-assisted instruction, the learner interacts with the computer so that the extent of material presented and the pacing are determined by the learner's rate of progress. A module of educational material is presented to the learner. When he or she has completed reading it, questions are asked to determine if the material had been mastered. If so, the learner "jumps" to a new module. If not, more details and examples are presented, with new questions following. Since a single lesson might conceivably be used by hundreds or thousands of students, great care can be taken to make the material both clear and interesting.

Some of the early work on computerized conferencing was based on a belief that communications would be built into these educational computer systems. It was recognized by those who followed this path that communications were an important part of the educational process; communications not just between teacher and student, but involving peer-group reinforcement among students as well.

New Problem-Solving Technologies

Both professionals and the general public have begun to give up the illusion that the computer is a perfect machine that can quantify and analytically solve all problems. Cases of computer error, misuse, and abuse are now common knowledge. Another aspect of changing attitudes is not so widely recognized. The early theories of operations research, as typified by former Secretary of Defense McNamara's brand of cost-effectiveness, have come into serious question, among both practitioners and clients.

The evolution toward the inclusion of nonquantifiable factors, reliance on the opinions of experts, and methodological emphasis on the problems of structuring communication among various experts or representatives of different points of view begin in the defense sector of the U.S. government and spread to the nondefense sectors. For example, the Corps of Engineers has moved from reliance on highly quantitative computerized cost–benefit models for evaluating construction projects to environmental impact statements and the need to involve the public.

There is growing recognition that complete quantification of problems is not possible. The more complex the problem, the more reflective of policy, planning, and decision issues, the more necessary are qualitative approaches

ISBN 0-201-03140-X, 0-201-03141-8 (pbk.)

and the use of informed human judgments or views. The risks and benefits of a major decision are just not completely quantifiable from "objective" data. As a result there has been a growing school of thought, outside the computer field, that what is needed are approaches that allow the coherent use of human judgments in the evaluation of complex problems. This conviction has been further supported by the growing recognition that our problems, whether organizational or societal, are largely impure. They do not fit neatly within the expertise of a single individual or even that of a few individuals.

What have evolved are techniques for structuring group communication processes so that it is more efficient for a group to pool and coevaluate their knowledge about a complex problem. Typical of these are the Delphi, using tailored paper questionnaires and feedback with anonymity for groups of 10 to 200 that are dispersed in time and space; and nominal group processes, which structure face-to-face meetings of small groups for the same objective. These methods share the concepts that any human group, as it grows in size, requires a structure in order to be able to communicate—a strong chairman or a parliamentary process or other set of procedures—and that this structure can be designed specifically around the nature of the problem. The first attempts at computerized conferencing were really an outgrowth of this activity and not of the computer field, as we shall see in the next chapter.

Summary

We have reviewed a complex of factors that has set the stage for the emergence of computerized conferencing. These factors will continue to interplay as we move through a transition period in which the new technology will become institutionalized.

After reaching out in so many directions, it is time to refocus. It is quite common in the introduction of a new technology for the innovators to first perceive it and utilize it in terms of their own problems or frame of reference. Therefore, as one would expect, the major use and evaluation of computerized conferencing has been by those scientists and technicians who have had the earliest access opportunities. In order to help you understand more fully what computerized conferencing is like, we will turn to a detailed examination of a particular system that was structured for scientific and technical communications.

EIES: Electronic Information Exchange System
for Scientific Research Communities

Once the computer is put into the communication loop, the potentials for structuring, facilitating, and augmenting the communication and information exchanged among members of a user group are virtually unlimited. With a

ISBN 0-201-03140-X, 0-201-03141-8 (pbk.)

dedicated minicomputer system, the design objective is to provide those communication features and computer-assisted computational functions that are most useful to the user community. We are going to describe the capabilities of EIES in some detail, to give you an idea of the nature and potential usefulness of a CC system.

The design philosophy is to start with the existing communication forms and functions of a user group and to build a system that accomodates or replicates such communication patterns at an overall increase in speed or efficiency and decrease in cost; and then adds communication and information-processing capabilities not possible without the computer. The specific nature of any particular system optimized for human communication would vary somewhat from one organization and application to another. Thus, we have to begin with an understanding of the existing forms and functions of communication in the target area.

Scientific Specialties and Scientific Communication

Scientific specialties consist of a set of scientists who engage in research along similar lines and who communicate often and intensively with one another (Hagstrom, 1970, pp. 91–92). As Chubin (1975, p. 1) has pointed out, "disciplines form the teaching domain of science, while smaller intellectual units (nestled within and between disciplines) comprise the research domain."

Such specialties have sometimes been called "invisible colleges" of scientists (Price, 1963; Crane, 1972) and have been seen as the social location of technical, cognitive, and ethical norms (Mulkay, 1972; Mitroff, 1974a) and as internally stratified on the basis of productivity (Cole and Cole, 1973).

Hagstrom summarizes the rather fuzzy overlapping definitions of scientific specialties and invisible colleges as follows:

> "These sets of groups are difficult to label or conceptualize. Considering their structure, one is led to call them networks, or more precisely, clusters in networks. Considering the intellectual content of their work, or their positions in encompassing disciplinary organizations, one is led to call them specialties or subspecialties. Considering the history of such groups, the most felicitous label might be "invisible colleges" [Hagstrom, 1976, p. 758].

Geographically dispersed networks of scientists working in the same specialty area can be viewed as the prototypical "production organization" of science, in which the "product" is scientific knowledge, and the social organization depends almost entirely on the communication system. Not only does the formal and informal communication system serve to direct and redirect efforts to "important" areas and the most fruitful methodological tools, but it also reinforces shared norms and theories and allocates rewards in the form of recognition.

Cole and Cole (1973, p. 16) describe the importance of communication in

ISBN 0-201-03140-X, 0-201-03141-8 (pbk.)

science as follows:

> Scientific advance is dependent on the efficient communication of ideas. The communications system then is the nervous system of science; the system that receives and transmits stimuli to its various parts.

The actual processes through which this crucial informal communication takes place have not changed in decades except that in many disciplines an exponential growth has slowed down the process and lengthened the time between the completion of a research project and its publication in a journal. Summarizing the results of a series of studies of scientific communication in the discipline of psychology, which is similar to patterns in many other disciplines, Garvey and Griffith (1971, pp. 354, 355) conclude that the scientist relies heavily on informal networks of discussion, small meetings, and exchange of drafts and preprints to keep abreast of current activities and of the current views of the community on the value and relevance of specific research problems. A simplified version of their findings is shown in Table 1–1. The journal article, by the time it is published, lags so far behind the research frontier that its functions are mainly to inform scientists in *other* specialties, and to allocate recognition for scientific achievement.

Increasingly, there have been calls for improving scientific communication and information dissemination. Many of these have focused on the information storage, processing, and network capabilities of the computer to provide assistance.

Some of the suggested innovations deal with the formal communication channels, the professional meeting, and the journal. There are predictions that there may be 100,000 journals published by 1980; something must be done to

Table 1-1
Dissemination of Scientific Information in Psychology[a]

Time from publication (months)	Event	Communication form
− 30 to − 36	Work starts	Discussion
− 24	Preliminary results	Drafts; informal meetings
− 18	Oral presentation (professional meetings, etc.)	Printed program; preprints
− 12	Technical reports; journal submission	Preprints
0	Journal publication	Circulated journal and reprints
+ 12	Abstracts	Reprints
+ 20	Literature reviews	Reprints

[a](From Garvey and Griffith, 1967.)

ISBN 0-201-03140-X, 0-201-03141-8 (pbk.)

decrease the costs and increase the efficiency of dissemination of "published" results. Selective dissemination of articles only to consumers who peruse computerized abstracts and order a copy of the full paper has been one answer; another has been more efficient, computer-assisted publishing procedures (see Rhodes and Bamford, 1976).

Another approach has been to make scientific information, particularly in the form of data bases and bibliographic files, directly available to researchers through an on-line, interactive computer system. One example of this is the NIH-EPA Chemical Information System, which stores both chemical data and chemical literature on a central computer that can be accessed from telephone-coupled computer terminals anywhere in the world. The user searches and retrieves information and performs data analyses on these files through conversationally designed computer programs (see Heller et al., 1977). Scores of abstracting services have been computerized and programmed to allow a person to interact with a computer to search these files using combinations of key words. However, the informal, prepublication communication within scientific specialties is also crucial to increasing scientific productivity. Recognizing this, the Division of Science Information of the U.S. National Science Foundation financed the building and field testing of a computer-based communication system designed specifically to meet the needs of networks of geographically dispersed scientists.

EIES Communication Structures

There are four main communications capabilities or structures within the EIES system, designed by Murray Turoff to replicate existing forms of scientific communications. These are displayed in Table 1-2. In addition, there are a multitude of available advanced features.

Table 1-2
EIES Communication Features

Structure and features	Replaces
Messages	Letters
	Telephone
	Face-to-face conversations, visits
Conferences	Face-to-face conferences or meetings
Notebooks	Sending or drafts or preprints
	Necessity for coauthors to be co-located
Bulletin	Newsletters (eventually, journals and abstract services)

ISBN 0-201-03140-X, 0-201-03141-8 (pbk.)

Each individual human member of EIES belongs to a research group, which has its own name and number and is just as much an entity in communication terms as is the individual. A *message* may be addressed to one individual, many specific individuals, or a whole group. The time and date of receipt of the message by each individual recipient is confirmed to the sender, and then about a week later it disappears from the system. The content, like that of the informal one-to-one communication channels that it is meant to replace, can vary from chatty "How are you? What's new?" messages to a request for assistance or information related to a specific problem in a person's research. Group messages are frequently used to introduce new members who are joining, to broadcast general requests for information on a research problem, and to announce the availability of drafts, preliminary research findings, or completed but as yet unpublished papers (preprints). As with all other entries in EIES, a user may choose not to divulge her/his identity by signing a message, but may instead have the message sent anonymously, or with a pen name.

The *conference* is a topic-oriented discussion in which a permanent tran-script is built up of the proceedings. A conference will typically last from a week to a few months, with participants entering and leaving the discussion at their convenience, and taking as long as they need to reflect on previous entries or consult references or data before responding. Each research group has one or more conferences, and every individual belongs to one or more groups. Since "intellectual mobility" is simple on EIES, it is a frequent occurrence for a member to join a conference of interest in another research specialty, as well as to participate in conferences set up by his/her own group.

A conference entry can be given an association number, noting the earlier items to which it is related. A participant can be helped in the review and organization of the proceedings by requesting all of the entries that are associated with a specific item. A user can also review the entries by asking to see, for instance, any that contain a certain word or phrase, such as "side effects," or "validity."

The *notebook* is the scientist's private on-line space for composing, storing, and reorganizing items on which he or she is working, with the aid of extensive computer-assisted editing routines. Designated pages of a person's notebook can be opened to others for reading; thus, for instance, after completing a draft of a paper, the scientist could send a group message inviting anyone interested to read it and comment. Parts of the notebook can also be opened to others to write in, thus facilitating remote coauthoring.

The *Bulletin* is a public space, like an on-line newsletter. Any user of the EIES system can request the titles of Bulletin entries and then the abstract and/or the full text, if something is of interest. The Bulletin is used to publish

ISBN 0-201-03140-X, 0-201-03141-8 (pbk.)

short papers or discussions, "letters," announcements of meetings or grants or job opportunities of interest to the specialty, etc. Eventually, it might carry full papers and abstracts of papers published elsewhere, thus becoming an on-line journal that facilitates the speedy review and publication of research results and their selective dissemination to interested persons. Using a computer to mediate communication does not necessarily result in "cold or impersonal" communication; nor does it obviate the need for human participants to play some very active roles. (See Table 1-3.)

Just as it takes a lot of work for the organizer of a session at a professional meeting to put together a group that is well balanced among different points of view and to help the session run smoothly, so too there is need for a human organizer of a computerized conference.

In order for a computerized conference to be successful, according to initial observations, the moderator has to work very hard at both the "social host" and the "meeting chairperson" roles. As social host she/he has to issue warm invitations to people; send encouraging private messages to people complimenting them or at least commenting on their entries, or suggesting what they may be uniquely qualified to contribute. As meeting chairperson, she/he must prepare an enticing-sounding initial agenda; frequently summarize or clarify what has been going on; try to express the emerging consensus or call

Table 1-3
Human Roles in the EIES System

Role	Functions
Bulletin editor	Solicit entries
	Appoint reviewers
	Write/edit contents
Conference moderator	Invite participants
	Set agenda
	Delete/edit
	Summarize; call for votes or new discussion items
Group coordinator (Administrative assistant)	Provide assistance to members
	Send group messages of general interest
	Act as interface to other groups
Console operator	
User consultants	Answer questions about how to use the system
System monitor	Add and delete members, groups, and conferences
	Send general announcements

ISBN 0-201-03140-X, 0-201-03141-8 (pbk.)

for a formal vote; sense and announce when it is time to move on to a new topic. Without this kind of active moderator role, a conference is not apt to get off the ground.

Since a group of scientists will typically have several conferences on different topics, each moderated by a different person, it also needs to have an overall coordinator. The group coordinator helps new group members learn to use the system, allocates additional time to group members who need it, sends messages to the group introducing new members who are joining, and answers questions addressed to the group by other groups. The role is something like that of a departmental administrator or manager.

The Bulletin cannot be published without an active, hardworking human editor and volunteer reviewers and writers, just as are needed for journals or newsletters published and distributed in the more traditional manner. Typically an author would submit a paper, located in his/her notebook, to the editor. The editor would appoint reviewers, who would read the paper by accessing the notebook; the reviewers would then confer with one another and with the author using pen names. When they reached consensus, they would message (communicate with) the editor regarding the decision and the entry could be transferred to the editor for copy editing and publication within a few days. High-speed printers at the EIES host computer can be used to generate periodic hard-copy compilations of the Bulletin for distribution to interested scientists who do not have access to the system.

One key question that will determine the success or impact of this system is, what are the possible rewards or motivations for scientists to assume these time-consuming roles? For example, being the editor of an established journal confers prestige, whereas being the editor of this new kind of journal may not be seen as having very many extrinsic rewards.

External to each group, there is always a console operator at the computer center to help users in trouble, either by answering a message addressed to "Help"! or over the telephone. There are also "user consultants" who have volunteered to advise or help other members of the system to use the available features.

The overall system monitor is responsible for managing the communication system as a whole, and is the only person who can automatically address messages or announcements to all members of the system.

Besides these roles, the EIES system itself has several built-in features designed to encourage informal discussion or to provide the equivalent of the coffee break at a face-to-face conference. One of these features is the Directory, which lists the name, number, address, telephone number, and a self-entered five-line description of each of the members of the system. Thus, if something written by a "stranger" interests a member, she/he can look up the unfamiliar person's affiliations and professional interests in the Directory.

ISBN 0-201-03140-X, 0-201-03141-8 (pbk.)

There are also several "public conferences," which are open to all members of the system and are meant for letting off steam or joking. By far the most popular in the experimental stage was "Graffiti," which was subsequently removed by National Science Foundation order because the NSF did not want to be publicly identified with sponsoring a way for scientists to have fun (an equivalent might be to ban all cocktail hours at scientific meetings). Among the other public conferences is Suggestions, where users can argue the merits of various modifications to the system.

Advanced Features

To the typical user the EIES system appears to be no more than what is described on the preceding pages. However, there are some aspects of EIES that have been incorporated in order to provide the ability to adapt to a wide range of research applications.

Text, the content of the communications on EIES, can be ordinary words or numbers, or it can contain any mixture of " procedures." These procedures provide a programming language power to the text in EIES. At the simplest level, procedures can be used to record any sequence of operations that a person frequently uses and establish that sequence as the personal command of the user's own choosing.

EIES advanced features allow the user to take advantage of the processing power of the computer, as an integral part of the communication process. (See Chapter 10 on design issues for more detail.) These include such capabilities as the user's developing his/her own personalized command or interface language, designing forms to collect and disseminate formulated information, writing text that allows the reader to indicate whether she/he wants other material (adaptive text), and performing various computations.

HAL: The Computer Who Talks to Computers

Another capability being incorporated into EIES indicates the role a CCS can play in the area of resource sharing. A fairly sophisticated microprocessor with its own computer-controlled telephone dialer has been programmed to engage in the conference system as a full-fledged member, with the same powers of interaction as any human member (Hal Zilog, as it/she/he is referred to). This entity may perform any of the following tasks:

1. It may enter EIES and receive or send messages or retrieve and enter items into the other components of the system.
2. It may exercise certain analysis routines or generate display graphics from data provided by other EIES members, and return the results to them.

ISBN 0-201-03140-X, 0-201-03141-8 (pbk.)

3. It may phone other computers and select data from existing data bases or obtain the results of a model to send back to any designated group of EIES users.
4. It may drop off and pick up communication items from other conference and message systems.

The microprocessor enables a conference system to become a central node to bring together and utilize a host of differing computer resources. An individual knowledgable about a particular computerized information resource now becomes, with the aid of the EIES system and Hal, the transponder for the group as a whole. That is, only one member of the group need be familiar with a particular computer model or data base. That person can utilize Hal as an agent to obtain information and/or data from the resource for the benefit of the group as a whole. The result is a mechanism for a group to produce a collective wisdom or knowledge base. (See Figure 1-2.)

Figure 1-2 Integration of information resources via computerized conferencing systems (CCSs) and human simulator via a microprocessor ("HAL").

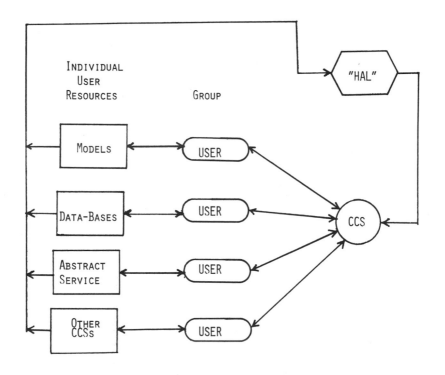

ISBN 0-201-03140-X, 0-201-03141-8 (pbk.)

As a result of the foregoing advanced features, the EIES system is a communication system that is inseparable from the capabilities we normally ascribe to a computer system. EIES provides a base within which individuals and groups can essentially program their own particular or peculiar communication patterns and integrate the analysis and adaptive capabilities they may need.

In the next chapter, we will describe a CCS with a similar user interface but a different purpose and set of communications capabilities. The main point for the reader to keep in mind is that designers and users of these systems have just begun to discover the power, efficiency, and ease of group communication made possible by the computer.

Social and Psychological Aspects

In Chapter 3 we will explore the social dynamics of this form of communication in detail. However, in order to understand computer-mediated communications at all, you must see them as a social process. In this introductory chapter, we will thus review two of its most striking characteristics, impersonality and autonomy. In the Appendix, a more systematic morphology is presented of the underlying differences between this form of communication and the other technologies that are available.

Impersonality and Freedom to Be Oneself

Communications channels are restricted in computerized conferencing, conducing to "impersonality" in serious discussions. Yet, some very warm and personal relationships can be fostered and maintained through this medium.

Computerized conferencing is much less intimate and self-exposing than oral modes of communication in the sense that only your words (which can be carefully considered and edited) are transmitted. Your appearance or other personal characteristics (or handicaps) or other nonverbal cues need not be known. The possibility of sending anonymous messages "legitimately" to other members of the conferencing group, so that even your identity need not be known, facilitates such impersonal communications. Another aspect of this impersonality is that the communicator is alone, rather than in the company of others. Thus, she/he can feel more free to express disagreements or suggest potentially unpopular ideas. In addition, statements may be considered on their merit rather than by the status of their proponents.

Impersonality obviously has potential disadvantages as well as advantages. The "latent functions" of face-to-face meetings, such as fulfilling the social and emotional needs of participants, are not as likely to be met. A person is likely to feel quite detached after several hours alone in a room with a

ISBN 0-201-03140-X, 0-201-03141-8 (pbk.)

machine. As a result, we can expect more reflection and/or introspection on the part of a conferee.

The emphasis in designing CCS's has been to maximize the amount of task-relevant information that can be shared among the members of a group, while still keeping the medium as "comfortable" as possible. Those who are most enthusiastic about the potential advantages of this form of communication tend to focus on this characteristic, as reported, for instance, by Johansen, Vallee, and Collins (1977, p. 3).

> Computer-based teleconferencing is a highly cognitive medium that, in addition to providing technological advantages, promotes rationality by providing essential discipline and by filtering out affective components of communications. That is, computer-based teleconferencing acts as a filter, filtering out irrelevant and irrational interpersonal "noise" and enhances the communication of highly-informed "pure reason"—a quest of philosophers since ancient times.

As the authors add, "Yet, whether computer-based conferencing actually does act as a filter, even in the majority of situations, is open to question. And where it does, some may object." Participants in a conference very soon "reach out" to one another through private messages of a social and emotional nature.

Although the medium seems inherently impersonal, there have been many cases observed or reported by the participants of the most intimate of exchanges taking place between persons who have never met face-to-face and probably never will. Revelations about personal inadequacies, deviant preferences, past love affairs, and serious personal problems that the sender may have told no one else except his/her psychiatrist have passed through the EIES system as private messages to "strangers" who were "met" on the system. What seems to be involved here is precisely the characteristic of the stranger that Simmel pointed out:

> The stranger is close to us, insofar as we feel between him and ourselves common features of a national, social, occupational, or generally human, nature. He is far from us, insofar as these common features extend beyond him or us, and connect us only because they connect a great many people . . . . Objectivity . . . is a particular structure composed of distance and nearness, indifference and involvement . . . . With the objectivity of the stranger is connected . . . the fact that he often receives the most surprising openness— confidences which would be carefully withheld from a more closely related person [1950, pp. 404–406].

Simply because the situation is somewhat impersonal and because one does not have to interact with the others on a face-to-face basis, people can feel free to be extremely frank and open with one another, whether discussing a topic such as a scientific or business problem, or in exchanging information about themselves and their feelings.

ISBN 0-201-03140-X, 0-201-03141-8 (pbk.)

Autonomy: User-Determined Rates and Topics of Information Flow
in Computerized Conferencing

Each participant in this form of communication chooses not only when and where to participate, but also whether to send or receive information at any specific time; at what rate the writing and reading (sending and receiving) will occur; and what topic this communication will concern. Everyone can "talk" or input whenever they wish, rather than having to "take turns" as in face-to-face communications. No one can be interrupted or shouted down. With no norms about "sticking to the subject," participants tend to develop several different topics or ideas at once and reading the transcript can be confusing. A question may be asked in, say, statement number 119, and an answer may not appear until entry 130 or even 150.

However, this general characteristic may also be modified by a combination of software and leadership structuring. For example, a conference moderator may force a vote or a response on a particular item before allowing a participant to read further into a conference, and may delete items that seem too far off the announced topic.

In the face-to-face communication mode, a person can and must respond immediately to communications received, or it is "too late." (A subject may be raised again later, by such a referent as "I just thought of something relevant to the matter we were discussing last week ... " but most of the context of what was said last week has been lost). There is immediate feedback of reactions or responses from the recipients, which in turn tends to stimulate further communication on the topic; or if disapproval or disinterest is indicated, to terminate the discussion or turn it to a new approach or topic. Each participant must take part at the same time and place and speed as the other members in order for this immediate feedback process to occur.

In the computerized conference, unless consensus-testing devices such as voting are used, a participant tends to be much more self-driven. Even if there is a synchronous (simultaneous on-line) conference, other participants may not receive and respond to a question or other communication for five or ten minutes or even longer, since they are not interrupted with incoming messages while they are composing something. Each person can read and fully review items at his-her own pace before responding or deciding to ignore them. However, there are tremendous elements of frustration involved in this lack of immediacy. Often, a person wants an answer or response *now*, and the minutes go by and nothing happens. Or hours are spent thinking about and polishing an entry into a conference. Then, after all this investment of time, it is ignored or misunderstood or criticized. Frustration can mount to anger in such cases, or to a resolution not to ever again make such a "useless" effort.

In addition, there are certain persons whose work style or workload has led them to be "interrupt-driven." That is, they do something only when an

ISBN 0-201-03140-X, 0-201-03141-8 (pbk.)

impending deadline or ringing telephone forces them to respond. Such persons may never seem to get around to participating in a computerized conference on a regular basis—since they are never forced to spend a *particular* period of time, as in a scheduled face-to-face meeting.

To some extent, rules and procedures established by a conference moderator can help to overcome a lack of self-pacing discipline. For example, deadlines for responses to items can be set; explicit expectations can be stated on frequency and amount of time devoted to participation, with a moderator working with a participant to agree to some regular time to be set aside for participation; synchronous conference sessions can be called once a week or so. The problem is that all such attempts at scheduling participation also take away from the advantages of the self-pacing aspects of the medium.

To summarize our observations, many persons who have not tried computerized conferencing shudder at the image of a Frankenstein's monster-like computer system conjured up by mad scientists to interfere in something so "human" as communication. The basic point of view we adopt is that what we have is more of a Dr. Jekyll and Mr. Hyde situation, in which the same basic characteristics have both a good side and a bad side, depending on the context in which they are used. Moreover, as will be explored further in Chapter 3, what may be seen as an inadequacy of this medium at first may later be overcome by participants' learning how to substitute for missing kinds of cues.

Summary

By computerized conferencing we mean any system that uses the computer to mediate communication among human beings. Changes in computer and communication technologies and the move to a complex information-based society have made this form of communication the cheapest, most convenient, and potentially most powerful option for geographically dispersed groups of people who must regularly exchange information and opinions.

For Discussion

1. How many different types of human group communication processes can you think of that might lend themselves to a specific form of computerized conferencing?
2. How well do you think a person can get to know another by communicating via computerized conferencing?
3. What initial impacts might be foreseen for the use of computerized

ISBN 0-201-03140-X, 0-201-03141-8 (pbk.)

conferencing by teenagers? What impact has the telephone had on this group, and how would you characterize the differences?

4. What problems can you foresee for the publishing industry—news-papers, books, journals, etc.?

5. What laws and regulations might need modification and what new problems might emerge in this area?

6. What might be the impact of the use of computerized conferencing on the use of language in written form?

7. What possible names would you propose for what this book is calling computerized conferencing? What connotations do those names convey to individuals of different backgrounds? What image does the word "computer" convey?

8. What are the consequences you perceive of a computer in every home?

9. How might members of a family utilize computerized conferencing?

10. Can you give some job descriptions for new occupations that might emerge as a result of the merger of communications and information technology?

11. What impacts have telephone and TV had? For those specific items, how is computerized conferencing likely to enhance, modify, or change the situation?

12. Should typing be compulsory in junior high schools in order to make sure that all adults in the future will be able to use computer terminals without difficulty?

13. What problem situations would you hypothesize as being difficult to computerize? In what situations would you be willing to rely totally on a computer model or analysis to make a decision for you?

14. Can you think of examples in which computer results have been utilized as a crutch for making a decision in order to remove the responsibility from the decision maker?

15. In what situations would you prefer to write to someone you had never seen or talked to?

16. In what situations would you like to use anonymity or pen names?

17. In what educational experiences is peer communication the significant factor, as opposed to things like drill, instruction, or reading reference material?

ISBN 0-201-03140-X, 0-201-03141-8 (pbk.)

APPENDIX

A Communications System Morphology

It is useful to establish a comparative framework that will serve to describe the unique physical properties and parameters of CCSs, since these differences in the physical parameters directly influence their impact on human behavior. The systems that are available for groups to employ in communicating are

Face-to-face meetings
Telephone (conference calls)
Video conferencing (TV or Picturephone-like)
Mail, telegrams, TWX (carbon copies)
Delphi (via paper, pencil, and mails)
Electronic mail (carbon copies and topic association)
Computerized conferencing
Citizens band (CB) radio

Table 1-4 depicts the differences in these forms of communication by specific physical parameters that appear to be the fundamental measures of the communication process. Many of these physical parameters are in fact associable into groups that serve more as a Gestalt measure. We will take each parameter separately in order to clarify this.

Medium

The first characteristic of these systems is the medium of communication, which basically breaks down to the written or the spoken word, with or without nonverbal cues and graphic or drawing capabilities. Most of the computer-based systems do not yet incorporate graphics because of the cost of graphic terminals. This then still provides certain advantages to the mail and specialized systems such as facsimile transmission. However, the costs for graphics is declining and it is really only a matter of time until this constraint is lifted.

It is very peculiar that we do not have a decent understanding, in a scientific sense, of the relative advantage in different situations of the written or the spoken word. However, anyone you ask is certain they know the answer. That any two people usually have two different views does not usually seem to perturb anyone.

ISBN 0-201-03140-X, 0-201-03141-8 (pbk.)

There is, though, an intuitive correlation between the complexity of the topic the group is handling and the need for written materials to support even a face-to-face environment.

Size and Structure of the Group

The technology governs the size of the group that can effectively communicate. Face-to-face is one of the most versatile situations since it can accommodate a few people to many hundreds of people. The difference in group size is really dependent on the protocols or "structure" agreed to by the group in order to conduct the meeting. Some of these alternative structures are:

Free discussion
Committee meetings with chairpersons
Parliamentary meetings (*Robert's Rules of Order*)
Nominal group processes
Seminars or formal conferences
Debates
Classes
Therapy and sensitivity sessions
Games

In general, as the group size increases, a higher degree of structure or regulation is needed in order for the communication process to proceed smoothly in a face-to-face situation. However, structure is also imposed on smaller groups in order to deal more effectively or efficiently with the objectives of that group, as is the case of nominal group processes or games.

Telephone and video conferencing, as well as CB radio (which is a form of audio conferencing because of the very limited structural variation allowed by the technology), impose very severe constraints on effective group size. This is also true for systems designed on the mail model, which although they can be very versatile in the electronic form, usually impose a single preferred structure. However, the introduction of the computer has the effect of removing the burden of imposing the structures from the group and essentially allows the computer to take over the bookkeeping on what is taking place. It also allows, in the CC approach, the ability to adapt as needed to a host of structurally different options made available by the logical processing ability of the machine. As a result there is the potential for much larger groups to communicate effectively via the computer than through the other media.

ISBN 0-201-03140-X, 0-201-03141-8 (pbk.)

Table 1-4
A Communications System Morphology

System parameter	System type			
	Face-to-face	Telephone	Video (TV) conferencing	Delphi
Medium of transfer	Verbal and nonverbal	Verbal	Verbal and nonverbal	Written word, graphics
Effective group size limit	Free discussion: few tens Structured: hundreds	Free discussion: Few Structured: Few tens	Few tens	Hundreds
Occurrence of interaction	Coincidence of all	Coincidence of all	Coincidence of all	Individual choice
Length of interaction	Minutes to hours	Minutes to hours	Minutes to hours	Minutes to hours
Frequency of interaction	Predetermined	Predetermined and chance	Predetermined	Individual choice
Speed of interaction limit	Talking rate	Talking rate	Talking rate	Reading speed
System delays	None	None	None	Days to weeks
System memory	Recordings	Recordings	Recordings	Hard copy
Memory modification	None	None	None	None
Memory retrieval	None	None	None	None
Memory size	Original	Original	Original	Book size
Transformations	Transcription	Transcription	Transcription	Copying
Structure	Varied but fixed	Single and fixed	Single and fixed	Varied but fixed
Skills	Speaking ability and mannerisms	Speaking ability	Speaking ability and mannerisms	Writing ability

ISBN 0-201-03140-X, 0-201-03141-8 (pbk.)

ISBN 0-201-03140-X, 0-201-03141-8 (pbk.)

Table 1-4 (*cont*)

System Parameter	System type			
	Mail, telegrams, TWX, mailgram, etc.	Electronic mail	Computerized conferencing	CB radio
Medium of transfer	Written word	Written word, graphics	Written word, graphics	Verbal
Effective group size limit	Few tens	Many tens	Free discussion: Many tens; Structured: To thousands	Tens (at one time)
Occurrence of interaction	Individual choice	Individual choice	Individual choice	Time coincidence
Length of interaction	Minutes	Minutes to hours	Minutes to hours	Minutes to hours
Frequency of interaction	Individual choice and change	Individual choice	Individual choice	Individual choice
Speed of interaction limit	Reading speed	Reading speed	Reading speed	Talking rate
System delays	Hours to weeks	None	None	None
System memory	Paper files	Electronic	Electronic	Recordings
Memory modification	None	Unlimited	Unlimited	None
Memory retrieval	None	Unlimited	Unlimited	None
Memory size	Personal files	Unlimited	Unlimited	Original
Transformations	Copying	Hard copy	Hard copy	Transcription
Structure	Single and fixed	Single and fixed	Dynamic and adaptable	Varied and fixed
Skills	Writing ability	Writing ability	Writing ability and typing	Speaking ability

The concept of structure is really at the heart of the distinction between the original computerized conferencing concept and the current popularity of the term "electronic mail." The working hypothesis for computerized conferencing is that the communications process should be tailored to the particular group and its application, if it is to be optimum for their use. The electronic mail concept is inherently based on the philosophy that there is one generalized system that is ideal for everyone's communications, and which should be a single uniform nationwide service utility. Later we will discuss the policy implications of these distinctions. In actual practice, there is very little distinction between a simple free-discussion CC system and an electronic mail system that allows messages to predefined groups to be stored and categorized by topic for later retrieval.

However, more sophisticated systems such as EMISARI (Emergency Management Information System and Reference Index) and EIES allow for a multitude of different structures for utilization by the groups involved and allow at least some degree of modifications to the structures to take place dynamically by user and group choice. One area that has received little attention to date, but which we expect to see more of in the future, is the concept of adaptive structures in which the computer can be programmed to make certain decisions on the nature of the structure, based on individual and group behavior. A trivial example of this decision making is not allowing a conferee to add any more comments to a particular conference if she/he has exceeded a certain threshold of authored comments relative to those put in by the rest of the group.

One must be careful not to look upon structures implemented in a CC environment as being completely analogous to their face-to-face counterparts, since the technology potentially modifies the rules of the structure. For example, in *Robert's Rules of Order* a "point of order" is often abused, since the audience can hear it before it is ruled out of order by the parliamentarian. However, when done through a CC version of *Robert's Rules of Order*, it could be passed to the parliamentarian before being allowed to go on to the group, and deleted if ruled out of order. Similarly, anyone in such an on-line parliamentary proceeding could be discussing any of many different motions or amendments, and the computer would deliver the items to the group only if the particular motion they were associated with had been brought to the floor. In other words, the computer would resequence the text items around the rules of order regardless of the order in which individual members of the group entered them. This then would provide a very different atmosphere from the one that exists in face-to-face parliamentary sessions. A similar example would be judicial proceedings where a judge rules, "Objection

ISBN 0-201-03140-X, 0-201-03141-8(pbk.)

sustained. The jury will ignore the witness's answer." (Sure!) In a CC version, a jury would never receive an "out of order" comment.

Size of the group and optimal communication structure are thus very intimately related, and computerized conferencing is very analogous in this sense to face-to-face and Delphi-like processes, whereas electronic mail is closer in objective to what we now perceive as mail and phone.

Space and Time

One very significant difference among these systems is the necessity for coincidence either geographically and/or temporally in order for the members of a group to be able to communicate. Obviously the telephone had its widespread influence as a result of being able to break down the geographical coincident constraint without the delays associated with mail. However, telephone still requires people to be available at the same time. The introduction of the computer maintains the lack of a system delay but allows individuals the freedom to choose their own time of interaction as well as the option to be there at the same time when that is felt to be desirable. This alone is a convenience factor that will lead to widespread use of both electronic mail and CCSs.

Control and Self-Activation

The written word systems allow individual control of the interaction in terms of when the individual interacts, at what rate (that is, how fast she/he reads or writes), and for how long. This factor introduces very noticeable differences in the behavior of individuals on these systems as opposed to participants in face-to-face or verbally oriented systems, where only one person at a time can talk and everyone else must at least appear to be listening at the same rate.

Those written systems that exhibit system delays, such as mail, do not allow the same degree of individual selectivity as computerized conferencing or electronic mail, since the person cannot choose the option of getting the material any faster than the length of the system delay.

There are many situations where the freedom of the individual to choose his/her own time and rate of interaction and what she/he wishes to do during that interaction can be quite crucial to the success of the communication process.

ISBN 0-201-03140-X, 0-201-03141-8 (pbk.)

Speed

Spoken word systems cannot move any faster than the average talking speed of an invidual in the group, whereas written word systems can move at the average reading speed of the individuals in the group. For the computerized systems this fact can represent anywhere from a two- to a tenfold increase in the throughput rate of the material being exchanged among the members of the group. For very small groups of a few people the rate will be more sensitive to the average typing speed, but as the group increases in size the throughput rate approaches asymptotically the reading speed or the print speed of the technology delivering the written word, whichever happens to be least. The impact of this increase of throughput for a human group can be quite startling for its members and sometimes a bit mind boggling. It also seems to impact on the sense of time passing, for at least some people who use these systems.

Memory

The only communication systems that allow a group as part of its group communication process to modify, update, reorganize, and reclassify what has transpired as an integral part of the communication process, with members automatically kept informed of such changes, are the computer-based systems. These systems introduce a "common" memory for the use of the group.* Although individuals may tape recordings of verbal discussion or gather mail correspondence files, there is a tremendous effort and significant delay involved in providing these materials to the other individuals in the group. The common or shared memory provided by the computer, its ability to manipulate this memory, and its potentially large size open the door to fundamentally new types of human communication processes. This is very evident in the EMISARI and EIES examples, where human communications can become associated with data and analysis tools provided by the presence of the communication system within the environment of a computer system. Of course, a very pragmatic aspect of the memory is that a new person can join an ongoing conference, and merely by reading what has taken place, be brought up to date. This includes, for instance, a person who has been away on vacation for two weeks.

The concept of memory begins to provide these systems with some of the same roles that libraries have played in society. Just as libraries represent the accumulated wisdom of the society, so computerized conferencing provides a

---

*A noncomputer analogy would be the *Congressional Record*, which is a "group memory" with revision privileges.

ISBN 0-201-03140-X, 0-201-03141-8 (pbk.)

Table 1-5
Summary Comparison Table

	Medium			
	Oral		Written	
Single structures		Telephone Video (TV)	Electronic mail	Mail, telegrams, TWX, etc.
Varied structures	Face-to-Face	CB radio	Computerized conferencing	Delphi
Control	Group	Group	Individual	Individual
Speed	Talking rate	Talking rate	Reading rate	Reading rate
Time coincidence	Necessary	Necessary	Not necessary	Not necessary
Geographical coincidence	Necessary	Not necessary	Not necessary	Not necessary
System delays	None	None	Some	Considerable
Memory	Separate	Separate	Integral	Separate

group the opportunity to accumulate a wisdom and/or folklore, and to be able to build from that point, as opposed to the time necessary to review and update that characterizes verbal processes. It is possible with such a memory for a person to enter a conference discussion late and not lose anything. Even if face-to-face discussions are transcribed, a newcomer who reads them loses all of the nonverbal content.

The result of these discussions allows us to formulate Table 1-5, which summarizes the principal physical parameters that characterize the differences among these communication systems.

ISBN 0-201-03140-X, 0-201-03141-8 (pbk.)

THE BOSWASH TIMES

Government Employees Edition

Boston, Mass. Mr. Michael Flattery, employed by the Department of Health, Education, and Welfare as a GS-8 Information Clerk, is suing the U.S. government for recovery of two weeks pay recently docked from his salary as a result of a Civil Service Commission hearing. Mr. Flattery is authorized to work from a computer terminal at home an average of nine out of ten days. In December Mr. Flattery took his terminal with him to the Virgin Islands and continued to function in his job, footing the added communication costs out of his own pocket. HEW claims that his job description uniquely specified the phrase "work at home," which is interpreted to mean that while in the Virgin Islands Mr. Flattery was in an unauthorized leave or vacation status.

Mr. Flattery's attorney has stated in his arguments to the Civil Service Commission that the phrase "work at home" is a colloquialism which does not legally pinpoint a specific location and that, furthermore, the regulations for work-at-home status do not follow the intent of Congress when they attempt to constrain the employee to perform the work at a specific location. Ethically the only issue should be whether or not the work is actually performed. Mr. Flattery claims he was better able to perform his job in the Virgin Islands, considering the oil heating rationing and the fact that his home temperature at that time was 60 degrees. In a separate petition before the Commission, Mr. Flattery has requested that he be allowed to retain his job with the Boston office of HEW, even though he is planning to move his "home" to Florida.

Needless to say many thousands of federal employees are anxiously following the progress of Mr. Flattery's struggles with the bureaucracy.

Washington, D.C. Congress is again debating a raise from 1-1/2 hours to 2 hours in the minimum time compensation law for those primarily employed via computer terminals. This law, of course, specifies the minimum number of hours of pay a person is entitled to receive for each hour spent on the terminal. While the president has stated that any increase at this time would be inflationary, the legislation has considerable union backing and is given favorable odds for passage during this session of Congress.

Washington, D.C. The Congress today passed into law the new Civil Service Act, which allows 30% of all civil service slots to become tied to individuals as opposed to agencies.

ISBN 0-201-03140-X, 0-201-03141-8 (pbk.)

Those lucky enough to obtain these slots will be free to work for any agency inviting them to do a job. Congressman Rutcku expressed the opinion that Congress was merely legalizing what had become common practice. He said: "With computerized conferencing in use on a government-wide basis, the significant work is really getting done by dedicated individuals in team efforts which cut across all agencies. It's about time we freed these individuals from bureaucratic red tape. We foresee a much more effective federal bureaucracy as a result."

Another Congressman expressed the view that certain agencies are going to suffer terribly because no intelligent person would want to work for them. However, in the long run this will force them to clean up their shops and give hardworking, dedicated people something meaningful to do.

ISBN 0-201-03140-X, 0-201-03141-8 (pbk.)

any of the standard terminals in use today. Within a few easy lessons you will be able to calmly and assertively engage a terminal in full view of your peers. At no additional cost, we supply our proven guide book, "Terminal Manners and Etiquette," which covers such topics as how to recover smoothly from an error; laying the blame on the system; computer jargon with which to impress people, etc.

ISBN 0-201-03140-X, 0-201-03141-8 (pbk.)

CHAPTER TWO

Development and Diversification of Computerized Conferencing

A computer-mediated communication system for dispersed human groups was first designed and implemented in 1970, at the Office of Emergency Preparedness (hereinafter OEP) of the Executive Office of the President of the United States. Since that time, a variety of computerized conferencing and related systems have been designed and implemented, and we have begun to understand the opportunities, limitations, and issues that are raised when the computer is used to facilitate and structure complex human communication processes. In this chapter, we will review the development of computerized conferencing as an idea and as an operational reality. A complete case history will be presented on the nature and evolution of the EMISARI system, which was the first full-scale conferencing system and is still a unique form of management information system for crisis management. This account will emphasize the organizational context that was an integral aspect of the development of the technological innovation. We will then briefly describe other notable efforts that have been completed at such organizations as the Institute for the Future and Bell Canada. Overall, some half dozen systems are in use today in tens of organizations serving a few thousand individuals.

The kinds of needs, applications, and technological developments that we feel will result in exponential growth of this new communications technology will be described in subsequent chapters.

Collective Intelligence?

To comprehend the basic purpose of computerized conferencing systems (CCSs), we must first raise what appears to us the philosophical or meta issue underlying the attempt to structure communications within a group:

Starr Roxanne Hiltz and Murray Turoff, The Network Nation: Human Communication via Computer

ISBN 0-201-03140-X, 0-201-03141-8 (pbk.)

Is it possible to conceive of a collective intelligence capability for a group of humans?

Is it possible for a group of humans utilizing an appropriate communication structure to exhibit a collective decision capability at least as good as or better than any single member of the group?

Though these are well-defined questions that can be addressed by suitable investigation, there have been no definitive answers. It is an example of a value or belief that underlies our faith or nonfaith, as the case may be, in things like the democratic or the committee process.

There are some experimental studies of group problem solving and risk taking which address the question, which can solve problems better, individuals or groups? The evidence indicates that the answer depends on the nature of the task, the social and communications structures that develop, and a number of other factors.

For so-called "insight" problems for which there is a single indivisible task and a correct answer, groups seem to perform at the level of their best member: if they contain a single member who can solve the problem, then they are likely to solve it. (See, for instance, Marquardt, 1955; Faust, 1959.) However, there is often loss: Even some groups containing such individuals may not reach a correct decision, because the individual either does not bring up the correct solution, or his suggestion is argued down. On tasks involving a great deal of division of labor and coordination in a single group effort, groups (especially large ones) often cannot "get it together" and end up being unable to accomplish the task at all, or performing at the level of their least able member. For example, McCurdy and Lambert (1952) found that on "problems requiring geniune cooperation," groups were inferior to individuals, because "the less alert and less interested individuals will always interfere to some extent with the progress of the group [p. 492]."

Looking over the many kinds of group or individual problem-solving experiments that have been conducted, we would agree with Davis (1969, p. 38) that

> The overall conclusion is that groups are usually superior to individuals in the proportion of correct solutions (quality) and number of errors, but somewhat less often are groups superior in terms of time required to reach an answer,

especially if we compute the number of person–minutes expended rather than the elapsed time from problem presentation to solution.

A basic factor at work in producing the general superiority of small groups over individuals for most kinds of problems was noted as early as 1932 by Marjorie Shaw. Whereas an individual is not likely to recognize and correct an error, group members are likely to recognize and reject errors made by others.

ISBN 0-201-03140-X, 0-201-03141-8 (pbk.)

Davis (p. 40) sums up the various processes and advantages working in favor of the group:

1. The group potentially can increase performance through redundancy. That is to say, if the problem requires that everyone work at the same thing and if individual performance is to some degree unreliable (i.e., some probability of error exists), then multiperson work by means of duplication provides a check on the quality of the group's output.

2. If each person possesses unique but relevant information, and the task requires the several pieces of information, then the pooling of this information will allow groups potentially to solve problems that an individual cannot attack successfully.

3. If the task may be broken into subproblems, then different group members may simultaneously work at different portions of the task. This strategy accelerates work and allows early responders to check the work of the slower persons.

4. In quite a different way, questioning and debating during social interaction may stimulate new or different intra-individual thought processes that the uniform environment of the isolated individual might not provide; thus other persons have a cue value in provoking new task approaches.

5. Finally, the mere presence of others is known to be motivating, and thus is an advantage for some tasks. Moreover, groups mediate a number of appealing by-products, ranging from status to plain fun, that have nothing to do with task performance, but which serve to keep one working.

The Delphi Method

Chapter 10 will discuss various structures or methods that have been designed to facilitate group processes in more detail. For the purposes of this chapter we wish merely to help you understand how computerized conferencing grew out of efforts to conduct Delphi studies, which are one form of structured communication.

The Delphi is a method designed to structure group communication in such a way as to attempt to capitalize on the strengths and minimize the weaknesses of collective problem solving. It derives its name from the Greek oracle at which the future was believed to be foreseen, because the method has most frequently been used to structure the discussion and forecasts of experts or representatives of different points of view who are trying to project the future impacts or likelihood of some specific developments. (For example, Delphis have been used to project the nature of the insurance industry in 1985, and of the desirability and impact of new life-extending technologies in bio-medicine).

The purpose of the Delphi technique is to generate and collate informed judgments on a particular topic. The participants should not be limited to "experts" in the usual sense. If the topic is welfare, for instance, then at least

ISBN 0-201-03140-X, 0-201-03141-8 (pbk.)

some of the persons included should be poor persons who are on welfare and poor persons who are not on welfare. As a communication structure, it has two key characteristics. The first is *feedback*, which means that the opinions and arguments of the participants are presented to the group so as to facilitate the exchange of information and views, and their input into a subsequent round of discussion and voting. The second is that all responses are usually (though not necessarily) reported anonymously.

The Delphi is thus a method for developing and reporting in an organized manner group judgments on a topic through a set of carefully designed sequential questionnaires, interspersed with summarized information and feedback of opinions derived from earlier responses. (See Linstone and Turoff, 1975.) One of the main disadvantages of a mailed paper-and-pencil Delphi is the amount of time that is consumed in administering all the rounds. The very first CCS was a Delphi conference, designed and implemented by Turoff in 1970 (see Turoff, 1972a). This conference involved 20 persons spread around the United States and was conducted over a period of 13 weeks. It provided a common discussion space that each participant could enter in order to:

- View all new discussion items entered by members of a conference;
- Vote on any or all of the discussion items according to evaluation scales automatically provided by the computer;
- View the vote on any or all items once a specified number of votes were in.

As a CCS, thus, the Delphi conference program was fairly rudimentary, but it did serve to speed up the Delphi process and it contained the kinds of general capabilities that might be modified to permit conferencing for another purpose. It is from this original Delphi conference effort that computerized conferencing springs, and therein lies a story.

EMISARI: A Case History of a Technological Innovation[1]

The development and application of a technological innovation is often a coincidence involving "the right people in the right place at the right time." Our history begins in the late 1960s when the Office of Emergency Preparedness in the Executive Office of the President formed a Systems Evaluation Division (SED), to conduct operations research type efforts.

---
[1]*Acknowledgments*

SRH is indebted to the following individuals associated with the Office of Emergency Preparedness and its successor agencies for interviews and written materials on the development and implementation of EMISARI: Robert Kupperman, Richard Wilcox, John McKendree,

ISBN 0-201-03140-X, 0-201-03141-8 (pbk.)

Over its six years of existence, SED's accomplishments included

1. The first nonlinear network optimization algorithms and programs for the construction of gas pipelines;
2. The first nonlinear network optimization of the Federal Telephone System;
3. The OEP Energy Conservation Study in 1972;
4. A major Delphi forecast of the steel and ferroalloy industry;
5. The first policy Delphi;
6. Delphi conferencing, computerized conferencing, and the EMISARI development.

It is the last three items that are primarily of concern here.

At this point the origin of the Delphi Computer Conference will be related from the point of view of Murray Turoff. Later, we will turn to accounts from other participants.

I joined SED in 1968 and was principally involved during the first two years with the Delphi work of the division . . . .

It was quite obvious to a number of people working in the Delphi area that a major improvement could be made if Delphis were conducted via computer terminals. At the same time, OEP's Computer Laboratory had hired Language and Systems Development Corporation to help out with system software problems on the Univac 1108. In 1968 I held a number of meetings with the Language and Systems Development people who were developing a higher-level language for the 1108 called XBASIC. Without any formal arrangements between LSD and OEP, I suggested a number of requirements and modifications to their language design that would allow for the programming of com-

---

Murray Turoff, Nancy Goldstein, Paul Bryant, Richard Scravette, and Joseph Minor. The principal contributors to the total EMISARI effort were (all affiliations with OEP unless otherwise noted):

Supervisory Member of Executive Staff	Robert Kupperman
Project Manager, System Architecture	Richard Wilcox
System Designer, Programming Manager	Murray Turoff
Operations Monitor	Nancy Goldstein
News File Manager	A. Lee Canfield
Exceptions/Exemptions File Manager	Albert Martin (Naval Research Lab.)

Programming Consultants

Thomas Hall, (Language and Systems Development, Inc.)
George Olmacker (Language and Systems Development, Inc.)
Ruth Anderson (National Bureau of Standards)

Programmers (OEP Mathematics and Computation Laboratory)

Rodney Renner	Robert Bechtold
Ronald L. Wynn	Charles Clark
David Marbray	

ISBN 0-201-03140-X, 0-201-03141-8 (pbk.)

munication structures, and LSD incorporated this into their design. As a result, in late 1969, there was a software capability available to automate a Delphi Process.

Since the previous Delphi work had been somewhat successful in terms of supplying results, and with a practical objective of potentially making quicker use of consultants for a short-fuse policy study, I proceeded to implement an automated version of the earlier paper and pencil policy Delphi structure. It was decided at the time to do this as an effort no one else outside of SED would be informed of, unless of course they asked. The reasons for this were manyfold:

1. There was no assurance that it could be done, and organizational politics are such that a failure no one knows about is the best kind of failure.
2. The computer shop in OEP felt that its function was threatened by SED, since people in that group could program. There were various regulations they had tried to promulgate that would assure that all programming had to be done by them, except very elementary things. And this entailed the usual protracted approval process as well as the difficulties of not being able to choose the programming talent.
3. Problems still existed with the 1108 software, one of which was that they could not tell who was doing what from any of their remote terminals. This was their problem, our advantage.

As it turned out, the system was implemented in about six months of unofficial effort, sandwiched in between tasks under way. At this time Dr. Kupperman approved an informal experiment with a few people knowledgeable about Delphi, computers and potential applications of the technology. I promptly invited 20 fairly well known people to participate in an on-line Delphi utilizing the 1108. Once again the informal experiment was not advertised outside of SED.

There were individuals in OEP who were sympathetic to what was going on, even though they were in that part of the organization involved with the management of the Computer Laboratory and the resource data bases that OEP maintained. Most notable among these were Stanley Winkler and Richard Wilcox, who have in their own right made notable contributions to the computer field, and were very quick to recognize the potential of what was taking place. When and where they could, they also ran interference, and Wilcox participated in that first "informal" experiment.

The experiment started in the Spring of 1970 and it went on for seven weeks of on-line discussion before the rest of OEP became aware that something was going on. This occured because the computer console operators began to talk about people from all over the country who would occasionally phone to find out some piece of information about hours and such things. This gradually filtered up the ladder until someone decided to ask what was going on. At this point the history becomes interesting, as the very first signs of overt organizational resistance emerged.

ISBN 0-201-03140-X, 0-201-03141-8 (pbk.)

The computer shop had not been informed about this use of its computers; none of the proper approvals had been obtained. After about seven weeks of the trial, they discovered that there were a lot of people accessing its computer from outside the organization and doing something with it that had not been programmed by them. They got a bit upset and conducted a complete investigation to see if government resources were being misused.

Kupperman was able to protect the experiment's continuation until the end, but as a consequence of the resistance encountered, Turoff had to be "punished" in a way that uniquely fit the "crime." His terminal was taken away for several weeks.

This seemingly bizarre incident was actually very symbolic of the internal strife that accompanied the development and implementation of the full scale conferencing system called EMISARI that eventually evolved from the Delphi conferencing program, under the impetus of a management information crisis set off by the wage–price freeze.

The Delphi conference and its descendant, the EMISARI computerized conferencing system, represented a highly innovative technique for using a computer to structure human communication for information exchange and collective effort to solve a problem. It is not enough to have the technological capability to design and use such a system. In order for this to happen, there first has to be an organizational *need* (that is, existing communication must be seen as severely inadequate for accomplishing an important task.) Second, there needs to be strong management support. It is this second crucial component, the organizational situation and the management support for a completely new approach in the face of inevitable resistance from the status quo, which will occupy a considerable amount of our attention in this chapter. This is because we believe that any attempt to introduce a computerized conferencing system into an entrenched bureaucracy will encounter similar kinds of resistance, and the manager who is providing the financial resources for the development and implementation of the system will need considerable organizational strength in order to ensure the survival of such an innovation through the development and implementation period. After all, we are trying to change one of the most habitual of all human activities, the way in which people communicate. By changing the communication channels in an organization, we are also threatening the power of those whose positions in the organization depend on control over the generation and dissemination of the information that flowed through these channels. Such an innovation is thus likely to be seen by many as a usurpation, a threat.

The Organizational Context

The attempt to halt the trial of the Delphi conferencing system was only the most manifest sign of the organizational resistance that surrounded the development of the Delphi and EMISARI conferencing systems. The entire

ISBN 0-201-03140-X, 0-201-03141-8 (pbk.)

effort could not have occurred without a kind of protective managerial support for innovation that is very unusual in a government bureaucracy. The Systems Evaluation Division of the OEP was such an environment. Its manager had been able to carve out an organizational mission to innovate that was legitimized by some tremendous successes at the top levels of OEP. Asked to describe his view of the purpose of the Delphi conferencing trial and the context in which it occurred, Robert Kupperman (personal interview, April 20, 1977) recalls

> In pre-wage–price freeze days, I felt that activities that were not directly related to coming up with instant analyses of one page or less, those which were traditional for White House Operations, were regarded with suspicion. As far as much of OEP's administration was concerned, we were an expensive addition, but they were fearful of getting rid of us. I think we were held at bay until the moment we could do something spectacular for them.
>
> Our status has to be looked at from the perspective of a number of the things the operation did before Turoff began dealing with Delphi and computer conferencing. We had a history which included saving six hundred million dollars in one fell swoop for an investment of only $100,000, because of an application of network analysis that we did. This gave us a considerable amount of notoriety at the tail end of the Johnson administration and annoyed everybody in the energy business, except those who were consumer advocates.
>
> At the same time, we had done some useful work in the strategic field. At the time, the head of OEP [General Lincoln] was one of the five statutory members of the National Security Council. I had helped Lincoln considerably at the very beginning, and so did Murray. There was a memorandum that made an impression on Kissinger. The details are classified, but it dealt with the question of should we or should we not have an ABM system. We had rather strong opinions. We had all come from defense, and we knew as much as anybody. We were able to put Lincoln in a useful intellectual position. Kissinger, commenting on Lincoln's memorandum, said that it was foresighted.
>
> *SRH*: To summarize, then, you had some good credits with Lincoln for being able to come up with unique and useful analyses; and suspicion from everybody else that you might upset the applecart. They were not quite sure what your organization would come up with.
>
> *Kupperman*: Yes, I set up the "schizophrenia" of the operation; the normal paranoia that affects all institutions. There was a significant intellectual rapport and fondness between Lincoln and myself. The palace guard who were around Lincoln were of my rank but not in a line capacity, and were jealous and very fearful of anything that I might do—a reputation of being willing to take on dragons. My fight was never with Lincoln, it was with the palace guard.
>
> We were able to get Murray computer time, etc. I was protecting Murray from the palace guard which was seeking to get him fired—a symbol of my vulnerability. Everyone questioned Murray's worth and I objected strenuously to the notion that a professional interested in doing some relevant, creative work

ISBN 0-201-03140-X, 0-201-03141-8 (pbk.)

should be blocked on petty grounds. Murray's work was valuable on personal, professional, and symbolic grounds. And so the fight went on vigorously.

What happened when the furor over the unauthorized use of the computer for the Delphi conference first broke was my plea of ignorance on procedural grounds, but I endorsed it thoroughly. One issue that they raised was improper use of computer time—I felt that it was being misallocated generally. If the issue were to be taken up as a flagrant violation, then I said that I thought that the Office of Management and Budget should be deeply involved in the matter and I would personally be happy to bring the GAO in as well.

SRH: So your reaction strategy was to call for a general investigation of everything, rather than just this one example?

Kupperman: That's right. The only infraction, if any, was a procedural one dealing with authorized account numbers. On the other hand, I argued the case that I felt that many of the issues were matters of personality, that there was jealousy and competition. And I argued the case of the computer division's incompetence in front of Lincoln and the palace guard, who were afraid of everyone on the outside.

On the other hand, the matter wasn't difficult to resolve if Murray were punished. I punished him summarily. I took Murray's computer terminal away for awhile—his lollipop. It was an important gesture to perform. And Murray scurried around—by the next day he had found another computer terminal to use. It really didn't matter.

This is the organizational setting that provides the context for the emergence of EMISARI: jealousy, power struggles, rival camps in the bureaucracy engaged in internal warfare.

Though the Delphi conference was designed for a specific, limited purpose, the software provided a basic conferencing capability that could be modified to fit other conferencing situations. A group of dispersed persons was able to add items to a permanent, mutually accessible written transcript, which could be accessed, reviewed, and added to at any time by any of the participants. The need for this capability arose with the assigning of the responsibility for monitoring the wage–price freeze to the OEP, and the result was the evolution of EMISARI as an interactive management information system.

To return to Kupperman's account:

Dick Wilcox and Murray and I spoke about it, and I said the hell with everything, go ahead. I don't want to stop the computer people from developing their own management system that I know will not be developed in thirty-five years, let's see if Murray can do something. We let him go ahead and steal what he could in programmer help, etc . . . .

All the bureaucrats were shuffling around to get a piece of the action. A very obvious thing occurred that everyone knew would happen but no one would ever admit. All of the efforts that went into planning to handle such a freeze—there was a whole division set up for it—disappeared in about a week.

ISBN 0-201-03140-X, 0-201-03141-8 (pbk.)

The extent of genuine fear was vast. We found ourselves in a totally pivotal position in the proverbial sense that "fools go in where angels fear to tread" . . . .

SRH: So you are the kind of manager who will take risks, who went ahead and told Turoff to design this new system without anybody giving you permission for the effort?

Kupperman: That's right.

SRH: And at the end of a week, somebody took what Murray had designed to General Lincoln and demonstrated it . . . .

Kupperman: I did. I told Lincoln we could manage the entire freeze by computer. He said, "How many millions of dollars do you need, I'll get it to you." And then everybody got kicked aside, because Lincoln was in deep, deep trouble. He had the whole nation upon him. I had the reporting functions . . . .

Our job, using the management information system, was to organize the government. That's what we did. It became the binding agent, the form, to allow feedback to go on between our regions and people who were administrators.

Richard Wilcox's version adds an interesting detail about the timing at this point:

Bob Kupperman told the director of the agency that he had some people who could put up a system in two weeks. We didn't know that he had said two weeks; all we knew was that he came back and told us that he wanted to set it up in one week. We did it by my sitting down and listing a set of requirements that had to be met, and giving it to Murray, who took a terminal home and worked around the clock. He came back in four days and he had the basic EMISARI system, and it ran.

Design of EMISARI: The Emergency Management Information System and Reference Index

The ten regional offices of OEP needed to be able to generate and share timely and useful data on the policies, problems, and progress related to the wage–price freeze. The regional offices had to respond to requests and inquiries from the public in a way that was consistent with the initial guidelines and emerging modifications and interpretations. The central OEP management had to make sure that policy interpretation was consistent from region to region, and that the Cost of Living Council was kept well informed of the progress and problems that occurred. One measure of the communications load is that the New York Telephone Company reported that the regional office there received 10,000 telephone calls in the first week.

The primary innovation in EMISARI was the ability to set up alternative communication forms, such as collections of numeric estimates, tables of numbers, and situation report forms, and have these assigned as a permanent responsibility to some member of the communication group who would supply the information on a regular basis. There was also considerable ability to reassociate items so that text could be used as footnotes to data and

ISBN 0-201-03140-X, 0-201-03141-8 (pbk.)

sequences of items of various types could be built up into reports. All this was under the control of a human monitor who could tailor the communication structure and the responsibilities as a function of the problem at any time.

As Wilcox and Kupperman (1972) describe the system,

> The specific form of EMISARI was strongly conditioned by our personal philosophies (biases?) concerning management information systems. First, a true management information system is much more than a mere reporting system to top management. Reporting systems only obtain information at lower organizational levels, collect and perhaps organize it, and forward it to potential users at higher levels. There are two serious shortcomings of reporting systems: (1) the people who are data sources seldom receive any feedback to find out whether their submissions are appropriate—or indeed, even used—and (2) failing such feedback or any other useful (to them) output from the system, the source personnel tend to become haphazard in the quality of data which they submit. In contrast, an effective management information system is a communications vehicle among managers at all levels. Thus, while it includes data reporting "up the line," it also provides for dissemination of feedback and policy guidance "down the line," and furthermore for coordination and cooperation among managers at parallel levels.

The basic unit in the system was the "contact," a person in a regional office or staff agency with a specific responsibility for reporting data or cases for which guidance was needed, giving interpretations of policy, etc. The system permitted the input and retrieval of data as either a single estimate or set of related estimates called a "program," or as a continuously updated table containing summaries of certain information for each region. Table 2-1 shows a sample table; the responsibility for continuously updating this information rested with a particular contact in region one.

Table 2-1
Sample Table from EMISARI[a]

#31 Telephone Queries—Region 1
Weekly Distribution of Queries
Received via Telephone

	Wages	Prices	Rents	Total
Citizen	10	246	145	391
Small business	157	205	42	404
Large business	50	96	13	159
Labor	47	T	T	T
Other organizations	16	25	5	46
Total	280	T[b]	T	T

[a]From Renner et al., 1973.
[b]T denotes "temporarily unavailable."

ISBN 0-201-03140-X, 0-201-03141-8 (pbk.)

Another important part of the system was seven text files, which included

1. Policy and guidance—contained the rules, regulations, and management policies issued by the Cost of Living Council. There were several hundred rulings and interpretations issued in all.
2. Actions—a description of exemption or exception actions, made at the regional or national level. By skimming the file of denied exemptions, a regional office would decide if an appeal was unique, or already covered.
3. Bulletin board—regional contacts could enter descriptions of requests for policy guidance or information forwarded to the National Office, in order to cut down on duplication by other regional offices; or describe a problem and ask if anyone else had run across it.
4. News—Abstracts of press releases or news stories that were pertinent to the wage–price freeze and might be particularly useful for briefing persons meeting with special-interest groups. (See Renner et al., 1973, for a complete description of the components of the system.)

Each of these policy guidance files could be searched on the basis of key words, used singularly or jointly, such as "insurance" or "cooperatives," in order to obtain all rulings or actions related to that area.

When EMISARI first started, the only text files were the rulings and the bulletin. Very early someone discovered that the Press Office was abstracting the five leading newspapers and preparing a daily summary for General Lincoln. A secretary was talked into typing these daily reports into the public bulletin. No more was thought about this until three weeks later, when the secretary was out ill and the daily summary was not entered. There were 20 calls from across the country before the day was over to complain. What was discovered was that people working 12 hour days did not have much time for reading newspapers, and a call that a labor leader or company president was coming in to see them would result in a search of the news file to see, for example, what a steel company executive might want to talk about. As a result a separate news file was established. Various other features of the system evolved in the same manner.

Commenting on the interactive nature of the design effort itself, Turoff says:

> A unique feature of EMISARI's actual use was that the designer and all the programmers working on the system were in fact members or contacts. Each programmer's description gave the elements of the system he/she was responsible for. Therefore, the typical user could message the direct source when a problem was encountered, or the designer when a new feature was thought desirable.

ISBN 0-201-03140-X, 0-201-03141-8 (pbk.)

In addition, there evolved three ways for the members of the system, or contacts, to talk to each other. One, called the "Party Line," was a simultaneous written conversation for up to 15 persons; it disappeared when the conference was over. The other was "Discussion," which kept a permanent record in the computer of all entries made by the participants, who were not on simultaneously. The third was messages among the contacts, which could also be attached to specific data items on tables as footnotes. The latter was extremely important in that both those reporting and those analyzing the data could augment the data with these footnotes.

Since there was no opportunity for personal training of terminal operators, and it was absolutely imperative that they be able to learn to use the system quickly, the user interface was designed as a menu approach. That is, the computer would print out a list of choices and ask which the person wanted to use. The whole system and the series of operations necessary to access each part of it was displayed on a one-page users's guide (see Figure 2-1). Later, a limited version was also reduced to a wallet-sized card, so that users could easily carry it around and have it available for reference wherever they might have something to report, and access to a terminal.

The EMISARI system was designed to be flexible and to evolve in terms of features, in order to meet the needs of users. It received heavy use during the ten weeks of the freeze that it was monitored. Overall, it was accessed about 900 times for purposes of entering data, the policy files were consulted about 1900 times, and individual estimates and text messages were accessed somewhat more than 2900 times (Wilcox and Kupperman, 1972). The number of terminal users was about 80, but many of these had responsibilities to service a whole group of people in their particular office.

The overall system underwent significant programming improvements, done in XBASIC by Language and Systems Development Corporation, and eventually evolved into a system called RIMS (Resource Interruption Monitor System), used by OEP for subsequent crisis situations, such as the voluntary petroleum allocation program and the truckers' strike. The RIMS package incorporated an improved EMISARI, conferences, notebooks for individuals or groups to store things in, and an Incident Reporting System called IRIS. (see McKendree, 1975). As with the tables in EMISARI, the nature of this information system was such as to allow discussion among the producers and consumers of data; and to use the computer to facilitate the organization and interrelation of data from multiple sources upon receipt. Table 2-2 shows an example of how the EMISARI component of the system could be used to pull out incident reports during a trucker's strike.

However, this eventual limited "institutionalization" of the EMISARI innovation was not without organizational resistances either, and was made especially problematical by the dissolution of SED and OEP themselves.

ISBN 0-201-03140-X, 0-201-03141-8 (pbk.)

Figure 2-1 Emisari User's Guide (Office of Preparedness Emergency Reporting System).

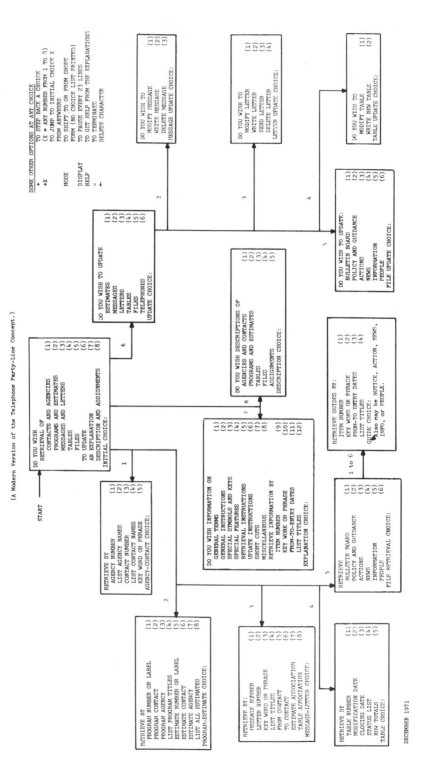

(A Modern Version of the Telephone Party-Line Concept.)

DECEMBER 1971

ISBN 0-201-03140-X, 0-201-03141-8 (pbk.)

Table 2-2
Example of the "Menu Approach" in EMISARI–RIMS[a]

@op. EMISARI
Emergency Information Exchange for Office of Preparedness
Date: 11/14/74
Access code please: ?   View

Do you wish
retrieval of:

Contacts and agencies	(1)
Programs and estimates	(2)
Messages	(3)
Tables	(4)
Notebooks	(5)
Update	(6)
Explanation	(7)
Current crisis arrangements	(8)

The menu approach
to using your terminal

Initial choice: ?   5

Retrieve

Bulletin board	(1)
Policy and guidance	(2)
Industry impact	(3)
News	(4)
State situation	(5)
People	(6)
Regional situation	(7)

Notebook choice?   7
There are 69 region items. The last is 1449

Retrieve region items by:

Item number	(1)
FRP to entry dates	(2)
Numerical values	(3)
Key word or phrase	(4)
List titles	(5)
Analyze usage	(6)

Pulling out the
Developing situation
in region 7

Region choice?   4

Primary key phrase? 2/11/74
Search titles only (Yes, No, L#, L# to L#)?No
Conditional key phrase? Reg.07
Search titles only (Yes, No, L#, L# to L#)?Yes
Hits by item number:
956   961   966   976   1086

Pulling out one
day's (Feb. 11, 1974)
report for Region 7

[a]From McKendree, 1977.

Table 2-2 (Continued)

Hit count:　6
Report hits (Yes, No, Copy(C), Analyze(A), Search on key word(s)? Yes
Print titles only (Yes or No)? No
#956　2/11/74 Truck work stoppage. Reg.07

> From R07. as of 2/11/74
> Morning: Reports from state policy/highway patrol
> Dispatchers indicate two acts of violence in Region
> 7 state: (1) Rock through truck windshield near　　(961, etc. would then be
> Lafayette, La., no injury reported. (2) Truck driver　printed.)
> shot near Beaumont, Texas, just after midnight by
> sniper. Injuries critical.

As of mid-1977, almost all of the persons involved in the original development and implementation of EMISARI-RIMS had moved on, and the OEP itself had been dissolved. In its successor organization, however, this kind of management information system had become part of the established "normal" (for a crisis) way of gathering and utilizing information for policy analysis. EMISARI itself is used as the weekly reporting mechanism from the ten regions of the Federal Preparedness Agency (FPA) of the General Services Administration (GSA), a successor to OEP. Whenever the GSA has a crisis to monitor, RIMS is revved up to full operational status. For instance, it has recently been used in the western drought and natural gas shortage situations. Somewhat ironically, the successor of the old computer division has coopted the conference system as its own. In addition, the Internal Revenue Service eventually took over management of the Nixon wage–price program, and had a special version, called IRMIS (Internal Revenue Management Information System), prepared for its use.

However, with none of the original design, implementation, or managerial staff to push for the continued development and expansion of the system as the state-of-the-art in computerized conferencing improved, it had become essentially a static system. There were proposals to improve and expand its capabilities but these were considered to be low-level priorities by FPA's current management.

Richard Wilcox describes the hard times of EMISARI as follows:

> On the ninety-first day of the wage–price freeze, memos came out rescinding all special authority, closing down the system, saying it was not to be used; stating that we had no authority to task people in the Computer Lab. Since the Computer Lab happened to be run by the Corps of Engineers, there was a memo put out charging that we had illegally tasked personnel of another agency without having the authority of the Commanding General of the Corps. Some of them wanted to put us in jail–that's literally true. Remember, the Computer Lab

ISBN 0-201-03140-X, 0-201-03141-8 (pbk.)

had originally been asked to set up a computer system to take care of information needs during the freeze, and they had said that it would take a minimum of ninety days to design the system . . . . It does not pay to be exceptionally successful in a bureaucracy . . .

If we had fallen flat on our faces, they would have forgiven us . . . . They couldn't forgive us for succeeding, especially when it was a job they had been asked to do . . . . Fortunately, Murray and his troops, about a third of the way through the freeze, had put in a monitor, which showed how many times different parts of the system had been used. This helped to refute the charges that nobody had used the system, and helped in its continued life . . . .

We had an assistant director of the agency who was smart enough to realize that there was a capability that we might need, so he wanted to make sure that it was not killed. The people in the Computer Lab could scream bloody murder and send all the memos they wanted, but he made sure that those charges, for instance, never got anyplace.

With the end of the freeze, all the special organization with ad-hoc assignments was swept out, and we reverted to the original organization. We thought of ourselves during the freeze as the architects, and the MCL (Computer Lab) as kind of the building contractors. And, you know, building contractors do not like architects.

Then, when the next crisis came along, the threatened Penn Central railroad strike, we weren't in the same carte blanche category we had been during the wage–price freeze, but he made sure that we could function in that role again. We brought all the personnel up to functioning again, and produced the first report, and then the strike was postponed. If they had needed it during a crisis, we would have been given whatever authority was needed and no more; and as soon as the crisis was over, we would have been turned right back, to keep us in our place . . . .

The assistant director always made sure that somebody was assigned to work on the project from MCL. It seemed to be used as a training ground. The programmers did not have any management supervision from within their own shop. They knew that it was not the way to get ahead in the agency, working on that program . . . .

During the entire 90 days, it always continued to evolve. After the Freeze, as the RIMS system, it continued to evolve. I happen to think that that is a fundamental requirement of any successful communications system. If you freeze it and try to make a set service out of it, it's dead.

At about this point, as Turoff and Wilcox and Kupperman and Goldstein began to leave the sinking ship of OEP, and as the dissolution of the agency itself was on the horizon, John McKendree took responsibility for EMISARI and its successors, working on a consultant basis for the Office of Preparedness in GSA (the successor agency). He relates the subsequent history:

Operationally, there were long periods of silence as far as the system was concerned. But on the implementation side, we always had people wrapping up

ISBN 0-201-03140-X, 0-201-03141-8 (pbk.)

the programming that was begun during the operational phases. For two years, there were at least six programmers working on it . . . .

In May of 1973, the voluntary petroleum allocation program began and OEP provided the core group to carry out the program. This program continued past the demise of OEP into GSA—EMISARI had never been busier.

The question of what OEP would do with EMISARI did not arise until the end of the Voluntary Petroleum Act. At that point, management had to examine EMISARI in terms of continuing the development work. So far, it had only been pressed into service because of specific emergencies. MCL was of a mind to terminate it. Then another emergency arose, the national truckers' strike in January of 1974. Emisari was used nationwide to provide the White House with twice-daily reports.

SRH: What were the reasons why they were going to essentially shelve EMISARI?

McKendree: My perception of the management thinking was that MCL was committed to FORTRAN as a programming language. They were also heavily oriented toward batch processing, and this was on-line interactive processing. So if they were to choose to embrace it as a philosophy for data processing support, they would have to be responsible for documentation and for always having personnel for maintenance . . . . They did not feel that it was the business of the MCL to provide this kind of support on a full-time basis. And there was no management commitment to use it on a day-to-day basis. It was viewed by MCL as not an appropriate use of their manpower.

I was in a position to recommend that EMISARI be maintained on a stand-by basis, not only in regard to the software, but also in terms of regional office secretarial proficiency. I proposed to use it for the weekly report from the ten regional offices, not because it was the most efficient way to get this information, but because it was a form of proficiency maintenance that was not an undue burden. That was early 1974. With that change of heart, the next step was to follow up on suggestions made earlier by Murray Turoff providing on-line data bases because that would give us a number of benefits. It would give us much earlier analysis of the data base—in hours rather than days or months. It would also develop a proficiency at the national office in the frequent use of the EMISARI system. So I made that proposal, which introduced what is now called RIPS on an experimental basis, still using the identical "front end" to EMISARI (sign-on procedures, protocol, etc.). This developmental period for RIPS began in 1974, and it is still under way, though it is also an operational system today . . . .

RIPS provides a version of a multiregional economic input–output data base as an integral part of EMISARI. It allowed the regional people to analyze for themselves the potential priority impacts regionally of a specific material or commodity shortage on other industrial components. As such it provided an anticipatory tool for estimating impacts of an impending crisis at the regional level. It still represents the first major integration of a large-scale data base directly into a computerized conferencing system.

ISBN 0-201-03140-X, 0-201-03141-8 (pbk.)

As for what might have been had OEP not been disolved, Wilcox adds this chilling final note:

> The regional directors were indeed interested in the potential application of the system to general war preparedness, and I lectured/demonstrated the possibilities to them subsequently in two locations. In essence, the full panoply of capabilities could be used for decision supporting/implementing communications between headquarters and the field. Access to the national resource data base and associated nuclear damage estimation models could support analysts attempting to figure what could be done with what was left after a nuclear attack. (Part of our role in OEP was to think about the unthinkable.)
>
> I still think a conferencing capability would be highly useful under such circumstances, particularly during the transattack and immediate postattack period when everybody would have to be holed up and traveling to face-to-face conferences would be impossible. But God save us from either EMISARI or EIES ever having to be used that way.

Our account of EMISARI thus far has omitted one crucial element. "Monitor" is the term used in EMISARI–RIMS for the human facilitation role in this computerized form of communication. From the very beginning, this was a designed-in component of the system, and the roles played by the human monitor were found to be crucial to the success of the communication form.

IMPORTANCE OF THE HUMAN MONITOR AND THE "HUMAN ELEMENT"

The original monitor of the EMISARI system, Nancy Goldstein, played a role that is user oriented rather than computer oriented, serving as the central point of contact among the users, the systems designers, and managers. She was also responsible for operational aspects of the system, such as keeping the data definitions up to date and entering contacts, codes, labels for tables, etc. Though the EMISARI system had a short one-page user's guide and some written instructional material, plus on-line explanations of the various choices and features available to users, there was also some need for in-person or telephone training of potential or confused users, which she carried out (see Renner et al., 1973, pp. 25–27). Finally, she served as a "public relations" person for the system, explaining and demonstrating it to potential users.

Nancy Goldstein reflects on her role as follows:

> The public relations role is, I think, the most important one for the monitor. The value of a system is dependent upon how it is viewed by those who use ("need") it and those who contribute to it. Many people in both categories were very nervous about EMISARI and using a computer in general. They had no prior computer experience, felt that the computer was "fragile" and that the computer and its technicians spoke a foreign language. It was very important for the

ISBN 0-201-03140-X, 0-201-03141-8 (pbk.)

monitor to be able to speak their language, to reassure them, and to break the operation of the terminal down into very elementary and easy to understand steps. I think traditional computer systems in other government agencies and private industry have ignored the importance of this "go-between"—the monitor who translates the users' needs to the computer experts and the steps necessary to attain their needs back to the users from the technicians.

Or, as John McKendree, who took over the monitor role for the successor systems to EMISARI puts it, "If you don't stroke the folks and make them feel that their problems are being addressed, they won't come back."

He took a very active role in seeking out persons who might need assistance, and in patiently teaching them the troublesome detail of use of the system.

If I saw a user had not been active for a day or two, or if I saw an entry that showed they did not understand how to do something, I would call them on the phone. We talked about our jobs on the EMISARI system as equals.

In the early stages of the implementation, the public relations, training, and helping roles are much more than a full-time job for one person, if done well. In subsequent systems such as PLANET, FORUM, and EIES, the order-keeping role of a systems monitor was separated out from the user-training role and from the assessment role (collation of user reactions and their feedback to designers for improvement in the system).

It is also significant that amid the barrage of serious policy pronouncements and tables of statistics transmitted over EMISARI, humorous items began appearing on the bulletin board, representing emotionally tinged communications about the feelings that the dispersed persons who were working together through the crisis wanted to share with one another. One example is the following poem, entered anonymously.

Ode to the Wage–Price Freeze

Egad! said Nixon,
Economy's a mess, needs fixing.
I think I'll have a wage–price freeze.
To OEP I'll resort.
Charge! said Lincoln, Report! Report! Report!

The people utilizing EMISARI at OEP were usually involved in a crisis environment and often called upon to work 12- to 16-hour days in certain situations. The humor that flowed through the system was an important factor in relieving the tensions and strain. However, there was a great deal more in terms of the democratization of roles and the feeling of all those in the operation that those on the system represented a peer group faced with the same common problems regardless of location.

ISBN 0-201-03140-X, 0-201-03141-8 (pbk.)

The one time the anonymity feature was used in EMISARI during the wage–price freeze was in the third week. OEP's staff had been augmented in the regional offices with persons who had been delegated on temporary duty from other agencies. Many of them had GS ratings of 10 to 13; those involved in the system at the national offices were GS 14 to 18. When a person at a high grade would send a message: "Having any problems?" to a low-grade regional worker, you could bet the answer would be "Everything is fine, sir or ma'am."

On the third week about 30 of the lower-grade regional people were ordered to join a conference at a designated time and enter a fake name for themselves. The first comment in the conference was an invitation to express any problems they were having. That conference went on for three furious hours of typing and sufficiently broke the ice so that grade levels of individuals began quickly to disappear and the operation and atmosphere became that of a group dealing with a common problem, where each had his or her contributions to make and his or her roles to perform.

For many of those engaged in the effort, this was a dramatic shift from their usual surroundings. When the wage–price freeze operation terminated and most of the individuals went back to the "normal" jobs in other agencies, there was for many a genuine sadness at leaving, despite the level of effort the operation had demanded of them.

Other Computer-Mediated Communications Systems

Another Forebear: Computer-Augmented Knowledge Production

In 1945, Vannevar Bush, President Roosevelt's wartime director of the Office of Scientific Research and Development, published an extremely foresighted article (Bush, 1945) in which he predicted, among other things, the Xerox machine, the Polaroid camera, and FORTRAN. One of the items predicted was Memex, a writing, reading, filing, and communication system contained in a desk and including a screen and keyboard. This device would allow an individual to accumulate and develop his own personal library of materials. It would also allow him to compose and edit his writings. Besides enabling a person to accomplish what then took card files, libraries, notebooks, typewriters, scissors, paper, copying machines, index tabs, graphic boards, and rolltop desks, Memex would also allow a high degree of nonlinearity in the structuring and association of text not possible with current linear forms such as articles and books. In essence, the electronic form of the printed word would facilitate our creation and use of text in manners more akin to our cognitive processes, which appear to be parallel and associative in nature.

ISBN 0-201-03140-X, 0-201-03141-8 (pbk.)

Needless to say there is no Memex machine available today. However, the pieces of Memex exist and it is now clear that it is only a matter of time, cost reduction, packaging, and human interface design before Memex becomes a reality. The current generation of word-processing equipment represents one crude but significant step toward the Memex machine described by Bush. Even more significant are the software developments that have emerged under the concepts of "hypertext" and NLS (ON-Line System); (see Nelson, 1965, and Englebart et al., 1976), as well as a significant number of large-scale text-editing systems (see, for example, Rice et al., 1971).

Among the most fully developed and widely available (through AR-PANET) of such systems is Englebart's NLS. As he explains (Englebart et al, 1976, pp. 228, 232),

> Over the past ten years, the explicit focus in the Augmentation Research Center (ARC) has been upon the effects and possibilities of new knowledge workshop tools based on the technology of computer timesharing and modern communication . . . .
>
> The core workshop software system and language, called NLS, provides many basic tools. During the initial years of workshop development, application and analysis, the basic knowledge work functions have centered around the composition, modification, and study of structured textual material.

NLS initially did not have any communication (conferencing) type features among users. It does have a "Send Message" form of electronic mail, and a synchronous form of discussion called "Linking," much like Party Line. In the past few years, there has been a "growing together" of these two types of computer systems, with NLS being made available on the ARPANET, which also has a sophisticated message system called HERMES; and EIES including both the computer conferencing components and the NLS-like components called "Notebooks."

Bell Canada's Computer-Mediated Interaction

Bell Northern Research experimented with an EMISARI-type system in the early 1970s, and then began building its own minicomputer-based conferencing system. One of the unique features of this system that is suited to its Canadian applications is that the user interface is bilingual. The use indicates a preference for French or English dialogue with the computer and all questions or choices are then presented or made in the indicated language. (The content of entries, however, remains in whatever language they were written in.) As Millard and Williamson explain this bilingual feature, (1976, p. 216),

> The two-language approach was carried through to the CMI user's guides and the accompanying summary charts, even to the open Problems conferences,

ISBN 0-201-03140-X, 0-201-03141-8 (pbk.)

with one being exclusively English, and the other French. Although only two languages are supported, it is easy to add any number of languages. A separate system message file is used for each language, thus adding a new language is as simple as having existing messages and commands translated, and put into another file.

The design emphasis has been on ease, flexibility, and "human-ness" of the interface, so that

> people can feel comfortable with the new system. It talks to users in either of Canada's official languages, French or English. It calls them by their first name, and prompts them with messages that are tailored to their familiarity with the system. Not only does the computer respond in two languages, but it also talks to you briefly or in detail, depending on how long an explanation you want. You can classify yourself as novice, intermediate or advanced to get the desired response. If a frustrated novice typed in "#!?!k$!" the computer would politely answer, "I do not understand #!?!k$! ... please try again," while for the intermediate, the reply is shortened to "I do not understand #!?!k$!". At the advanced level, the computer tersely echoes the user's command with "#!?!k$!"[Millard and Williamson, 1976, pp. 214, 216].

CMI has been tested in a set of trials including conferences among Canadian government bureaucrats and a business organization as well as internally within Bell Canada. In addition, Bell Canada has made extensive use of NLS.

The Institute for the Future: Extensive Field Trials with PLANET/FORUM

The Institute's original FORUM computerized conferencing system in the early seventies derived from the Delphi studies and EMISARI tradition, and included some unique components such as a supplemental computer-controlled voice channel (see Vallee et al., 1974, p. 11). It allowed a very structured Delphi-like form of activity in which the organizer elicited answers to specific questions, with the answers not distributed to participants until and unless the organizer requested. Subsequently, however, a simplified system called PLANET was focused upon, which provides a common conference and also allows a one-to-one private message. The main focus of the Institute's efforts has been real-world trials of computerized conferences, and the development of basic methodologies for monitoring them, understanding their strengths and weaknesses, and exploring their social effects.

One of the more interesting of the conclusions to come out of 28 FORUM conferences with scientists, engineers, and other "experts" has been the delineation of at least five different "styles" or types of computer conferencing that had been observed, even with the fairly limited capabilities and user types that were investigated. These are described as the "notepad" and the "questionnaire," which are noninteractive styles, and the "seminar,"

ISBN 0-201-03140-X, 0-201-03141-8 (pbk.)

"assembly," and "encounter." There are no clear-and-fast distinctions among these styles; rather, they indicate a range of purposes, size of group, amount of structuring by conference leaders, and degree of interpersonal interaction and of synchronous (simultaneous on-line) sessions.

A short description of the distinctions among these styles has been given by Andy Hardy (in the Applications and Impacts Conference, EIES, comment 65, February 1977):

> The "notepad" is a multiple topic discussion with an unstructured group. It lasts several weeks to a few months. It is almost completely asynchronous with little interpersonal interaction.
>
> The "seminar" is a conference addressing a specific topic. It usually involves asynchronous usage, with periodic synchronous interaction. Message sending is moderate. The structure of the interaction is still fairly informal.
>
> The "assembly" usually has multiple topics, and a larger number of participants. All the topics are related to a single theme. To deal with the problem of increased topics, interaction is somewhat more structured. It is mostly asynchronous, with some synchronous interaction. Message sending rates are low to moderate.
>
> The "encounter" is a conference of short duration, with an intense degree of interaction. Discussion is topical and task-oriented. The amount of structure is low. The message rates in this type of conference are very high. Interaction is almost totally synchronous.
>
> The "questionnaire" involves an unlimited number of conferees in an extremely structured conference. The goal is response elicitation according to a prescribed format. Interpersonal interaction is almost nonexistent in this style.

Another good description of the five styles has been presented as a series of short "scenarios" which characterize them in terms of extrapolations of IFTF observations to future conferences that might take place (Vallee et al., 1976, pp. 211–212; a fuller description of the styles and complete scenarios may be found in Vallee et al., 1975).

The "Notepad": When a Scientific Discipline Goes Network

In 1979, an Astronomical Resource Network (ARN) was established to link astronomers at all major universities as well as observatories and a number of research institutions. It is estimated that approximately 100 astronomers use it daily to communicate with colleagues at remote sites, to access an astronomical data center, and to monitor flash announcements of important phenomena such as comets and novae.

The "Seminar": Conferencing Unlimited, Inc.

Conferencing Unlimited, Inc. is a commercial computer conferencing service. It provides access to computer terminals and takes care of all the formalities of computer networks for short conferences and meetings among people who do not regularly use computer conferencing . . . . It has serviced conferences

ISBN 0-201-03140-X, 0-201-03141-8 (pbk.)

ranging from committee meetings to topical debates by scholars to psychotherapy sessions. Even the college player draft of the National Basketball Association was organized in 1982 by Conferencing Unlimited, Inc . . . .

The "Assembly": The Meeting without Travel

From July 3 to July 6, 1983, the World Future Society held its general assembly using computer conferencing. This first major worldwide meeting of scientists to use this medium was arranged in response to complaints about large conferences, stuffy panels, lack of participation, and general frustration. The chairman of the assembly noted afterward that the conference was both more authoritarian and more democratic than any previous assemblies. It was authoritarian in those sessions in which leaders exerted strong control over the proceedings, sometimes limiting comments to three lines per message or limiting the total number of questions each person could ask. It was more democratic in that more complete information was available to more people. Access to well-known persons was higher in this medium, and private messages were exchanged freely among all participants. The assembly demonstrated that a satisfying form of communication is possible without travel and that this medium should be taken seriously for more regular meetings.

The "Encounter": An International Crisis Management Network

On June 2, 1983, the United States officially joined the International Crisis Management Network. The ICMN is a worldwide network of statespeople as well as economists, scientists, military strategists, psychologists, and other resource people who are connected to a 24-hour-a-day computer conferencing service. It has been uniquely successful in dealing with a range of worldwide crises. This success rests in the skill of ICMN members in manipulating its wealth of information—including on-line models and data bases—to support their proposals for solutions to crisis, to evaluate the proposals of others, and to adjust their own proposals quickly . . . .

The "Questionnaire": Alaska Mineral Wealth Forecasts, Inc.

Alaska Mineral Wealth Forecasts, Inc., is a mineral forecasting service which resembles the "Delphi" concept of the 1960's. A highly trained group of mineral geologists, located all over North America, participate in weekly computer conferences to review information records about mineral properties in the files. These "conferences" are highly structured iterations of quantitative forecasts in which the probabilistic estimates supplied by each expert are mathematically aggregated to produce graphic feedback for the group. The results are supplied directly to government and private subscribers engaged in mineral exploration.

On our Way to a Wired World? Teleconferencing in Socialist Sweden

One of the most significant experimental programs was launched in 1976 by the National Swedish Board for Technical Development (see ERU/STU, 1976, 1977). At the time this book was written this was the only sponsored

ISBN 0-201-03140-X, 0-201-03141-8 (pbk.)

research effort placing a high priority on experiments in the area of public use and socially conscious applications. Among the applications undergoing experimentation with a version of the PLANET system were

1. Use of this technology by the deaf from their homes;
2. The sharing of information and consultants among small-scale business organizations;
3. Regular group discussions (required by law in Sweden) between labor and management, prior to any management actions that would affect working conditions;
4. Planning conferences on the northern regions of Sweden, to examine the impact of current excessive travel requirements on those involved;
5. Health care consultation and information distribution in rural environments; this involves tying "mobile nurses" and "psychiatric first aides" who work in these areas into consulting networks and the "networking" of regional hospitals for resource pooling among specialists.
6. High school level educational courses in subjects such as English, and some university courses as well.

As of 1977 about 300 users were engaged in these various experiments. Interestingly, a much higher percentage of Swedish users are reported to tend to become fanatical (very frequent) users than American users engaged in trials to date. In discussions one of the authors had with some Swedes involved in these experiments, it was pointed out that many of the frequent or fanatical American users tend toward some degree of introversion. The Swedes very quickly rejoined that nine months of the year their whole society tends to be introverted, due to the weather and lengthy periods of darkness. Unfortunately, this is only an intuitive impression gained from informal observation, and remains to be confirmed by more careful experimentation. The Swedish concern was that very extensive use of these systems might intensify introversion in their society.

Although Sweden is a well-developed country, it has extensive underpopulated rural areas with very harsh travel conditions in winter. The Swedish government is encouraging decentralization as much as possible, which in turn produces severe travel and communication problems for many organizations. In 1977 the Research Institute of the Swedish National Defense was split into three different locations. They estimated, in evaluating conferencing, that on the average a face-to-face meeting would cost $28 per hour with $20 representing lost travel time. This is compared to $10 per hour for the conference system. These factors taken together tend to imply a high potential use and rapid growth of computerized conferencing within Sweden.

ISBN 0-201-03140-X, 0-201-03141-8 (pbk.)

Other Systems

Among the other systems developed for human communication via computers in the 1970s are CONFER, used at the University of Michigan for trials of citizen involvement in public issues, among other applications; ORACLE, programmed at Northwestern and seen as an adjunct to CAI (see Hough, 1977, pp. 74–75); CONCLAVE, a message and conference system developed in England that gives the chairperson in a conference a high degree of power; and General Conferencing System, Ltd., developed in Canada by Bert Liffman. (See Hough, 1977, pp. 71–75 for more detail on these specific systems.) The General Conferencing System is interesting because it started with the Delphi conferencing design, and then was rewritten in APL for an IBM machine, which is very different from a Univac or a minicomputer, whose internal operating systems are more amenable to handling a conferencing system. The program was developed under funding by the Canadian government, through Liffman's organization, called "Memo from Turner." It is also noteworthy because it was an operational system used by the Non-Medical Use of Drugs (NMUD) Director to tie together field offices around Canada, with over 21,000 messages passing among the NMUD offices in the initial three-month period (Panko, 1975). Reports on this project are proprietary, but informal discussion with those involved indicates that the following sorts of outcomes resulted from the increased communication made available by the conferencing system:

The separate provincial organizations became a more unified single national organization.

The regions felt more informed about what was happening in other regions.

There was a new openness to discussion of fundamental issues, especially between the national office (oriented toward research) and the operational field offices in the provinces.

Morale was improved.

In addition, there are a variety of store-and-forward message systems (used for internal mail by such organizations as IBM, Bell Telephone), and Bolt, Beranek and Newman's HERMES and Scientific Timesharing's MAILBOX, which attempt to replicate a post office system for the user. Panko (1976) states that the level of use of message systems has been much higher thus far than experimental conferencing systems. What is interesting is that the recent designs for message drop systems have added capabilities that begin to make them look like conference systems. What happens in an electronic message system is that a set of individual users discover there is some subject they wish to address as a group, and that they desire to collect all the discussion

ISBN 0-201-03140-X, 0-201-03141-8 (pbk.)

about that topic in a single file that now represents a common proceedings of the discussion the group is having. It is soon discovered that the ability of the computer to aid in editing and reviewing the discussion is useful for further development of the topic. Once these capabilities are incorporated, there is little distinction possible between a simple conference system and a sophisticated message system.

It is widely recognized that all of the existing systems are very rudimentary. It is also generally accepted that current costs for the computer and the communications network will drop substantially in the next few years. As a result, cost factors alone should stimulate the widespread proliferation of such systems over the next decade and their attempted application in many kinds of organizations and for many kinds of communication purposes.

Summary

Within seven years, computerized conferencing has developed from an idea to a wide range of operational systems that have been utilized in government and business and experimentally applied to many different services. Several of the systems are available for purchase or lease. Large-circulation journals in the business and computer fields are beginning to carry articles and features on the characteristics and possibilities offered by this new medium.

This and the preceding chapter have provided a perspective on the factors that have influenced the development of computerized conferencing and its establishment as a new and unique form of human communication. In essence, the threads that come together or the roots that have sprouted the tree are

1. The technological forecasting reaction to pure computer decision modeling, and the emergence of Delphi and other group communication processes;
2. The declining cost of computer hardware and the growing recognition of indirect user costs;
3. The natural evolution of text-processing computer systems to provide forms of on-line communication;
4. The evolution of simple message systems to provide richer and more useful features;
5. The evolution of computer-assisted instructional systems to incorporate direct communications;
6. The growth of digital oriented communications networks;
7. A growing recognition within the telecommunications community of the alternatives offered by the computer;

ISBN 0-201-03140-X, 0-201-03141-8 (pbk.)

8. The new understanding of communications as a major problem in the organizational use of information;
9. A need to balance analytical and quantitative approaches with informed and qualitative subjective approaches;
10. An understanding that computers can augment group as well as individual cognitive processes;
11. An anticipation by those in this area of the potentials offered by incorporation of computers into human communications;
12. A realization of the potential for expanding our knowledge of how to deal with complex problems.

The research and development work to date (as reflected in the Bibliography for this book) emphasizes that individuals have come at this area from many different starting points: Delphi and technological forecasting; word and text processing; computer-assisted instruction; telecommunications; management information systems; message systems and communication networks. Regardless of where it started we believe this area is now unique in its own right and our objective in this book is to tie these threads together under one perspective.

The probable spread of the new communication mode into a variety of applications with many purposes will be the subject of Part II of this volume. First, however, we will focus our attention on some of the social and psychological aspects of communication via this medium, and its weaknesses or limitations as well as its possible advantages. These characteristics are likely to affect the outcome of any of the applications explored in Part II.

For Discussion

1. What are the situations and conditions under which an organization might be forced to introduce a computerized conferencing system? Give specific examples.
2. What impact would such systems have on job functions, the role of the boss, relationships among employees, etc?
3. Assuming a "collective intelligence" for a group of people could be established through the use of computerized conferencing, what particular groups might most benefit from such a capability and why?
4. What are some situations in which one might employ the Delphi technique? To what extent would computerized conferencing be more or less useful than the paper and pencil Delphi in the given situations?
5. In what employee situations would anonymity be useful?

6. In what type of situations would one not want the democratization that occurs in these systems?

7. What uses of humor can you imagine in a computerized conferencing environment?

8. What is the role of the secretary to be in the computerized conferencing world?

9. How would you propose introducing computerized conferencing in a particular bureaucratic environment with which you are familiar?

10. What are the alternative roles for computer professionals and/or computer shops in organizations, and how would this impact on evolving the types of user capabilities we have been discussing?

11. What are the psychological factors that make for a good monitor in these systems?

12. What are the conditions necessary in an organization to foster innovative users of computers in the open as opposed to the use of subterfuge?

THE BOSWASH TIMES

The Computerized News Summary Service of the Megalopolis

All the News
Fit to Display

ISBN 0-201-03140-X, 0-201-03141-8 (pbk.)

Oral Denial Syndrome Described at APA

Spokane, Washington,

September 16, 1993

Dr. Slip, in speaking before the tri-yearly meeting of the American Psychiatric Society, has offered evidence of a new syndrome affecting an alarmingly increasing number of Americans. Referred to as "oral denial," it appears to be a form of frustration and ego degradation afflicting those whose prime talent was domination of small group meetings. Dr. Slip believes this has been brought about by the increasing use of computerized conferencing for most significant meetings in government and industry.

As Dr. Slip describes it, most people with a tendency to this syndrome are extremely assertive and even aggressive in face-to-face situations. They all tend to low verbal latency scores. The growing lack of opportunity to dominate meetings is usually too much for their inflated egos to handle, and they sink into a severe state of despondency, with little motivation to try to do anything as a substitute. Dr. Slip has suggested that the high rate of manic depressives reported over the past few years may be a case of large-scale misdiagnosis of this new syndrome, since the end product is very similar.

For those who have the talent, he encourages extensive retraining in the use of the English language. In other cases all that can be done is to attempt to get individuals to move out of the jobs they are in and into either blue collar work, direct sales, or academia, where for some reason little or no use is made by faculty of computerized conferencing.

Washington, D.C.

September 17, 1993

Battle over CC Builds in Congress

Distinct rumblings can be heard behind the scenes in Congress about the recently installed computerized conferencing system. While general agreement exists on the convenience of transacting a large proportion of committee business in this manner, a number of unexpected side effects seem to have occurred.

Recently Senator Gallstone and 20 other senior Congressmen introduced a resolution to have major modifications made in the current system. Among these are elimination of the pen name and anonymity features, as well as incorporation of

73

a review procedure for all comments added to committee records by the committee chairman before they can be transmitted to the rest of the committee.

Senator Mavrik, representing an outspoken group of junior Congressmen, felt that the current system has had a beneficial effect on loosening the reins of control over the committees by the senior Congressmen. In an off-the-record interview, this reporter was informed that many of the senior members felt they had been hoodwinked into a technology which is destroying the established customs and procedures of the Congress.

We have also been informed by a high-ranking source that OMB is strongly backing the resolution because of some of the recent problems involving career civil servants entering anonymous comments into the system for various Congressional committees. This has led to a number of embarrassing disclosures on current administration practices.

There is little doubt that this will be one of the major issues facing Congress during this session.

ISBN 0-201-03140-X, 0-201-03141-8 (pbk.)

changed to protect individual privacy. In our program you will be under the tutelage of a select staff of our nationwide network of language experts, who will be picked by the computer to match your unique needs and situation. Just message us at Writemasters, 301 HAL to receive free our on-line questionnaire and a resulting report of just what we can do for you and your communication problems.

Writemasters is licensed by the Informations Systems Regulatory Commission as an accredited on-line business, and by the Association for On-line Education to award credits for course work.

ISBN 0-201-03140-X, 0-201-03141-8 (pbk.)

CHAPTER 3

Social and Psychological Processes in Computerized Conferencing

Computerized conferencing as a social and psychological process differs markedly from other modes of communication, such as face-to-face meetings, telephone, or letter writing. In this chapter, we will explore computerized conferencing as a social process that has some unique characteristics that may be strengths or weaknesses, depending on the nature of the group using this medium, the level of skill they have attained, and the purpose for which it is applied. We will see that there is an extensive learning and socialization process as participants learn to take part in "electronic group life."

As illustrative material, for the first part of the chapter, we will use actual messages and conference entries from Group 72's Applications and Impacts of Computerized Conferencing conference, held on EIES for several months in 1976–1977. These entries will serve the dual function of making substantive points and of illustrating the variety of actual forms and contents of communications on this medium. To introduce the group members quoted, we have included self-entered Directory descriptions (Figure 3-1). More quantified data come from controlled experiments and from statistics on user behavior during the first nine months of EIES operation.

Psychological Differences; The Narrowing of Communication Channels

In face-to-face communication, a person simultaneously receives information through many channels,* which may be broken down into the audible channels and the visual channels. In turn, the audio channels contain both

---

*The classification used here draws on that presented by Dittman (1972) in his chapter on channels of emotional messages. As he points out, there have been many definitions and classifications of communications. A fairly simple and nontechnical classification scheme seems best suited to our purpose of qualitative comparisons among media.

Starr Roxanne Hiltz and Murray Turoff, The Network Nation: Human Communication via Computer

ISBN 0-201-03140-X, 0-201-03141-8 (pbk.)

Figure 3-1 The cast of characters: Self-entered directory descriptions of EIES members quoted in this chapter.

ELAINE KERR (ELAINE, 114):
As a "user-consultant" I am available to answer questions and complaints about the use of this system. As a sociologist, I am interested in the applications of computer conferencing, especially with the handicapped. Other interests include research methodology and complex organizations.

CHARLTON PRICE (CHARLTON, 116):
Sociologist and management consultant. User consultant for EIES particularly interested in applications of computer conferencing.

ROBERT BEZILLA (ROBERT, 213):
Public opinion research, information processing and communications research. Group communications research. Studies of elites, experts, scientists, etc. Epistemology

RONALD L. WIGINGTON (RON, 235):
Electrical engineer, member of management, interested in computer and information systems—design, application, and use.

W. DAVID PENNIMAN (DAVE, 245):
I am associate section manager of the information systems section at Battelle and am responsible for the group that develops and maintains our on-line data/information retrieval system (BASIS). In addition we do a variety of studies in information science for sponsors such as NSF, NLM, industry, etc.

RA3Y PANKO (RA3Y, 705):
Specialties: organizational communication, computer mail, public policy analysis, applied epistemology.

ROBERT JOHANSEN (IFTF, 706):
I am a sociologist doing research relating to various teleconferencing media, as well as some general futures research.

RONALD UHLIG (RON2, 708):
Manager and Computer Scientist. Responsible for computer support to scientists and engineers in almost all of the Army's labs and test activities. Responsible for implementation of computer-based office support systems to provide computer-based tools, including conferencing and message systems to 100,000 employees.

BARRY WELLMAN (BARRY, 720):
I am interested in studying urban social networks. Presently analyzing Toronto survey data on such matters as density, effects of spatial distance on in-person and telephone contact, and the use of nets to control urban contingencies and access resources. Several papers available to those interested.

ROBIN CRICKMAN (ROBIN, 730):
I teach basic research methods to library students at Drexel and work for Belver Griffith. I hold a Ph.D. in urban and regional planning from Michigan (dissertation with Manfred Kochen), an M.S. from Illinois in library science, and a B.A. in linguistics. I used the Michigan system of conferencing extensively for the last two years for class and research.

the actual words used and their arrangement, as well as what might be called "vocalizations." The visual channels may be broken down into facial expression, clothes and other aspects of general "appearance" that give status cues, body movements, and psychophysiological responses.

Language content includes the direct, manifest content of the language used to convey information. Besides the "manifest" content of the words and their arrangement, there is also "latent" content given by slips of the tongue, *double entendre*, etc. In addition, the richness of the vocabulary and the quality of the grammar tell much about the educational, social class, or intellectual level of the presenter.

Vocalizations are all the "nonwords" or sounds that accompany the delivery of the words or are used as interjections rather than for their manifest meaning. There are sounds, like "um" or "you know," that are sometimes used merely as habit and sometimes to indicate that the speaker is pausing to find the right words, or checking to see if the receiver understands. There are emotion-laden sounds like laughter or tongue-clicking. The speed, loudness, and tone or pitch of voice give cues about emotion, age, and sex. Accents convey cues about nationality or social class, or ethnic group of the speaker.

Visual Information

General appearance, such as height, weight, and other culturally determined aspects of "attractiveness" and the clothes, makeup, jewelry, and other props used by persons to present themselves to others, provides an important filtering context for face-to-face communication. So do the visibly apparent cues that are provided by sex, age, and race and by visually apparent physical handicaps. In general, those aspects of self that are devalued by a culture—such as being black, female, old, "ugly," or disabled—have the effect of acting as a general stigma. In Goffman's sense, they pervade the social interaction and discredit the speaker in the eyes of the recipient of a communication. The stigmatized person, in turn, is likely to become aware of the prejudice of others and avoid aggressive participation in a group discussion, out of a feeling that her or his contributions will not be well received by other members of a group.

This largely visual information about "social category" or "social type" is important in group communications when the group is "skewed," so that there is a numerically dominant category (approximately 85% or more) and a "minority," whom Kanter (1977) has designated the "tokens." "They are often treated as representatives of their category, as symbols rather than as individuals" (p. 966). For stigmatized minorities, or tokens, undue focus on their social category tends to blot out other aspects of their performance. For

ISBN 0-201-03140-X, 0-201-03141-8 (pbk.)

example, Kanter's study of the first women sales trainees in a Fortune 500 company found that

> the women had to put in extra effort to make their technical skills known, to work twice as hard to prove their competence. Both male peers and customers would tend to forget information women provided about their experiences and credentials, while noticing and remembering such secondary attributes as style of dress [Kanter, 1977, p. 973].

Facial Expression

Smiles, frowns, raising of the eyebrows, and the many other expressions of the face are for most people the primary source of information about the feelings or attitudes of the person behind the face. They show whether the recipient of communication is pleased or angry or amused by it; they show the degree of seriousness with which a statement is made. Whether a person "looks you in the eye" or not is also used as a cue about honesty or sincerity or warmth; though, of course, these cues may be directly manipulated by the actor. In other words, they may be consciously or unconsciously manipulated sources of "misinformation" as well as of "correct" information about the emotions or the intentions of the person behind the face.

The Importance of Eye Contact

The eyes have been identified by many scholars who have studied face-to-face interpersonal communication as the single most important channel for communicating emotion and other subtle messages (see, for example, Fast, 1971; Knapp, 1972). Among the assertions and conclusions about the role of eye contact are the following:

1. The entire pattern or interaction in a group discussion may be derived from the records of duration and direction of gaze among members (Borden et al., 1969). In Bales interaction coding, for example, the person "to whom" a statement is directed is most often detected by direction of eye gaze. It is much less frequent that a person says, "Ken, I agree with you," let alone, "Ken I agree with you and I want you all in the rest of the group to say what you feel and really I want you to agree with us." In other words, a person would indicate that she or he wishes the rest of the members to express an opinion by intently gazing at all of the group members in turn while saying some of the sentence.

2. The duration and frequency of direct eye-to-eye gaze seems to constitute one of the greatest sources of social pressure in interaction. It is a demand by the speaker-starer that the listener actually listen, and eventually respond. It is a sign by the listener that attention is being paid and a response

ISBN 0-201-03140-X, 0-201-03141-8 (pbk.)

can be expected (see, for instance, Argyle and Dean, 1965; Allen and Guy, 1974).

3.   Kendon (1967) referred to the kinds of processes just described as expressive and monitoring functions:

> "Expressive" means a signal of the degree of interest, involvement with the speaker, or arousal.

Monitoring refers to a checking of the partner's reactions. In addition, he identified two other functions of eye gaze during face-to-face interaction. The "regulatory" function refers to what can be termed "ocular breaks" as a source of controlling the beginning and end of verbal sequences. The "cognitive" function refers to the breaking of eye contact to reduce sensory load at those points in the verbal stream when intensive mental processes are involved.

A recent test of Kendon's theory found that it was indeed true that there was a higher incidence of ocular breaks in the speaking mode than in the listening mode and at points where alternative judgments or other demanding mental processes were involved (Allen and Guy, 1977).

We are jumping ahead of ourselves a bit, but the question becomes, if eye gaze does all these things to help to regulate social interaction in the face-to-face condition, then what are conversational partners going to do without any of these cues? Are they going to flounder and feel frustrated and misunderstand one another? Or are they going to find available, written channels for new kinds of cues within the computerized conferencing mode, to substitute for the functions normally performed by eye gaze?

Body Movement

This includes gestures that are substitutes for words. For instance, a nodding of the head "yes" to indicate approval or agreement with the speaker is a frequently observed communication in both everyday life and laboratory experiments on group discussion processes. Pointing and other arm and hand gestures are used to give emphasis to speech or to try to illustrate meaning or denote emotion or feeling. "Nervous habits" such as foot shaking or finger tapping convey information about the speaker's emotional state or degree of attention. In addition, there are cues of posture and distance that indicate that a person likes the other or is paying attention, such as standing or sitting close to or leaning toward the other. Termination of an accompanying gesture is often a "turn-yielding" cue in a conversation.

Psychophysiological Responses

These have to do with "uncontrollable" cues about emotional state, such as blushing, heavy or rapid breathing, yawning, and eye blinks, which can indicate intense states of nervousness or excitement or other aroused emotional states.

ISBN 0-201-03140-X, 0-201-03141-8 (pbk.)

Available Channels of Communication and Cuing Processes in Computerized Conferencing

The beginning user of a computer-mediated written form of communication is usually most conscious that all the kinds of cues just listed are *missing*, and has an initial tendency to try to ascertain some of these cues by supplementing the written channel with a visit or a telephone call. As the user becomes more practiced, however, there is the development of skills peculiar to this medium and the recognition that it has many potential advantages.

The following reactions are fairly typical:

5C W. DAVID PENNIMAN (DAVE, 254) 11/15/76 3:44 PM
I think there are two major categories of non-verbal communication that concern me here. First, there are the broader underlying meanings of verbal statements which are conveyed by overall context. In this conference setting that context is certainly restricted. We may never see each other "face-to-face" so our context will be what we say to each other or hear about one another from others. Secondly, in the narrow sense, I can't watch your eye contact (to see how shifty you are). Your body language, or hear your tone of voice to help me interpret those underlying meanings. In fact, I'm having to resist the urge to call you up on the telephone to reply to your question instead of sending this message. Your receipt of this in the conference will prove that I resisted the temptation.

8C CHARLTON PRICE 11/20/76 4:00 PM
This is in answer to Roxanne's request for comment on early use experience.
It took me about two days to get over my fumbling. Now the only problem is to discipline myself to stay off the line and get some work done. EIES is potentially a uniquely effective means to intensify and upgrade just about everything I do in my kind of work. It gives me a kick comparable to first learning to drive. I just would like to get more facile and adroit . . . .

What happens to new users, without their conscious recognition in most cases, is that they are faced with a kind of "culture shock" in which all of the very complex "rules" for combining the various kinds of communications channels described earlier do not work, because the nonverbal channels are missing, there are some new channels or means of communication available, and the rules or possibilities for using the written equivalent of the spoken verbal channel work differently. For example, among the basic regulating mechanisms for face-to-face conversations are "turn-yielding cues," which are a complex set of language content (finishing a sentence), vocalizations, body language, etc. As Duncan (1972, p. 299) points out, turn taking is one of a number of communication mechanisms operating in face-to-face interaction that serve the function of integrating the performances of the participants in a number of ways, such as regulating the pace at which the communication proceeds and monitoring deviations from appropriate conduct. New users get frustrated when they do not receive an instantaneous answer to a message they have sent, since they are used to having someone respond immediately when they pause or otherwise show that their "turn" is over.

ISBN 0-201-03140-X, 0-201-03141-8 (pbk.)

Stage 2 of learning to use this new medium begins very quickly. Within 20 to 30 minutes, people are able to learn the basic mechanics; within 2 to 4 hours, they report that they feel they have "learned to use the system reasonably well." Soon they learn to send and receive in "batches" with perhaps a half dozen conversations being conducted simultaneously.

Goffman commented on integrating mechanisms in general, and on turn taking in particular, as follows.

> In any society, whenever the physical possibility of spoken interaction arises, it seems that a system of practices, conventions, and procedural rules comes into play which functions as a means of guiding and organizing the flow of messages [Goffman, 1955, p. 266].

In computerized conferencing, a new system of "conventions and procedural rules" emerges among participants, who must master them as well as the mechanics of the system before they feel comfortable and accomplished in this medium. Learning is a continuous process. Almost all of the 45 subjects we used in experiments were able to learn to send and receive items in less than an hour. On the other hand, the more a person uses the system, the more skilled she/he becomes. (See Table 3-1 for an illustration of improvements in effective input rates on EIES.)

Among experienced users, the "written equivalent" of the language content tends to be somewhat better organized and more fully thought out than comparable statements recorded from a face-to-face conversation. This is because the participant has a chance to take as long as desired to think about a response or comment, to reorganize and rework it until it presents the idea

Table 3-1
Effective Input Rate by Cumulative Time On Line[a]

Usage class (hours on line)	N	Effective input rate (words/minute)
1–2	8	6.4
2–4	17	8.0
4–8	22	8.6
8–16	31	10.7
16–32	15	14.4
32–64	19	19.3
64 +	17	27.7

[a]From a sample of 129 users during the first year, obtained by eliminating EIES programmers, special service roles such as system monitor, and those whose total access to the system was for less than two months. The basic data were collected by a computerized monitoring routine.

ISBN 0-201-03140-X, 0-201-03141-8 (pbk.)

as fully and succinctly as possible. Not all will take advantage of this opportunity for revision and review, of course, but on the average the written channel will tend to have a somewhat richer and better-organized content than spoken conversations, in terms of topic-related information.

Visual Cuing and Expressiveness

One reason for this better-organized content is that written communication introduces an additional visual channel. The use of indentations and numbering can help a person in following a complex argument. The computer can be used to generate statistical summaries and to line up and present tables of the data. Pictures can be drawn using terminal characters, or using a computer graphics routine to produce histograms, two-variable plots, or other visual means of information display. For example, take a look at the two conference comments in Figure 3-2, related to the nature and necessity of humor in CC systems and in other group communication modes. Note that in this exchange:

> The original "question" and the answer were several days apart.
> The statements are much longer and fuller explanations of a set of ideas than would normally be elicited in any face-to-face discussion.
> Numbering or outlining cues are used to help the recipients of the comments to receive the information in an organized manner.

As conferees become more familiar with the visual presentation techniques that may be utilized, they tend to become quite skilled at adding visual components to their text. Figure 3-3A is an example of a holiday greeting. Trees, stars, and other elaborate pictures composed of computer-terminal-typed characters seem to proliferate during the Christmas and holiday season and are used in "sociable" messages rather than those concerned with scientific information exchange. In Figure 3-3B, we see the use of graphics to succinctly present a set of relationships within a scientific conference on structural modeling techniques.

Computerized Conferencing as a Goffmanesque Establishment

The following quote from Goffman (1959, p. 249) describes the inadequacy of cue-searching in any medium, while the second quote below (Goffman, 1969, pp. 4–8) shows his later concern with the fact that cue searching can go on in written communication, as well as face to face.

> When one individual enters the presence of others, he will want to discover the facts of the situation. Were he to possess this information, he could know,

ISBN 0-201-03140-X, 0-201-03141-8 (pbk.)

Figure 3-2   Use of outlining to provide cues in computer conferencing.

C 72 CC194 MURRAY TUROFF (MURRAY, 103) 4/16/77 11:39 AM
   I would like to bring up what may seem like a trivial problem in terms of some of the other items you have been discussing. However, since I started working on systems like this in 1970, it has always been a very personal nagging sort of issue that never goes away.

   My intuition leads me to feel that humor is a key element in promoting group cohesion on systems of this sort. I think in other communication forms we take it very much for granted and are not aware that we have to look carefully at it to determine its actual degree of necessity for any group communication process.

   As opposed to a face-to-face meeting, humor gets documented in systems of this sort, and in practically every system that has ever been built there have been specific conferences set up to collect humor and conventional folklore.

   There are two types of reactions that this produces that lead to difficulties.

   1. That Congress might make hay out of government funds being used to support people having fun in such a manner.
   2. That work is not supposed to be fun.

My concern really comes down to the following two items:

   1. Systems of this sort appear to raise issues we have never really viewed as significant for study, and I believe that the role of humor in the group process is one that deserves considerably more attention than it has received.
   2. The ability of the computer to process means that there are a host of auxiliary services and structures that can be incorporated in a communication environment to foster the use of humor and other tension relaxation methods (games, e.g., bridge; producing funny remarks by having the computer modify a human's comment, etc.)

C 72 CC209 ROBERT BEZILLA (ROBERT, 213) 4/20/77 11:15 AM
   (In reference to C72 CC 194-Turoff-Humor) I can't suggest any experiments on the use of humor in CC, but the following areas of use of humor may be of help to anyone who is working on the problem.

   1. Humor as an assessment tool
      A.   As a means of assessing the intelligence, sophistication of a new acquaintance in a socially acceptable way.
      B.   As a means of assessing group membership, i.e., does the new acquaintance understand the in-humor of the group.
   2. Humor as a socializing mechanism
      A. Previous suggestions about its use as an "icebreaker"
      B. Previous suggestions about its use as a means of creating multi-stranded relationships
         Note: Both, however, may be closely related to the assessment function.
   3. Rhetoric—The use of satire, irony, sarcasm, etc. to underscore, vivify, call attention to the point one is making.

ISBN 0-201-03140-X, 0-201-03141-8 (pbk.)

Figure 3-2 (Continued)

4. Catharsis
    A. The suggestions that graffiti-type outlets are necessary for programmers
    B. Observations that most hostile msgs. usually include a stab at humor at the end to help reduce the tension
    C. Comic relief, perhaps, to relieve tension in extended sessions or when the atmosphere becomes unusually tense.
5. Use as a reductio ad absurdum in testing hypotheses generated in a conference.
6. Induction—Mental excursions into the absurd, use of humorous analogies, etc., in the search for cohesion in seemingly unrelated ideas, in attempting to solve knotty problems.

ISBN 0-201-03140-X, 0-201-03141-8 (pbk.)

86

Figure 3-3A Use of Graphics

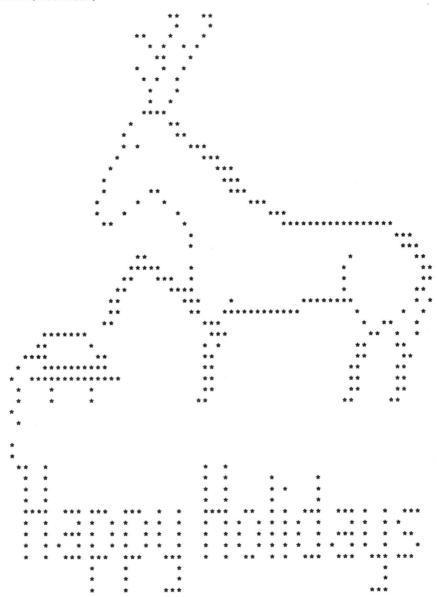

ISBN 0-201-03140-X, 0-201-03141-8 (pbk.)

Figure 3-3B Use of Graphics (Continued)

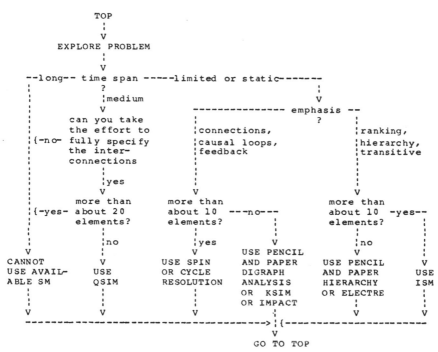

```
KEYS:/FLOWCHART FOR SELECTING STRUCTURAL MODELING METHODS/
                     TOP
                      ¦
                      V
             EXPLORE PROBLEM
                      ¦
                      V
      --long-- time span ----limited or static------
      ¦               ?                              ¦
      ¦               ¦medium                        V
      ¦               V             --------------- emphasis --
      ¦            can you take                      ?
      ¦            the effort to  ¦connections,      ¦ranking,
      ¦:{-no- fully specify       ¦causal loops,     ¦hierarchy,
      ¦            the inter-      ¦feedback          ¦transitive
      ¦            connections     ¦                  ¦
      ¦                            ¦                  ¦
      ¦               ¦yes         ¦                  ¦
      ¦               V            ¦                  V
      ¦            more than    more than          more than
      ¦:{-yes- about 20        about 10  ---no---   about 10  -yes--
      ¦            elements?    elements?   ¦        elements?    ¦
      ¦               ¦            ¦         ¦          ¦          ¦
      ¦               ¦no         ¦yes       V          ¦no        ¦
      V               V            V       USE PENCIL   V          ¦
    CANNOT                      USE SPIN   AND PAPER  USE PENCIL   V
    USE AVAIL-       USE        OR CYCLE   DIGRAPH    AND PAPER   USE
    ABLE SM          QSIM       RESOLUTION ANALYSIS   HIERARCHY   ISM
      ¦               ¦            ¦       OR  KSIM   OR ELECTRE   ¦
      ¦               ¦            ¦       OR IMPACT     ¦         ¦
      V               V            V         ¦          V         V
    ------------------------------------------------>¦{----------------------
                                                      V
                                             GO TO TOP
```

Figures 3.3A and 3.3B are discussed in text on page 83.

and make allowances for, what will come to happen, and he could give the others present as much of their due as is consistent with his enlightened self-interest. To uncover fully the factual nature of the situation, it would be necessary for the individual to know all the relevant social data about the others. It would also be necessary for the individual to know the actual outcome or end product of the activity of the others during the interaction, as well as their innermost feelings concerning him. Full information of this order is rarely available; in its absence, the individual tends to employ substitutes—cues, tests, hints, expressive gestures, status symbols, etc.—as predictive devices . . . .

Any contact which a party has with an individual whether face-to-face or mediated by devices such as the mails, will give the party access to expression. Immediacy, then, does not mark the analytical boundary for the study of expression. Nonetheless, face-to-face interaction has a special place, because whenever an individual can be observed directly a multitude of good sources of expressed information becomes available. For example, appearance and manner can provide information about sex, age, social class, occupation, competencies, and intent . . . .

. . . The behavioral and technical process through which information is COMMUNICATED, like all other human activities, will naturally exude expressions . . . . The least the communicating can express is that the sender has the capacity and apparently the willingness to communicate. Further, for those familiar with the sender, the style of a mediated communication is likely to be sufficiently expressive to tell them whether or not the claimed sender is sending it.

What is involved is the question what kinds of cues are useful for task accomplishment and for individual comfort and group cohesion, for different communication goals and periods of interaction. In some situations, the "narrow-banded" nature of a computerized conference allows more information to be contributed and rationally judged and synthesized than would any other form of communication. In other situations or for other purposes, it may seem desirable to supplement this task and information-oriented narrow band by a combination of:

Software (such as joke conferences and on-line games to provide diversion and a coherent way for conferees to relate to each other and get to know more about one another);

Conference leadership (such as a request that group members might exchange information about their various personal and professional interests or biographies);

Supplementing the written computerized conference channel with telephone exchanges, face-to-face meetings, or even picture exchanges.

It is with what Goffman terms "the presentation of self" and what social psychologists term "person-perception" that much of the subjectively experienced social awkwardness occurs. An adult has spent years learning to emit a complex set of cues in face-to-face situations, designed to present a certain

ISBN 0-201-03140-X, 0-201-03141-8 (pbk.)

impression or image to others. All of the learned behavior and set of cues that make up what Goffman terms "personal front" are missing:

> It will be convenient to label as "front" that part of the individual's perfor-mance which regularly functions in a general and fixed fashion to define the situation of those who observe the performance . . . . As part of personal front we may include: insignia of office or rank; clothing, sex, age, and racial characteristics, size and looks, posture, speech patterns, facial expressions, body gestures, and the like . . . . It is sometimes convenient to divide the stimuli which make up personal fronts into "appearance" and "manner" . . . . Ap-pearance may be taken to refer to those stimuli which function at the time to tell us the oncoming situation. Thus a haughty, aggressive manner may give the impression that the performer expects to be the one who will initiate the verbal interaction and direct its course [Goffman, 1959, 22–24].

The cue-emitting capabilities available in computerized conferencing are both more limited than and different from those in the well-rehearsed face-to-face situation. Certainly, the pen name that a person chooses gives some cues as to the status and manner of the person using it; to use "Super-Person" or "Socrates" is meant to create a definite impression. How-ever, the participants are quite aware that this alter ego may be deliberately false; "Socrates" may turn out to be a young female, and "Super-Person" a middle-aged male, both of whom have deliberately chosen to mislead. A great deal about personality and degree of literacy and intelligence are also conveyed by the language chosen. What seems to happen is that the par-ticipants pick up these cues but are not able to have confidence in them, at least at first, because they are missing the confirmation of additional kinds of cues.

What may seem an inadequate set of cues in computerized conferencing for novice users may later be overcome by participants learning how to substitute for missing kinds of cues.

For example, if someone does not respond to a message or comment directed to her/him, how is this to be interpreted without any other cues? Is the recipient disinterested? Does he/she disagree but not want to say so? Is she/he away on a trip or busy? Ill and unable to communicate? Is the terminal broken? Consider the following self-report from EIES:

M 7111 ROBIN CRICKMAN (ROBIN, 730) 4/26/77 11:18 AM
ELAINE AND CHARLTON,
   I am struck by a missing cue on conferences. I often make allowance for colleagues whom I see regularly if I can see they are feeling ill or depressed or looking harassed. I can't make allowances on conference for the same reason because I don't know if someone is sick or down or busy. I can't see them to tell.

ISBN 0-201-03141-8 (pbk.)   0-201-03140-X,

A colleague, Carl Drott, suggested that you might want to see if someone has studied how blind people manage the same problem among themselves. He suspects they may verbalize it more explicitly than the sighted do. I observed it when I realized I hadn't heard from Roxanne for several days and asked if anything was wrong. It turned out she had been ill. I do not see how the forgiving capability of face-to-face meetings can be incorporated. Maybe C.C. folks will just be altogether more forgiving of lapses in one another.
cc: Elaine, Charlton, Roxanne

Norms tend to develop in ongoing computerized conferences, so that participants will sign in and send excuses or explanations for periods of nonresponse, such as "I will be away all next week folks, and will not be able to participate"; or "I am down with the flu this week, but intend to catch up with your deliberations as soon as I can" (types of accounts observed on EIES) or "Wife having baby, will not be around for the next 24 hours" (a memorable message from EMISARI).

An experiment by Reid (1970, reported in Short et al., 1976, pp. 122–123) provides indirect support for this interpretation of the subjective aspect of the limited range of cues. He compared face-to-face with an audio-only condition in terms of the objective degree to which persons were able to tell if another was lying, and the subjective feeling of confidence about being able to make this determination. The speakers were asked to lie or tell the truth in answer to questions in a prearranged sequence for 30-second intervals. There was virtually no difference in the ability of listeners to make correct judgements: 59.9% in the face-to-face condition and 58.7% in the audio-only condition. However, there was an unjustified higher degree of *confidence* in judgements in the face-to-face condition, when rated on a three-point scale from "very confident" through "fairly confident" to "not confident at all."

The "private message" represents in some ways the equivalent of "whispering" to another person during a group meeting to carry on a side conversation or negotiation. However, whereas other persons can hear a whisper or see a note passed, no one who does not receive a private message knows that any communication has occurred. What is a "missing cue" for other group members is a new capability for the sender.

The private message can be sent to one or more individuals, whereas a person cannot "whisper" to two others at once. Moreover, the sender may choose to show to each the names of all those who received the communication, or to "blind copy" it, so that each recipient does not know if it went to others. This has become the source of some emerging normative force on the EIES system, with the pressure generally being to get senders to list "who else" received a message, rather than leave each person without this cue and unsure of who else shares the information.

ISBN 0-201-03140-X, 0-201-03141-8 (pbk.)

Written Vocalization

It is thus possible for people to use substitutes to replace some of the cues and information normally transmitted by vocalizations or by nonverbal channels. For instance, instead of nodding the head, a message "I agree with you" might be sent. Accomplished users often insert written descriptions of cues or emotions to provide the proper context or make an emotional point. For example, among the equivalent of vocalizations that have been embedded in messages sent on EIES are "Ha! Ha!," "(giggle)," and "Boo Hoo."

Unless supplementary forms of communication are used, however, most of the nonverbal content is lost, for better or for worse. On the positive side is the consideration that it is the *content* of the communication that can be focused on, without any irrelevant status cues distorting the reception of the information, especially if anonymity makes even the sex of the contributor unknown. On the negative side, there are situations in which people want to know how the other person looks or feels, in order to know how to interpret the meaning or significance of a communication or lack of it. Consider, for instance, the excerpts of a fairly heated exchange in the Applications and Impacts conference held on EIES, shown in Figure 3-4.

Computerized Conferencing as Several Different Communication Forms

One reason for the lack of consistent findings about communication behavior on a computer-mediated system is that it is not one communication form demanding a single set of skills, but rather many forms. Though the systems are designed to allow some kind of communication by messages and conference comments after only a half hour or so of practice and instruction, mastery of the different kinds of interaction with others facilitated by different communication structures within the system takes a great deal more trial and error. In addition, the "norms," or expected rules for participation, have not fully emerged. When users feel a great deal of confusion about how best to utilize the system, and which part to use for which purpose, and what is socially correct or likely to be considered in poor taste or deviant, then the communication behavior observed is likely to show some irregularities. With time, it can be expected that users both individually and as a kind of collective "subculture" will develop much more skill as well as some shared norms and understanding about etiquette and level of participation, such that the observed behavior will be much more "regular" or "predictable" than has occurred in field trials thus far.

Of all the communication forms and conditions permitted by computerized conferencing, the synchronous discussion seems to cause the most difficulty and feelings of confusion. After all, other forms can be likened to written

ISBN 0-201-03141-8 (pbk.), 0-201-03140-X

Figure 3-4 Opinions on using picture exchange to provide general appearance cues.

C 72 CC210 (ANONYMOUS) 4/22/77 4:38 PM
A fellow conference member has semi-seriously proposed that one way to increase broad-band communications is to exchange pictures, so that our images of each other as we communicate would be more than verbal. How could this be designed into future cc's, if feasible?

72 CC211 ROXANNE HILTZ (ROXANNE, 120) 4/22/77 8:15 PM
One proposal is that a conference coordinator would collect the pictures (old, retouched, or whatever) and make a composite page and send it out to all members of the group. Personally I think this may be a good idea for a group that is going to be interacting and working together over a long period of time, but best avoided for a short-term conference of only a couple of weeks or so. But obviously, this is yet another area where there are no data.

C 72 CC213 (ANONYMOUS) 4/22/77 9:47 PM
The silliness about exchanging pictures shows some people's lack of social finesse or social illiteracy. No sense of what kinds of norms or reciprocal expectations are essential/desirable to be system norms and which are "private" (i.e., off-line business). I for one have been amazed at the amount of communication and cuing and sense of the other is possible on EIES systems, without the need for auxiliary or broader-band media. Social learning or socialization includes the growing ability to customize or fit types of communications to types of media. In the electronic age, the amount of social learning required may be too much for some. Hence there will be more instances of "inappropriate" (and hence unreciprocated and/or negatively sanctioned) communications, in streets, at work, and in the home too. The dimension of error seems to be that of intimacy—how much is deemed appropriate where, when and with whom—and via what communications medium.

C 72 CC215 RA3Y PANKO (RA3Y, 705) 4/23/77 11:32 PM
I'm confused. Is the idea of exchanging pictures part of the proposed experiment on humor?

C 72 CC217 BARRY WELLMAN (BARRY, 720) 4/25/77 12:24 PM
My picture ⟨trial balloon⟩ seems to have generated more interest than anything else recently in the c.c. Possibly because this seemingly trivial issue hits at some fundamental concerns: should all computerized conferences be narrow-banded in information and single-stranded relationships? In favor of the latter, I might note that some of the highest-powered, most "instrumental" groups about which we have information, such as British merchant bankers or the phage /DNA group, depend heavily on broad-banded multistranded information in order to evaluate each other's work, to have sanctions for the enforcement of trust, etc. The picture idea was a minimal version of this, and perhaps ill thought out.

C 72 CC218 ROBERT BEZILLA (ROBERT, 213) 4/25/77 3:32 PM
REF: C72 CC210 (PICTURES)
Pictures and other cuing devices might be helpful if the mission of a CC group is is socializing. From non-CC media there may be some contrary evidence: one could hypothesize that part of the success of CB is that it affords people opportunities to communicate without reference to physical appearance; the ultimate in use of

ISBN 0-201-03140-X, 0-201-03141-8 (pbk.)

Figure 3-4 (Continued)

picture conferencing might be video telephone conferencing—which has been a dismal failure.

If the mission of a CC group is scientific inquiry and communications, the pictures and other cues could distract more than enhance. The elimination of pictures and cues could serve also to eliminate ad hominem and authority fallacies . . . .

For that matter, in some cases should pen names be chosen randomly in order to prevent cuing? (Are we influenced to any extent by the act of someone choosing to call herself "Wonder Woman"?)

How should we be applying Occam's razor to CC? What are the necessary components of our communication and what are the components that merely produce distracting noise?

---

letters or notes or papers, since the sending and receiving are separate in time, and it seems easier to transfer one's usual writing habits to the new form. However, a real-time "conversation" presents the expectation of interaction, tends to lead to the sending of many short unreviewed comments, and makes the participants most conscious of the missing kinds of cues, as compared to a face-to-face or audio situation. Unless such interaction is well structured by a leader, the participants are likely to feel that they do not know what they are supposed to do or how to behave. For example, Figure 3-5 shows selected entries in an unstructured and disorganized on-line "party" that was attempted without the necessary kinds of clear instructions as to exactly what everyone was supposed to do. (It did not help matters that as many as 17 persons were participating in the interchanges at one time, whereas only six or seven had been "expected"; thus, not enough computer core had been allocated, and the system slowed down to the point where it would pause for 30 seconds or so before printing each new line.) The participants had all taken part in exchanging messages and conference comments for several weeks, but only two of them had ever tried a simultaneous interaction with several other persons. It was actually rather amazing that despite the system delays, most participants "stayed at the party" until the end.

Figure 3-5 Reactions to missing cues during a synchronous "party."

14530 M BARRY WELLMAN (BARRY, 720) 4/1/77 5:13 PM
  Hi, Barry Wellman here ready to join real-time party.
  How do I get in???

14541 M ROBIN CRICKMAN (ROBIN, 730) 4/1/77 5:25 PM
ROXANNE
  Like Barry, I am confused exactly how to proceed at this point. Am I supposed to respond to specific items. Which ones? Maybe an item or group message from you would be timely just now. RC

ISBN 0-201-03140-X, 0-201-03141-8 (pbk.)

Figure 3-5 (Continued)

14551 M ROBERT BEZILLA (ROBERT, 213) 4/1/77 5:38 PM
This seems to be one of those parties where everyone stands around waiting for someone to say something.

162 C JACQUES VALLEE (704) 4/1/77 5:42 PM
I just realized I would not be able to participate if I listed the messages waiting, so I had to bypass the earlier discussion. I hope what I am going to enter will not be too out of context.

14570 M ELAINE KERR (ELAINE, 114) 4/1/77 6:05 PM
Private message: Roxanne, this is awful, please be dictatorial, seems nothing's being accomplished.

14577 M ROXANNE HILTZ (ROXANNE, 120) 4/1/77 6:18 PM
I am simply not going to try to tell you what to do or say, so please stop asking me, people!

14620 M ANDY HARDY (719) 4/1/77 7:09 PM
Roxanne when I said I wanted to see who was there I meant in conference 72 or with Group 72. I also by mistake discovered that everyone was interacting through the message system, not the conference system as I had assumed. Kind of like going to the wrong house when you're trying to get to a party.

M 14797 ROBIN CRICKMAN (ROBIN, 730) 4/4/77 12:01 PM
An open letter to Roxanne on the "cocktail party" experience
I have one observation I wanted to try out on you to see if you also think this would help group activities. When I was using a computer conference in Michigan to collect the opinions of citizens on recombinant DNA research, I found it necessary to structure the interaction highly. This was initially resented as being constricting and stifling creativity, but was later accepted (in retrospect) as the only way to get a group of neophytes into the activity quickly and to keep the discussion active.
Just as the hostess of a cocktail party must see that most guests are introduced to people and that new topics are suggested when conversation lags, I think it becomes necessary for a CC organizer to accept a heavy burden when simultaneous or mission-oriented activity is contemplated. The subject needs to be broken into some series of questions and each person must understand how they are to contribute to those subjects. The organizer must undertake to summarize positions and suggest what each person is to do next. Some amusement or other diversion has to be suggested for participants who are waiting until the system is ready for them to take the next step. Some sort of closing convention is needed . . . .
I know we are all grown people who don't need to be led by the hand but I also think that when a new technology is evolving, more structure should be evoked for interaction. I learned how to greet people at f-t-f cocktail parties several years ago, when to break off conversation and when the party was over and I should go home. These were carried as implicit cues of the environment and interaction. I don't have the cues in a CC cocktail party, and so I hope for structure to replace them . . . .

CC: ROXANNE   (GROUP 72)

ISBN 0-201-03140-X, 0-201-03141-8 (pbk.)

Letting Go of Face

Participants in a computerized conference are physically alone. While they learn to concentrate more and more on sending and receiving typed cues, they almost immediately drop the usual "face-work" or maintenance of "front" that would occur in a face-to-face meeting. (See Goffman, "On Face Work," 1967.) For example, consider the following observation of a first-time participant, made through a one-way mirror (see Hiltz et al., 1978, for a further description of the experiment):

> Subject chews gum, blows bubble, while reading intently. Laughs at receipt of one message. Pops a bubble. Leans over and re-reads items just received. Typès a reply; touch types but checks accuracy every few letters. Checks watch. Chews gum while reading messages as they print out. Goes back and re-reads one just received. Shakes head, no; again, shakes no. Then deletes scratchpad; pauses and looks back. Then moves chair in; sighs slightly; begins typing.

Would the subject have felt as free to chew and pop bubble gum, or to shake her head vigorously "no," in a face-to-face group? What seemed to be happening was that subjects did not worry about controlling overt behavior that would be discrediting or result in loss of face in a face-to-face meeting.

Other participants have reported experiencing a certain amount of delight about not having to worry about what to wear to a computerized meeting, or at being able to eat and drink freely while communicating, or to carry on side interactions simultaneously. In other words, what would be deviance in other group communications contexts is not communicated and not perceived as such in a CC situation.

Pen Names and Anonymity

A pen name is like a mask or a costume; it helps people to play a role in a conference, and to say things that they would not particularly want attributed to them or to their organization. The most frequent use is in conferences in which all the participants agree to use pen names. Sometimes participants begin to feel too type cast and boxed in by the name they have chosen; then they simply adopt a different pen name and play a different role. The word "play" has been purposeful; the use of pen names often signals a "play form" of conferencing; something that is not meant entirely seriously. The advantage of pen names over total anonymity is that participants can address a reply to a pen name, whereas you cannot send a message to "anonymous."

Anonymity seems to be used to avoid embarrassment. Figure 3-6 shows some comments sent anonymously by our experimental subjects. In the first, the person was sending her/his first entry, and was obviously embarrassed by the lack of skill in using the medium that it revealed. In the second, a rather sharp criticism of an unknown latecomer is made; in the third, there is

ISBN 0-201-03140-X, 0-201-03141-8 (pbk.)

Figure 3-6 Examples of anonymous comments.

C 910 CC8 (ANONYMOUS) 6/24/77 1:17 PM
It is suppose to rain later on today, but Ishhh
It is suppose to rain later on today, but I hope that is does not.
What do uou hhh
What do you thind about that?
Nothing
;hhhhhhhhhh
What does this mean hhhhhhh?
It mean that I do not know hos to work this machine yet.
Do you?

C 910 CC21 (ANONYMOUS) 6/24/77 1:34 PM
Well I am glad that this project is going to begin soon because
I am tired of waiting
you people that were late should try harder next time to be on time
it is not nice to keep people waiting. I could be doing something
else

C 910 CC24 (ANONYMOUS) 6/24/77 1:36 PM
I agree its not nice to be kept waiting, but sometimes
there are unforseeable circumstances

C 910 CC26 (ANONYMOUS) 6/24/77 1:38 PM
No one is out here
This is all a figment of your imagination

---

disagreement with the second comment. These are the kinds of situations in which experienced users also tend to sign things anonymously.

The motivation of the sender of an anonymous message or conference comment is self-protection. However, anonymity can have some very important social consequences for the groups. As Robert Bezilla points out (1977, p. 5) the use of anonymity can promote interaction, objectivity, and problem solving:

1. Interaction. The free exchange of ideas or reporting of matters without the threat of disclosure . . . .
2. Objectivity. The masking of identity can serve to suppress distracting sensory cuing or *ad hominem* fallacies; that is, the matter being reported or suggested can be considered on its intrinsic merits without regard to personal origin or aspects of origin.
3. Problem solving. The total subordination of the individual ego to the group task . . . e.g., one would not have to worry about peer relations, risk ridicule, advancement of unpopular ideas, etc.

Norms and Sanctions

As an illustration that groups can form and develop norms and sanction deviants and engage in the social processes that normally characterize the face-to-face condition, consider the excerpts from an EIES conference in

ISBN 0-201-03140-X, 0-201-03141-8 (pbk.)

Figure 3-7. This is a conference group that formed spontaneously among EIES members who had belonged to several different task-oriented conferences. They had come across one another in the Directory or been introduced electronically by third parties who thought they had something in common, or had otherwise "met" fortuitously. As far as non-EIES ties are concerned, "female one" has met "male three" face to face; Hiltz and Turoff have met all except "the couple"; but the other parties have met only electronically.

What happens is that "male one" uses obscenity. He is then sanctioned by most of the other participants, defends himself (offers an account), and

Figure 3-7 The emergence of social processes (norm formation, sanctioning, coalition formation) in "electronic group life" (excerpts from an EIES conference).

C 117 0021 (*MALE ONE*) 7/26/77 12:41 P.M.
ACTION PLAN: COMPUTERIZED CONFERENCING AND SOCIAL DECISION MAKING
  THE SETTING OF SOCIAL POLICY DERIVES ORIGINALLY FROM THE PEOPLE.
The theory of democracy is based on the doctrine of personal choice. Even the patriarchal kings referred to the royal WE. Democracy has never been more than an ideal. Because of the severe economic and time constraints, it has always been assumed that true democracy, government by the people, is impossible. Under this constraint, the legislative process is a fair approximation of the ideal. Therefore, for my current purposes, I will claim that the setting of social policy is properly the domain of the elected legislative bodies.

Social Engineering is the art of making practical application of the social sciences. The setting of social policy (objectives) is an act of social engineering. It is important that the portion of social engineering which is policy formation be in public domain i.e. legislative bodies.

C 117 CC26 *MURRAY TUROFF* 7/26/77 9:52 P.M.
Reference 21, you are making a lot of assumptions implicitly in your comments. Social Engineering is a two-edged sword, the Germans practiced it as a fine art. Who is to say if you are manipulating values for the better or the worse. At least to me social engineering is the design of human systems and such systems will alter behavior and values.

As for legislatures you have two underlying problems. Too many politicians believe the public cannot grasp anything but the simplest of rationales, and refuse to raise issues above that level. And an educational system that still seems to produce a public which cannot always deal with complex issues and have really delegated authority to the political body.

C 117 CC38 (*THE COUPLE*) 7/27/77 4:44 P.M.
Regarding Murray's comments in 0026:
You bet social engineering is a two-edged sword, just like any other technological/design problem where we take into our hands that which has been "outside" our control—some times it helps things, some other times it makes them much worse. Are we really intelligent enough to design a humane society when we can't seem to be humane in the first place?

ISBN 0-201-03140-X, 0-201-03141-8 (pbk.)

Figure 3-7 (Continued)

As for legislatures—we completely agree that most politicians have virtually no faith in the public and will generally act in ways to assure that the public has little or no information with which to take charge. Our primary interest in cc is as it relates to enabling and empowering people who now do not have rich access to information.

C 117 CC45 (*FEMALE ONE*) 7/27/77 7:45 P.M.
various comments on the proceeding entries:
   CC 26 (Murray) and CC 38 (the couple): I'm glad that others here share my skepticism of that-which-now-is (politically and otherwise?)
   CC 21: I'm not at all sure that "the setting of social policy is properly the domain of the elected legislative bodies." More education of the public is necessary, yes, but combined with the political cynicism exposed in the above paragraph, I for one am not at all comfortable with the current situation of having elected politicians set social policies and bureaucrats implementing them. My vote is to have sociologists (and their friends, of course!) set social policies . . .

C 117 CC48 (*MALE ONE*) 7/28/77 10:11 A.M.
   I have a considerably different perception of a state legislator than Murray, (the couple) and (female one). ((Female one) is an elitist!)
   A legislator is just as human as the rest of us. They are grossly underpayed to do a very unpleasant job with little or no staff. They have no job security.
   Most of their work consists of servicing constituent complaints. Most of the rest of their work is spent in a polite fight with colleagues over money for street lamps (etc.).
   Within all this mess they are supposed to be writing law, and at least in Massachusetts, being a legislator is a part-time job.
   As for (female one's) opinion that POLICY should be set by the sociologists, what the fuck right do you have to dictate your opinions to the rest of socitey? I am very much afraid of giving anyone the right to set policy, but with interdependence (and all that) someone must have the power of social control. I think that the majority of the people are the only people who have an inherent right to dictate their opinions. It is currently impossible to manage society by the will of the majority, so I'm willing to go with the closest available alternative. If the sociologists gain control of the country, I'm leaving.

C 117 CC49 (*MALE TWO*) 7/28/77 10:51 A.M.
   Please do continue to copy any items on the pen-name stuff. I am very much interested in the ethical issues conference.
   I cuss like a trooper, but I see no point whatsoever to using four-letter words in any conference.

C 117 CC50 (*MALE TWO*) 7/28/77 10:59 A.M.
   Ignore the first two lines in CC49. (I became so irate that for the first time in about eight months I forgot to erase the scratchpad.)

C 117 CC51 (*MALE ONE*) 7/28/77 11:18 A.M.
   This is an interesting difference. I swear very little, but I'm not particular where I use such words. The reason I swear very little is that otherwise they have no impact. The reason I used one here is precisely to have some impact—this I seem to have done.

ISBN 0-201-03140-X, 0-201-03141-8 (pbk.)

Figure 3-7 (Continued)

C 117 CC63 (*THE COUPLE*) 7/28/77 4:43 P.M.
Here's our contribution of bits and pieces about seven of you—not exhaustive—
just a few impressions. No offense meant by any of the words. Change, modify, add,
whatever. Join the psychodrama!

(Female One):
— always thinks in small letters
— likes to help people
— sophisticated EIESer
— knows that sociologists are wiser than politicians
— feminist

(Male One):
— refuses to use American Standard English
— likes to be involved in the action, wherever it is
— his religion is democracy and (yet?) sees social engineering as the domain of
  legislatures
— brief

Murray:
— knows the secrets of the undocumented system commands
— waits patiently while we discover what he already knows
— changes diapers
— somewhat sardonic sense of humor about computers
— likes word games and intellectual pranks
— hints at things

(Male Two):
— ALWAYS THINKS IN CAPITAL LETTERS
— helps clear up brainstormer's overload in conferences
— has a good garbage detector
— philosopher by training
— intellectual wit
— skeptical about participation and legislative cc
— a cues and clues man

C 117 CC65 (*MALE THREE*) 7/28/77 5:09 P.M.
I'm sure we're all loving your vignettes, (couple). You adroitly have cited the
things I suspect most of us would all most like to be perceived by all the rest of us.
You also have freed us from the rather heavy and nasty mood induced by (male
one's) semantic faux pas. I agree with (male two) and some others who did not say
so in the conference that it was worse than a faux pas, by the way.

C 117 CC66 (*ANONYMOUS*) 7/28/77 8:41 P.M.
A note to (male one) on his insensitive use of words.
To a woman, the use of obscene language is like being called nigger if you're
black or other kinds of degrading epithet. It seems like a male power play
deliberately meant to assert superiority and put the female "in her place" as some
kind of subhuman object. At least, I THINK that is why obscenity directed at me

ISBN 0-201-03140-X, 0-201-03141-8 (pbk.)

Figure 3-7 (Continued)

by males makes me very angry and upset. Any other psychiatric interpretations in the group?
Actually, I changed my mind and I'm going to sign this. Roxanne

C 117 CC68 (*MALE ONE*) 7/29/77 10:26 A.M.

I've been bit. Actually my——should have been directed at Murray, as he started that line of discussion. I don't know about the rest of you, but I come from a lower class background. I am the only one in my family who has gone to college. I am going to stick by my guns on this one. There are a number of things I consider truly obscene, the VN war, the U.S. educational system, other wars, etc. Words are words and cannot in themselves be obscene. I'm not going to try psychiatric interpretations, but I'll try a few words on semantics and connotation. Words have meaning only in a sociological context. The context in which I have existed for most of the last few years defines fuck as something like messed up. This is a much better meaning than the more traditional one, and if you don't like the connotation of a word then I think the best thing to do is step on it. I am truly sorry that I appear to have hurt some feelings, but at the same time I don't feel that the feelings had any reason to be hurt. To Roxanne in cc66, the use of such language may offend some women but I know a large number who swear much more than I do. You used the word insensitive, in my opinion I made very sensitive use of language in this context. (Female one) had said something which I consider morally obscene. I wouldn't continue on with this line of discussion if I didn't think talking about it could serve a purpose. (Male three) mentioned T-groups. I'm ready.

C 117 CC74 *MURRAY TUROFF* 7/29/77 1:33 P.M.

Some feminists use 4 letter words to express that they are equal, the movement should print up buttons or signs so the rest of us know which type we are dealing with.

C 117 CC75 (*MALE ONE*) 2:02 P.M.

As long as this conference keeps me amused I promise not to use anymore of those words some of you are so upset about. If the conference begins to drag I promise an item consisting entirely of those words.

C 117 CC89 (*FEMALE ONE*) 6:09 P.M.
various comments on the preceding entries:
    Last time I tried this, the results were negative, but I'll be brave and type ahead:
    My CC 45 which enraged (male one) to obscenity was meant as a quasi-tongue-in-cheek cynical view of the american political structure. I tend to be optimistic and elastic in my human one-to-one relationships, but pessimistic and skeptical when it comes to societal levels. This may be part of my sociological bias. In any case, my training, professional and otherwise, has taught me that there are more normatively acceptable ways of disagreeing than that chosen by (male one). This issue seems to have been resolved, and I'm saying so for the record.

C 117 CC93 (*MALE THREE*) 7/29/77 7:05 P.M.

Although I said (the couple) had moved us to a higher level, and although (female one) says the matter is resolved, I have to dump on (male one) some more because I think there are some points to be made along the way that may be useful to our common enterprise.

ISBN 0-201-03140-X, 0-201-03141-8 (pbk.)

Figure 3-7 (Continued)

(Male one), you baffle me. Sometimes you are cogent and engaging, and other times you are the bad little boy who will make a mess on the living room rug unless you can have everyone's undivided attention.

In cc 68 you claim lower class origins and the "obscenities" of current history make letters referring to sexual and bodily functions ok because they are much less obscene. As to the first, lower-classness may be an explanation (which I doubt, because it's such a denigration of the dignity of all persons including those low on the social scale) but not an excuse: it's just the "Officer Krupke" argument (West Side Story): "I am ass I am I am because of my background"—I think I will leave the typo in the previous line, even though I have at last learned the edit command to fix it.

In cc75 you threaten us with making a mess on the rug if we don't amuse you. I predict that if you try that you will get the deepfreeze treatment. So, since we would like you to stay around, please don't.

C 117 CC94 *MURRAY TUROFF* 7/29/77 7:40 P.M.

(Male three), get down off your high horse, you have your modes too.

C 117 CC146 (*MALE ONE*) 8/2/77 1:35 P.M.

(Male three) wishes to continue on this, and I'm not going to let Murray's dumping on him stop us.

I'm not going to take your etiquette any more than I'll take American Standard. I'm talking to your mind, (male three), not your body. If you expect me to hide behind social walls you've set up you can just go away. I'll admit that I can be a bull in a china shop, but in this conference the one and only thing I intend to accomplish is break ALL the china.

C 117 CC161 (*THE COUPLE*) 8/2/77 7:48 P.M.

We have been watching the conversations here about four letter words and the rules/etiquette/whatever for communicating together in this game/real-life situation. First, we have no particular objection to four letter words; we use them ourselves on occasion in appropriate situations. Our comments below are in no way based on negative feelings toward (male one) for his choice of words. In fact, we are mildly surprised that there was such a strong reaction from several others.

What we seek to communicate in this comment is the depth of our currently negative feelings towards (male one). We are bored by most of what he has said recently. We resent his taking up time in this conference with what seem to us to be mini-tantrums directed at those of us perceived by him to be parents. We further resent the fact that he seems incapable of responding in a fairly direct, open manner without making completely vacuous threats. There is virtually nothing (male one) can do to tear down walls or break china here. This is a completely open communications space in which any of us can say whatever we choose, including (male one). This openness is balanced by the complementary fact that each of us can also choose to ignore anything that anyone else says. According to our understanding, these are the only limitations (walls?) that exist here. Clearly (male one) can continue to fill this conference up with comments until he is blue in the face. Clearly we can all ignore him. In cc we all have power, and yet we all have none.

By making such threats, (male one) is acting in a way which makes us feel quite walled off from him. (Male one), we feel you are building at least as many walls as

ISBN 0-201-03140-X, 0-201-03141-8 (pbk.)

Figure 3-7 (Continued)

you are tearing down. Walls develop between people when they stop trying to communicate openly. Your tirades make us so burned out that we don't even care to communicate with you beyond this last-ditch effort. Is this the winning you seek? If you want to break down the walls you must use LOVE. We suspect that you want the same from us. Maybe that's why you have been acting so weird. Hate, accusations, defensiveness, threats, and things like that may indeed give you proof that you are responsible for what you say and are free to say what you will, but they will not tear down any walls. If we trusted you a little more, we might let our walls down. If you were not so threatened by this situation, you might be willing to give and take a little more.

We expect you can find some points to pick at here. Do what you will. But we ask, quite seriously, that you consider the spirit of what we are trying to say. Come on in, join us in working together, and please, please, stop making it so difficult. We know you have as much power here as anyone else. Let's get on with it.

C 117 CC162 (*MALE ONE*) 8/2/77 8:05 P.M.
Thank you very much (couple). I have cried. That is what I really wanted out of this . . . . The final paragraph was totally unnecessary . . . . I am crying . . . . I knew it but I could not stop it. The way to put me down is to talk to me not at me . . .

---

eventually promises to behave himself. There is also a great deal of emotion expressed; "male two" gets so upset that he makes an elementary error (forgetting to erase his scratchpad). "The couple" play the role of peace-makers, offering love and forgiveness to the deviant. He cries, and tells the others that he has done so.

This transcript is evidence not only that group social processes emerge in computerized conferences, but also that the exchanges are often very personal and highly emotional. Some participants come to feel that their very best and closest friends are members of their electronic group, whom they seldom or never see. That this level of intensity of involvement with one another can be generated and sustained has important implications for possible applications and impacts of the medium.

Phobias, Deviance, and Computerized Conferencing

There are some persons who dislike face-to-face social interaction, or who are not skilled at interaction rituals, who can nevertheless seem particularly suited to communication in this new medium. One example is the hypochondriac, who is afraid of being contaminated by the germs of sick persons. The sterile isolation afforded by conversation by computer is perfectly suited to such a personality. Another is the person who is constantly interrupting others, talking over them, interjecting comments that are "off the subject," or otherwise aggressively deviant from the norms of polite face-to-face conversation. What might seem to others to be obnoxious behavior in

ISBN 0-201-03140-X, 0-201-03141-8 (pbk.)

the face-to-face condition is apt to seem scintillating or original in computerized conferencing since no one is actually interrupted by such out-of-order contributions. Another is the person who cannot remember names, and who is constantly insulting others by calling them the wrong name. In a computerized conference, the computer remembers the proper and complete name for the sender, who need only pick it from a list.

Approximately a year's observation of behavior on the EIES system has yielded the following kinds of situations:

- Several dyads (two-person relationships) in which the persons involved dislike one another when talking face to face, to the point that the interaction often terminates with overt hostility, whereas the two work together very well over the conferencing system;
- At least one case of a dyadic relationship in which two persons are fairly friendly and cooperative in face-to-face meetings, but in which disagreements and hostility soon surface when they communicate by computerized conferencing;
- A young woman who exhibited signs of schizophrenia or other severe personality disturbance, communicating in grunts, nods, and monosyllables in a face-to-face condition; within ten minutes of being introduced to CC, she was sending a constant stream of long messages, all signed "anonymous."

These observations and speculations are offered to suggest our extreme ignorance in this area. We know that there are personality factors that are very strongly related to the amount and style of use a person is likely to make of computerized conferencing. At present we have only the skimpiest of insights into what these factors are.

Stage Three: Addiction and Dependence

Among experienced participants in computerized conferences there emerges a strong urge to check in several times a day to receive any waiting messages and to see what is happening in various conferences. One EIES user in Boston formally had access only once a week, when he worked in the office building that contained a terminal. He reports that he found himself making special trips to that building several times a week, just to sign on the system. This "addiction" is the source of considerable antagonism from the families of EIES members with terminals at home because members are prone to sneaking off after dinner for a session, rather than talking or helping with the dishes.

One place that conscious addiction manifests itself is in the somewhat sheepish or apologetic requests for more time on the system, when a person's

ISBN 0-201-03140-X, 0-201-03141-8 (pbk.)

allotted quota is threatening to expire. Here is an example of a request that explicitly recognizes the tendency toward dependency on the system:

OK Murray,

You invented this damned system. You know fully what the potentials are for trapping young innocent minds like ours into a vicious addicted circle of sending messages and hungrily waiting at the keyboard for the response. So you shouldn't be surprised to find us knocking at your terminal once again, asking for more computer time—please, please more computer time. We have only a few more hours left before we fall away into the cold abyss of "your allocation has expired." Quick, we need our August fix so we can continue collaborating with Barry Wellman on the INSNA network, participating in the C117 therapy group, participating in the NCSL demo tomorrow, showing EIES to the emerging new regional science resource network, and contriving more seductive phrases and paragraphs to entice even more young innocent minds into the perverse practices of CC. Have you ever read the article titled, "In Each of Us a Monster Dwells"?

Your hopelessly addicted friends,

Of the members of EIES during the first year we would consider most of those who had spent in excess of 64 hours on line to be addicts. Table 3-2 shows how their communication behavior differs from that of other members. They sign on over 17 times a week (since the system was down on Sundays,

Table 3-2
Comparative Parameters for Profiles

Usage class/Total hours	1–2	2–4	4–8	8–16	16–32	32–64	64 +
N	8	17	22	31	15	19	17
Mean interaction time (minutes)	11	9	10	12	15	18	20
Interactions per week	0.6	1.1	2.5	2.0	3.2	4.4	17.4
Hours on per week	0.11	0.17	0.42	0.40	0.80	1.32	5.80
Comments received: comments sent	84	56	63	28.5	18.5	8.4	5.8
Messages received: messages sent	12.9	8.9	3.9	4.4	4.5	2.6	1.7
Total items received: total items sent	20.0	12.7	7.6	7.7	7.2	4.5	2.3
Items received per interaction	7.1	3.8	2.8	3.9	5.8	5.8	5.5
Items sent per interaction	0.36	0.30	0.37	0.51	0.80	1.30	2.40

[a]Data gathered by statistical monitor, sample of EIES users, 1976–1977. The user sample was obtained by eliminating EIES programmers, special service roles such as system monitor, and users who had access to the system for only a very short period (two months or less).

ISBN 0-201-03140-X, 0-201-03141-8 (pbk.)

that is about three times a day). Each individual session is somewhat longer than for more casual users of the system.

One thing that seems to be able to explain addiction, in theoretical terms, is exchange theory. In its simplest form, as stated by George Homans (Homans, 1958, 1961), no person will continue to engage in any behavior that is not profitable. "Profit" is defined as rewards for engaging in an interaction minus costs. Costs are, essentially, the value of other activities that have to be foregone in order to continue to engage in a particular interactional exchange.

If we look at rows referring to total items sent (further broken down by messages and comments), we see that in computerized conferencing, even the most active users "profit"; that is, they receive back considerably more items than they send. How can this be? It is because each message can be multiply addressed, and each conference item is received by all members of the conference. This is not possible in any one to-one form of communication, such as the telephone call or the personal letter. In addition, the system is rich enough so that there are always new tricks and features to be learned, each of which makes communication quicker or more effective. Thus, time on the system is continuously rewarded not only in terms of receiving more communications than are sent, but also in terms of large, observable gains in communications skill.

The data on the total of items sent plus received for each interaction are also of interest. Information theorists have tended to find seven to be something of a "magic" number. The short-term human memory seems to be capable of dealing with seven plus or minus two items (see, for example, Martin, 1973, p. 337). Note the very interesting fact that the total number of items per interaction (sent plus received) tends to average about seven regardless of how active the user is. Could it be that unconsciously the user tends to sign on just frequently enough so that the number of items to be dealt with will stay in the psychologically comfortable range of about seven? One observation that would tend to confirm this is the moans and cries of "mock" distress ("Oh, no!") from users who have been away from the system for a much longer time than is usual for them and who sign on and receive notification of something like "waiting: 23 confirmations, 27 private messages, 13 group messages." Their vocalizations can be interpreted as signs of genuine "information overload."

Social Dynamics: The Impact of Computer Mediation on the Group Discussion and Decision-Making Process

Face-to-face group discussions tend to have many dysfunctional aspects. We will begin our comparison of the potential contrasts in social dynamics by reviewing the experimental literature on the nature of the social interaction

ISBN 0-201-03140-X, 0-201-03141-8 (pbk.)

that occurs during group problem solving processes, focusing on phases of interaction, inequality of participation, and leadership or dominance. This review will be interspersed with hypotheses about how group processes using computerized conferencing tend to differ from the face-to-face mode. Where there is preliminary evidence available that supports or refutes these hypotheses, it will be summarized.

One of the most important of the potentially dysfunctional aspects of face-to-face group problem solving is the tremendous pressure on participants to conform. The experimental work of Robert Bales relates to this problem—that is, the tendency to agree with other members of the group, rather than disagree. His work is also related to domination of the face-to-face interaction process by one or a few people; and, third, to the "problem" that a large portion of the interaction that goes on in a face-to-face meeting is not related to the task at hand, but to the personal needs or motives of the participants and the social needs of the group itself.

Bales presented groups of from two to seven members with a standard task. Each member was given a three-page description of a human relations problem in business, with no clear solution. They then met face to face to discuss the problem and tried to reach a group decision. All interactions in these meetings are coded by Bales' Interaction Process Analysis (see Table 3-3), a series of 12 categories or types of statements.

Observers who have been trained in this coding technique can reach very high levels of reliability in their observing and recording. What is involved is that for every statement or action, such as nodding of the head in agreement, the observer notes, on a form on which these categories are listed, who made the statement, to whom it was directed, and the category (1 through 12, Table 3-3) into which it falls.

Through hundreds of replications of group problem solving using Interaction Process Analysis, Bales and his colleagues have established that for small groups (two to seven members) asked to discuss such real-life-type problems and reach a decision, there emerges both a fairly standard distribution of types of contributions and clear "phase" movements and regularities. One of the most important findings is the documentation that agreement is much more prevalent than disagreement in such problem solving discussions. Specifically, for these standard task discussions, 16.5% of all statements were in agreement with the suggestions or opinions of others, whereas only 7.8% were in disagreement. There is also a fairly clear shift in types of statements over the course of a face-to-face problem solving meeting. For instance, over a third of all statements during the first third of a meeting tend to be information giving, and this declines in the next two thirds. Rates of giving opinion are usually highest in the middle portion of the meeting. There are marked increases in positive and negative reactions in the last third of a problem solving conference, and these are said to represent the need for a group to deal with the internal problems generated by the problem solving

ISBN 0-201-03140-X, 0-201-03141-8 (pbk.)

Table 3-3
Categories in Interaction Process Analysis[a]

Social–emotional areas: Positive reactions	A	1	*Shows solidarity*, raises other's status, gives help, reward
		2	*Shows tension release*, jokes, laughs, shows satisfaction
		3	*Agrees*, shows passive acceptance, understands, concurs, complies
Task area: Attempted answers	B	4	*Gives suggestion*, direction implying autonomy for others
		5	*Gives opinion*, evaluation, analysis, expresses feeling, wish
		6	*Gives orientation*, information, repeats, clarifies, confirms
Task area: Questions	C	7	*Asks for orientation*, information, repetition, confirmation
		8	*Asks for opinion*, evaluation, analysis, expression of feeling
		9	*Asks for suggestion*, direction, possible ways of action
Social–emotional area: Negative reactions	D	10	*Disagrees*, shows passive rejection, formality, withholds help
		11	*Shows tension*, asks for help, withdraws out of field
		12	*Shows antagonism*, deflates other's status, defends or asserts self

[a]From Bales, 1950a, p. 258.

effort (see Bales, 1955). Manifested by these shifts is an overall phase movement between task-oriented problem solving attempts relating to the external environment (adaption and goal achievement, in Parson's terms) and the social–emotional internal needs of the group and its members to resolve the tensions generated within it—integration and pattern maintenance, in Parsons' (1951) terms.

Inequality of Participation

Bales found that in face-to-face discussion there usually emerges a "top man" who sends and receives a disproportionate number of messages and who addresses considerably more remarks to the group as a whole than he addresses to specific individuals (Bales et al., 1951, p. 465). Commenting on the processes that produce this dominance, Bales (1955, p. 34) has written

ISBN 0-201-03140-X, 0-201-03141-8 (pbk.)

This tendency toward inequality of participation over the short run has cumulative side effects on the social organization of the group. The man who gets his speech in first begins to build a reputation. Success in obtaining the acceptance of problem-solving attempts seems to lead the successful person to do more of the same, with the result that eventually the members come to assume a rank order by task ability.

Many other studies have also documented strong correlations, running around 0.95, between amount of talking in a face-to-face discussion group and the likelihood of being perceived as a "leader." (See, for instance, Norfleet, 1948; French, 1950; Bavelas et al., 1965.)

What, then, causes a person to do most of the talking? This tendency for an individual to be slow in responding or jumping into a conversation, or prone to speedy replies and interruptions, was noted by Chapple and Arensberg in 1940 and has come to be recognized as a fairly stable individual characteristic (the LVR, latency of verbal response, measured by response time on sentence stub completion tasks). In a task that minimized differences in competence (moral dilemmas, such as whether a man with a wife dying of cancer should steal some expensive drug that might save her), Willard and Strodtbeck (1972) found that a participant's LVR was the strongest predictor of participation, compared with measures of IQ and personality (correlation of -0.60; the correlation between IQ and personality, for instance, was only 0.12).

What is interesting here is that the evidence indicates that persons who happen to be "fast on the draw" in a face-to-face verbal situation, and who may not be particularly intelligent or correct, tend to dominate the discussion and decision-making process in small groups. Computerized conferencing as a mode of communication would pretty much suppress LVR as an operative variable, it is hypothesized, since all participants can be "talking" at once. Moreover, the relative verbosity of a person in written communication is much more likely to be resented than unconsciously deferred to. Thus, it is quite possible that intelligence and correctness might be much more highly correlated with the leadership and dominance processes in decision making that developed in a CC group. Specifically, it is hypothesized that in computerized conferencing, there will be less tendency for a single dominant individual to emerge, and that this contrast in degree of dominance will increase as the group size increases. The hypothesized reason for these anticipated contrasts is that one participant making a statement in no way interferes with the ability of another person to be making a statement that overlaps it in time; those with slower (more latent) verbal responses will not be shut out by the faster reactors in the group.

A second hypothesis is that in computerized conferencing there are more likely to be multiple leaders, each specializing in, and deferred to for expertise in, a particular aspect of the problem. Among the reasons for this, besides the fact that speedy verbalization (LVR) is not operative as a factor, is that there

ISBN 0-201-03140-X, 0-201-03141-8 (pbk.)

is no pressure created by a large number of participants for a single leader to emerge and keep social order by recognizing speakers, etc. The computer substitutes for this order-keeping function and removes the need for a single leader. There may well be a "division of leadership," with different individuals playing moderating, compiling and editing, and policymaking or "expert" roles.

On the other hand, it may be that groups without a single leader who focuses discussion and dominates the others may more frequently fail to reach any decision at all within a specified time.

Available evidence from conferences does tend to show that this form of communication can support fairly heavy participation in a discussion by many persons, rather than dominance by a single leader. For instance, here is the distribution of the number of text lines contributed by each participant in EIES conference 72 (which we have used as an example in this chapter) during the three months through February, 1977 (for the first 109 conference entries, 2345 lines of text).

1–99 lines	5 persons
100–299 lines	6 persons
300–399 lines	3 persons

Two of the five "low" participants were actually in the conference for less than a month. What these figures show is that the majority of the participants made fairly substantial contributions; it seems unlikely that a face-to-face conference of 14 persons would have resulted in such a relatively equal participation pattern.

On the other hand, some conferences have seemed to be virtual monologues, with the moderator putting in comments but no one else responding. What seems to happen in these cases is that the "participants" felt able to "leave the meeting" if it did not interest them; whereas such an act would be considered inexcusably rude in a face-to-face meeting.

Replicating Bales: Evidence from a Pilot Study

Bales experimental problems and procedures were used in a pilot study in order to compare directly the "Interaction Profiles" and amount of inequality of participation for face-to-face and computerized conferencing. Altogether there were only 12 groups, and the face-to-face and computerized trials were conducted sequentially, rather than having groups randomly assigned. There were also several other uncontrolled sources of variation (see Hiltz, Johnson, and Agle, 1978), so the observed differences cannot be said to be conclusive evidence.

ISBN 0-201-03140-X, 0-201-03141-8 (pbk.)

DIFFERENCES IN BALES INTERACTION PROFILE ANALYSES

The mean proportion of statements recorded in each of the 12 Bales categories for the three face-to-face groups for which full coding was reliable and eight computerized groups are shown in Table 3-4.*

The main differences are in categories 3 (agreement) and 10 (disagreement). There was a lot more overt agreement communicated among the members of the face-to-face groups than was typed into the conferencing system. Included in "agreement" was anything that was understandable on the tape recording as a symbol of agreement. In other words, the participants did not have to say "I agree with you"; an "uh-huh" or "yeah" would suffice. It appears that at least with new users of computerized conferencing who have not become socialized to the need to make everything explicit in this medium, there will be less overt cuing of the extent of agreement with the statements of others.

Most of the other differences that are statistically significant are substantively so small that we hesitate to say that they might mean anything. They very well might be due to lack of reliability in coding. On the other hand,

Table 3-4

Mean Responses (%) in Bale's 12 Categories for Face-to-Face ($N = 3$) and CC ($N = 8$) Groups: Mann–Whitney U Values, and Probabilities

	Category	Mean (%)		M–W U	P
		Face to face	CC		
1	(Solidarity)	0.75	2.45	19	>0.10
2	(Tension release)	3.17	0.56	0	<0.02*
3	(Agrees)	22.51	5.42	0	<0.02*
4	(Suggestions)	5.24	9.33	18	>0.10
5	(Opinions)	41.16	55.64	21	<0.10
6	(Orientation)	14.91	11.86	7	>0.10
7	(Asks orientation)	3.20	3.16	9	>0.10
8	(Asks opinion)	2.06	5.73	22	<0.05*
9	(Asks suggestion)	0.64	0.94	13	>0.10
10	(Disagrees)	4.34	0.70	0	<0.02*
11	(Shows tension)	1.14	1.60	15	>0.10
12	(Antagonism)	0.94	2.99	16	>0.10

*Significant at 0.05 level.

---

*The nonparametric Mann–Whitney U-test was chosen to compare the two conditions (face to face and computerized) because of the nature of the samples. The sample size of three groups for the face-to-face condition was too small for use with parametric techniques, and the unequal size of the two sets of trials also makes the assumption of homogeneity critical for parametric techniques. Thus, the nonparametric Mann–Whitney U-test was chosen, since it requires only independent selection of samples and an ordinal level of measurement. With the number of groups used, the 0.02 level of probability is the lowest value that can be obtained.

ISBN 0-201-03140-X, 0-201-03141-8 (pbk.)

there is a fairly substantial larger amount of "giving opinions" in the computerized conditions. This just failed to reach the 0.05 level of significance (the U value was 21; 22 is the 0.05 level in this case). We do think that there is a good chance that a larger sample of groups would produce a statistically significant tendency toward more people giving more opinions in the CC condition than in the face-to-face condition.

INEQUALITY OF PARTICIPATION

An index of inequality of participation in a group was generated using the same approach economists use in constructing a Lorenz curve to measure inequality of distribution of income in a society. It compares the cumulative percentage of statements made, starting with the least active participant, with the cumulative percentage of the number of participants. This index is constructed in such a way that it yields a value of 0 if there is total equality of participation, and 1 if there is total inequality, regardless of the size of the group. The numerator represents the observed differences between the proportions of statements made by each of the participants and the proportions they would have made if each contributed an exactly equal share. The denominator consists of the maximum value that this sum of observed differences could possibly reach in a group that size in which there was total inequality, with one of the members making all of the statements. Thus, the index compares observed inequality with the maximum possible for a group that size.

A graphic representation can be made using histograms in which shaded areas show the difference between observed participation and the amount that would have occurred had there been total equality. The index itself is computed as follows: Let I be the index of inequality, N the number of members in the group, O_i the observed cumulative proportion of statements, and E_i the expected cumulative proportion if there were total equality of participation (E_i equals the cumulative proportion of the number of members of the group). Then

$$I = \frac{2N^{-1} \sum_{i=1}^{N} (E_i - O_i)}{(1 - N^{-1})}.$$

The value of the indices for each of the groups is shown in Table 3-5. There is a statistically significant tendency for there to be more inequality in the face-to-face discussion mode. It should also be noted that the only reversal was a face-to-face group that used a different problem than all the other groups. Thus, the relatively lower amount of inequality in that discussion could have been due to the difference in the problem being discussed.

ISBN 0-201-03140-X, 0-201-03141-8 (pbk.)

Table 3-5
Index of Inequality of Participation

	Index value
Face-to-face groups	0.4155
	0.5527
	0.2560
CC groups	0.2464
	0.2965
	0.1237
	0.2286
	0.1648
	0.2867
	0.0974
	0.2184
$U = 2; P < 0.05.$	

THE PUSH TOWARD SOCIABILITY

Knowing that computer-mediated communication seems "cold" or "machinelike" or "inhuman," most new users seem to compensate heavily for this by engaging in very strong efforts to be warm, friendly, and personal. Some evidence for this arises out of the replications of the Bales experiments. In the face-to-face condition, there is usually a brief period when the participants exchange names, but no extensive socializing among strangers who were brought together for this single group meeting. In the CC condition, however, we observed very overt attempts to be personal and friendly. In Figure 3-8 are some examples of messages sent among participants who had never met face to face and who were 5 to 20 minutes into their period of learning and practicing the use of the system.

Leave-taking also tends to be more explicitly friendly than the verbal (spoken) segments recorded among participants at face-to-face meetings. For

---

Figure 3-8 The Push towards Sociability

C 963 CC3 STEVE (2,962) 6/28/77 3:44 PM
   My name is Steve. What's yours?

C 963 CC5 JOANNE (5,965) 6/26/77 3:48 PM
   Hello Steve. I find this type of machinery interesting.

C 963 CC9 STEVE (2,962) 6/20/77 3:59 PM
   Hi Joanne. What year are you in?

C 963 CC11 JOANNE (5,965) 6/26/77 4:03 PM
   I am a biology major hopefully going on to physical therapy afterwards. I am there for the summer taking organic chemistry which find difficult if you do not do the work entail. Hello again Steve, I am in the third year.

ISBN 0-201-03140-X, 0-201-03141-8 (pbk.)

Figure 3-8 (Continued)

C 904 CC7 JUDY (5,905) 6/21/77 1:22 PM
   Hi my name is Judy. I am a part time student at Upsala and a full time mother of three. I am also employed in an alcoholic rehab. center. My typing is terrible so please excuse my mistakes. Thank you. Hi Cindy. Hi Anne.

C 904 CC9 CINDY S (6,906) 6/21/77 1:24 PM
   I am Cindy and I am a mother of a 6 year old boy. I am here today for a group experiment that I am sure we will all enjoy.

C 904 CC11 ANNE (2,902) 6/21/77 1:27 PM
   The weather today is partly cloudy.
   Hi, my name is Anne and i am a mother of four, two daughters in college, one boy in high school and one boy in junior high. i am presently enrolled in the paralegal program here at Upsala. i used to teach chemistry and physics in the high school, but i am changing my career. My husband is an attorney.

C 904 CC12 CINDY S (6,906) 6/21/77 1:28 PM
   Is there anyone there who also speaks Spanish or Italian? Anyone there who is a music buff?

C 904 CC15 DAVID (3,903) 6/21/77 1:33 PM
   Is this how we are all going to spend the next two hours of our time? Well, I say that if they want to pay for it, than by all means it is okay with me.

C 904 CC18 ANNE (2,902) 6/21/77 1:36 P.M.
   I once took a short course in conversational Italian because I took a trip to Italy two years ago, but I cannot speak it. Does onyone play tennis?

C 904 CC19 CINDY S (6,906) 6/21/77 1:36 PM
   I still didn't get any answers folks, how about it? Dear David you are right. This is exactly how we are going to spend the next two hours. Have fun. Cindy.

C 904 CC21 JUDY (5,905) 6/21/77 1:37 PM
   Sorry Cindy. I do not speak any languages. I do enjoy good music: when I have the time. Anyone out there sports orientated?

C 904 CC25 CINDY S (6,906) 6/21/77 1:41 PM
   Sorry Judy, I am not into sports. Anne would you consider teaching me to play tennis? That is about the only thing athletic I might be able to handle.

C 904 CC27 JUDY (5,905) 6/21/77 1:42 PM
   Yes Anne. I play tennis every chance I get. Lucky me. I have a tennis court one half block from my house.

C 904 CC32 ANNE (2,902) 6/21/77 1:46 PM
   Wow. We all could get together for tennis at my house. I persuaded my husband to pave our backyard and we put in a makeshift tennis court. It is a lot of fun. So Cindy, I'll be glad to show you some of the fundamentals. But, I'm still working on my backhand and serve.

C 904 CC34 CINDY S (6,906) 6/21/77 1:48 PM
   Thanks Anne. I would like that. By the way folks, I am a senior here and a music major. So if anyone takes an intro. course to music and has problems, I will gladly help out.

ISBN 0-201-03140-X, 0-201-03141-8 (pbk.)

example, consider these final comments from conference participants in another two-hour experiment.

C 903 CC165 MARJORIE (3,963) 6/20/77 5:37 PM
  I agree. Joanne, Kyle, Steve and Dolores. Fondly, Marge B.

C 903 CC106 KYLE (6,906) 6/20/77 5:37 PM
  I agree with Steve. It has been fun working with you guys. Bye bye.

C 903 CC107 JOANNE (5,905) 6/20/77 5:39 PM
  It has been fun discussing this problem with you. Thank you.

C 903 CC106 DOLORES (4,904) 6/20/77 5:39 PM
  It was fun. Have a nice summer.

Remember, these persons had never met face-to-face and had been interacting for less than two hours. The level of positive comments and feelings appears to be very high.

Among participants in long-term conferences, fairly strong feelings of friendship and colleagueship often arise. During and following conference 72, a very significant portion of the participants altered their business and vacation travel plans so as to include a face-to-face meeting with one another.

This is because for a group of persons to continue to cooperate on an extended basis, there must be a certain amount of communication that is directed at maintaining social ties and providing a context of "good feelings" within which the task-oriented aspects of the communication can flow. The degree to which any particular individual engages in social–emotional categories of communication via this medium may be somewhat dependent on personality structure and the proportion of her/his communication that takes place over this medium.

Sociologist Barry Wellman made a similar kind of observation after noticing how "shocked" some of the nonparticipants in the "on-line party" were at the amount of joking and personal exchanges among those who did take part:

15120 M BARRY WELLMAN (BARRY, 720) 4/5/77 5:10 PM
THIRD IN A SERIES OF OPEN LETTERS RE THE COCKTAIL PARTY
  There is clearly a lot of hung-over ambivalence re our recent party. I'd like to tie it into fundamentally different styles of using cc—and by extension, of structuring interpersonal relations. Some people in this conference prefer what is called in network terms, "single-stranded" ties—that is, their utterances are narrowly defined, businesslike and to the point. Others (myself included) press toward a more "multi-stranded" relationship—they press toward having a wide variety of points of contact (and more multiplex role relationships) with fellow cc'ers. That partially explains references to children, spouses, drinking preferences, personal requests (such as asking a cc'er to look a same-city friend up), "bad taste" remarks and puns. The explanation for this press toward multi-strandedness is often made in network terms—the more connections you have with someone, the greater the likelihood that you'll have a successful interaction with him/her in any role relationship.

ISBN 0-201-03140-X, 0-201-03141-8 (pbk.)

Some Hypotheses

STATUS EFFECTS

An earlier series of experiments provides evidence as to how high-status persons can easily convince a group to make a "wrong" decision by dominating the discussion in face-to-face communication. In 1952, Maier and Solem reported a study in which individuals and groups were asked to solve a "horse trading" problem, which was adopted as the task in several subsequent studies:

> A man bought a horse for $60.00 and sold it for $70.00. Then he bought it back for $80.00 and sold it for $90.00. How much money did he make in the "horse business"? [Maier and Solem, 1952, p. 28].*

In Torrance's (1954) version of the experiments with groups of three members of B-26 crews, pilots had the highest social status, navigators medium, and gunners lowest. Using three members of intact crews, Torrance found that among gunners who knew the right answer, 63% were able to convince their associates to accept this correct solution. Comparable rates were 80% for navigators and 94% for pilots. Of course, the pilots were also more successful in getting groups to accept their wrong opinions, too. As Steiner (1972, p. 25) summarizes in his review of these experiments, especially in groups with a history and a future, the opinions and suggestions of higher-status members are likely to be accepted *even when they are wrong*.

It is hypothesized that the "chilling effect" of rank on disagreement with a bad idea or decision advocated by a high-status person will not be as operative in computerized conferencing. To the best of our knowledge there is no experiment that directly tests this hypothesis, however.

GROUP SIZE

Motivation seems to be a key process mediating the effect of group size on group performance. Shaw (1960) found that ad hoc groups of college students with two to five members were more willing to work harder on a group task than were members of groups with six to eight members. Similarly, Wicker (1969) found that members of large churches reported spending less time and energy on their organization's programs than did members of smaller churches. Shaw interpreted his results as evidence that group members who are responsible for a large share of a task will be more strongly motivated to work hard than will members of larger groups, whose work represents a smaller part of the total output. Other investigators have concluded that

> members of large groups report less opportunity to contribute freely and to influence the course of events . . . [and] are more inclined to complain that

---

*The answer is $20.00, but the majority (55%) of subjects in the Maier and Solem population thought it was either $0 or $10.00.

ISBN 0-201-03140-X, 0-201-03141-8 (pbk.)

activities are poorly organized and that their group does not function very well [review of "Effects of Group Size on Actual Productivity," Steiner, 1972, p. 85].

On the other hand, a group that is "too small" in terms of resources to perform the task is likely to become so demoralized that it gives up completely.

It seems quite possible that organization of discussion and problem solving through computerized conferencing might enable a large, diverse group to tap the resources of all the members without losing the ability to communicate freely or experience other negative effects of large group size. In any case, a problem solving experiment with small and large groups would seem worth replicating. With the ability of the computer to allow structured subconferences, it may also be possible to make a large group feel it is really a collection of small working groups and to retain the small-group motivation.

LESS "RISKY SHIFT"

Beginning with Stoner (1961), a number of experiments have presented individual subjects with problems that involve a series of dilemmas entailing various degrees of risk versus possible payoff. The subject is asked to choose what the odds for success would have to be before he should advise someone to attempt the riskier opportunity (1 in 10, 5 in 10, 9 in 10, etc.). Then there is a period of group discussion, and group consensus is reached on the items. Finally, there is an individual posttest.

As Teger and Pruitt (1967, p. 189) summarize the findings:

> Many studies have shown similar results over many tasks and conditions . . . . The difference between the mean level of risk taken initially and the mean of their later group decisions is termed a "shift." If there is a change toward greater risk, it is termed a "risky shift." A risky shift is almost always found.

One hypothesized explanation is that the group causes a "diffusion" of responsibility, as in the following conclusions by Kogan and Wallach (1967, p. 51):

> Failure of a risky course is easier to bear when others are implicated in a decision; . . . . Consider a homogeneous group composed of test anxious individuals, that is, individuals uniformly fearful of failure . . . [such people] might be especially willing to diffuse responsibility in an effort to relieve the burden of possible fear of failure.

If this is truly a strong factor, then changing the decision-making mode to computer conferencing should not have much of an effect.

A second type of explanation is that the very type of individual who tends to choose the riskiest decisions is also the "take-charge," persuasive, leader

ISBN 0-201-03140-X, 0-201-03141-8 (pbk.)

type of personality, who therefore tends to dominate the group discussion and influence the low risk takers to accept his/her position. (This explanation is advanced by Collins and Guetzkow, 1964, among others, but rejected by several subsequent experimenters, such as Wallach, Kogan, and Burt, as unconvincing and not supported by direct testing.) To the extent that this factor is operative, the risky shift would be lessened by computer conferencing, because the personality attributes determining leadership and discussion dominance in the face-to-face group are not operative. Anonymity ought to especially decrease the dominance of an established leader for real groups.

Another hypothesis is that something about the social nature of the group discussion process itself is involved—perhaps the emergence of the norms of American society that people (especially men) are supposed to take risks in order to achieve success, and the consequent desire of individuals not to appear deviant from commonly accepted norms in publicly announcing their choice. A key experiment along these lines is that of Wallach and Kogan (1965), who contrasted the amount of risky shift in the three following situations:

1. Discussion until consensus was reached;
2. Discussion and revoting before consensus was reached;
3. "Consensus without discussion," in which subjects communicate their risk preferences to each other by written messages without face-to-face discussion.

The risky shift occurred for both face-to-face groups, but not for the written communication group.

Teger and Pruitt (1967) used a written successive ballot technique similar to a Delphi technique, and found a small risky shift.

To the extent that groups have such a normatively induced tendency to generate riskier decisions than individuals would make on their own, the experiments suggest that computer conferencing may cut down the likelihood of imprudent or risky decisions being made.

SATISFACTION OF PARTICIPANTS

The final variable we will explore in relation to experimental data is the effect of communications medium on interpersonal attraction and satisfaction of participants, and the question how, in turn, this alters task effectiveness. The evidence is very skimpy here, and obviously more comparative experiments need to be done even on "older" media than computerized conferencing. The pilot replication of Bales reported earlier found no statistically significant differences in satisfaction of participants. There are, however, situations where differences would be expected.

ISBN 0-201-03140-X, 0-201-03141-8 (pbk.)

Mehrobian (1971, p. 11) has pointed out that "in terms of the immediacy that they can afford, media can be ordered from the most immediate to the least: face-to-face, picturephone, telephone . . . " (and below this, synchronous and asynchronous computerized conferencing and letters or telegrams). He states that the choice of media in regard to intimacy should be related to the nature of the task, with the least immediate or intimate mode preferable for unpleasant tasks.

To explore this hypothesis, Williams (1975) used two tasks, supposedly differing in "intimacy," for two-person conversations utilizing face-to-face discussion, closed-circuit TV, and telephone communication. The conclusions were that

> Significant media effects on evaluation of the conversation and (less strongly) of the conversation partner have been found . . . . For the less intimate task, the most immediate medium, face-to-face, leads to the most favorable evaluations; and the least immediate, the telephone, leads to the least favorable. For the more intimate task (of the two used), a medium of intermediate immediacy, closed circuit television, leads to more favorable evaluations . . . . This would suggest that with tasks of very high intimacy, perhaps very embarrassing, personal or conflictful ones, the least immediate medium, the telephone, would lead to more favorable evaluations [Williams, 1975, p. 121].

Obviously, these results are suggestive of greater participant comfort and satisfaction with a low-immediacy and low-intimacy mode, such as computerized conferencing, for some kinds of communication tasks and for heterogeneous groups with hostile factions. So little experimentation has been done in this area that there is a great deal of room for further research.

Satisfaction with Teleconferencing Media for Specific Aspects of the Group Process

Figure 3-9 shows how users of three different kinds of teleconferencing systems (audio, video, and computerized) rated the media for specific aspects of the group discussion process. The scales were developed by the Communications Studies Group (CSG).

The data in Figure 3-9 must not be taken at face value, because there is no comparability among the various data sources. The experienced EIES users were not engaged in any particular task at all, but had all spent at least eight hours on line. The subjects, tasks, and all other conditions for the four studies were not at all comparable. In addition, the scales for the Bales replications reported here and for the Institute for the Future's studies of PLANET users were reversed, with the "unsatisfactory" on the left and "satisfactory" on the right. (Corrections were made for this reversal in Figure 3-9 by reversing the

ISBN 0-201-03140-X, 0-201-03141-8 (pbk.)

118

Figure 3-9A Comparisons of first-time users on CSG scales: Video conferencing, audio conferencing, and computer conferencing.

Completely
satisfactory

Completely
unsatisfactory

Giving or receiving information

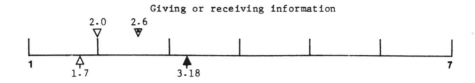

Problem solving

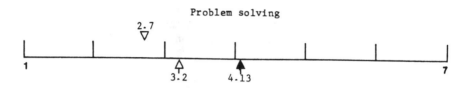

Bargaining

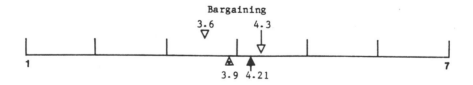

Generating ideas

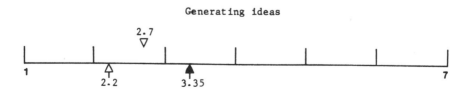

ISBN 0-201-03140-X, 0-201-03141-8 (pbk.)

Figure 3-9B Comparisons on CSG scales (continued)

Completely
satisfactory

Completely
unsatisfactory

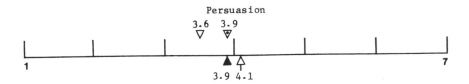

Persuasion

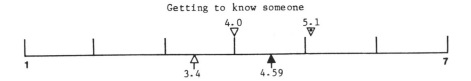

Getting to know someone

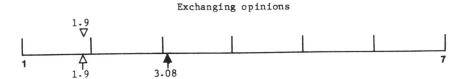

Exchanging opinions

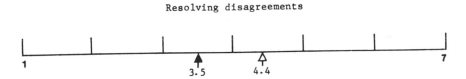

Resolving disagreements

<u>Key</u>

△, 13 experienced EIES users (*N*=13);

▲, computerized conferencing: Bales replication with students:
(scale reversal performed) (*N*=40);

△, Confravision: Champness (1973a), as reported by Pye and
Williams (1977);

▲, Audio conferencing ("Remote Meeting Table"): Champness
(1973b) as reported by Pye and Williams (1977).

ISBN 0-201-03140-X, 0-201-03141-8 (pbk.)

direction of the deviation from the neutral 4.0 point; but the different order may affect answers, since there tends to be a slight response bias on scales toward answers that appear first.) Thus, the only totally valid comparisons are differences in suitability within each medium, not across studies.

New users of computerized conferencing do not make very large distinctions after only an hour about the users for which it is most and least satisfactory. Overall, they find it most satisfactory for exchanging opinions and giving or receiving information, and least satisfactory for "getting to know someone." The range of mean ratings is only 3.2 to 4.6.

There is a greater distinction made by the more experienced EIES users about the specific kinds of tasks for which the medium is more and less satisfactory. First impressions or esthetic responses are not the same as the opinions of experienced users. For example, the experienced users find it almost completely satisfactory for giving or receiving information (1.7) and for exchanging opinions (1.9). (These are the same tasks for which it is rated most highly by the new users; but the ratings have shifted much farther toward the completely satisfactory end of the scales). It is on the satisfactory side of neutral for "getting to know someone." It is rated least satisfactory for persuasion (4.1, just slightly on the unsatisfactory side of neutral) and "resolving disagreements" (4.4).

As stated earlier, since the data were gathered for noncomparable tasks and subjects, any comparison between the CSG data and the EIES data can at best be suggestive of things worth studying in future controlled experiments. However, it is very interesting that of all three media rated in any of these studies, the highest rating for "getting to know someone" occurs for experienced users with computerized conferencing. Even the new users rate it higher for this task than did the CSG users of the audio conferencing system. Therefore, we cannot by any means dismiss computerized conferencing as cold or impersonal or low in "social presence."

Pye and Williams (1977, p. 233) conclude that "numerous carefully conducted experiments on information transmission, problem solving, group decision making and interviewing have found all vocal media to be very similar in effectiveness for these tasks" (face to face, audio, and video are included). However, as we have seen from their data, people *perceive* audio to be less satisfactory than video (and both to be less satisfactory than face to face). We hypothesize that part of the explanation is that they simply have not become adept at using the unfamiliar communication channels.

Computerized conferencing also seems competitive with audio or video conferencing for giving and receiving information, exchanging opinions, and generating ideas. Since it is less expensive and more convenient, it would therefore seem to be preferable for these functions. For functions on which it is not rated as highly satisfactory—such as getting to know someone, bargaining, and persuasion, audio and video conferencing are not rated satisfactory,

ISBN 0-201-03140-X, 0-201-03141-8 (pbk.)

either. People will therefore, be likely to want to continue meeting face to face for such purposes. On the other hand, it may be that face-to-face meetings are just as unsatisfactory for resolving conflicts; it is an oversight of the existing experiments, including our own, that face-to-face groups were not asked to make such judgments. In this case, the most promising direction for future development of telecommunication systems would be in the incorporation of structured group processes explicitly designed to facilitate outcomes such as getting acquainted, resolving disagreement, or reaching consensus.

EIES as a Social System: Social Stratification and Social Mobility

Most working groups have a single leader and a fixed hierarchical order of manager, project leaders, and team members for all projects. Thus, a person has a single, commonly recognized group membership and position in its social hierarchy.

The unique thing about EIES or similar conferencing systems is that the same person may play many different roles in many different conferences that involve different subgroups of people. In one, he or she may be an ordinary member. Since a person is free to browse through the Directory to find compatible groups conducting conferences in related areas, and to request admission to such conferences, a person is quite likely to have the role of outside expert in some conferences; and since every member has the privilege of setting up and acting as moderator on a temporary conference on any topic of his or her choosing, every member of the system has the opportunity to play the lead or moderator role in at least one conference. Thus, we have an extremely fluid social structure, with "electronic migration" from one group to another according to interests and ability a real possibility.

Another unique feature of a CC system is that two or more individuals can communicate using the same identification (ID) number.* In some cases, this has happened by accident on EIES, because it was bureaucratically difficult to get National Science Foundation (NSF) approval to add a new member, so an old member in the same location shared access to an existing ID code. In most of these cases, there was little ambiguity, because the multiple users would all explicitly indicate who authored each item. In two cases, however, a "group identity" was purposeful. One "member" of EIES has been "the gang" at *Creative Computing* magazine, which sometimes enters an item as a group opinion, without indicating a specific individual as author. Then there is a couple who purposely never indicate which of them contributes a specific item.

---

*The identification number is an account on the system, corresponding to a membership for a communicating entity.

ISBN 0-201-03140-X, 0-201-03141-8 (pbk.)

What we have, in other words, is a social mechanism for "collectivities" or groups, rather than discrete individuals, to participate in social interaction. As Barry Wellman puts it,

M 22801 BARRY WELLMAN (BARRY, 720) 6/17/77 6:14 PM
CCSs FACILITATE
The participation of collectivities as nodes. While this happens in FTF meetings— as when a group has the right to send a delegate, and different ones show up at different meetings—the substitution is highly noticeable. But in CC, any number can play at being "Barry Wellman." Or, it could be a membership in the name of the "Toronto Network Group."

Certain kinds of roles are easier to play in a CC system than in a face-to-face group. For instance, an attendee at a face-to-face group who is merely an "observer" is likely to make the other members feel resentful or self-conscious. EIES conferences can have "observe-only" members who are specifically permitted to read comments but not to write. This is a mechanism that allows nonexperts or students to follow a conference topic in which they are interested without intruding on the discussion.

This fluid world is not an environment that everyone will find easy to adapt to or swim through. It very well can produce an individual "future shock" situation and may be difficult to digest by those who prefer a rigidly structured social order.

Why Conferences Fail

In this chapter we have concentrated on the individual and group processes involved in successful adaptation to this new medium of communication. However, by no means all of the people given the opportunity to try this new medium survive through the learning period and become addicts; nor are all conferences successful in generating a large and heated discussion and in reaching their goals.

The data from our pilot period with EIES can only be generalized with caution, given that the system was so unstable and the user groups more just a collection of people than groups with a shared purpose. However, our preliminary conclusions are that there are a number of individual and group characteristics that are associated with failure. The most important include

1. Lack of convenient access to a terminal.
2. Lack of a need or desire to communicate with other people on the system. This is related to a condition in which the group itself lacks a shared goal to which all members are willing to contribute several hours a week of their time.
3. Lack of adequate training material in an acceptable medium. Some

ISBN 0-201-03140-X, 0-201-03141-8 (pbk.)

ISBN 0-201-03140-X, 0-201-03141-8 (pbk.)

people simply will not go through a 100-page printed users' manual. Alternative training, such as in-person sessions, videotapes, or on-line lessons, is necessary in order to motivate them to learn more than just the basic mechanics of how to send and receive a message.

4. Lack of strong or adequate leadership. As mentioned in Chapter 1, unless a moderator takes hold and sets an agenda and keeps the group working toward its goal, nothing much will occur within a conference.

5. Lack of a "critical mass" within a conference or group. This has to do with both a minimum number of active participants and a minimum number of geographic locations. The minimum seems to lie somewhere between 8 and 12 active participants in three or more geographic locations. Below this critical mass, there are not likely to be enough new messages or conference comments entered, so that there would not always be new items to receive and respond to when a member signed on. Above the minimum size and dispersion, enough activity and controversy is generated to motivate members to sign on frequently and to participate actively in the exchanges.

The critical mass hypothesis arose as the result of observation of three groups with only four or five active members. Those persons were highly motivated in the beginning when they tended to sign in several times a week. After four to six weeks, however, there was too often "nothing new" when they signed in. Some of these members of too-small groups stopped using the system. More interestingly, some engaged in "electronic migration" to larger and more active groups and conferences.

Summary and Conclusions

Our observations of users of CC systems can be summarized as a list of hypotheses about individual communications and learning behavior. Most of these have been discussed in detail in this chapter; for more data, see the final technical reports on the FIES field trials (Turoff and Hiltz, 1978) and the experiments which were performed (Hiltz, Johnson, and Agle, 1978).

New Users

A new user is usually passive, receiving much more than he or she sends relative to more experienced users. A new user is more likely to send messages than engage in conference activity. A new user's motivation is likely to depend on the availability of individuals he or she desires to talk to and the availability of interesting topics in ongoing discussions (attractions). Barriers are the other aspect of new user motivation, in terms of access to terminals, inadequate user training materials, or system problems.

Regular Users: Continued Learning and Changes in Behavior

1.   Users evolve specialized norms with respect to the use of the facilities and communications and writing style. The acquisition of these norms by individual users and groups appears to be an important learning process on such systems.

2.   User participation in conferencing in an active sense of contributing items seems to require some degree of usage above the basic level of learning the mechanics. This may be a second-level learning plateau involving the acquisition of norms established by the user communities.

3.   Users will gain facility as time passes, so that their input rates become higher than usual typing rates. For large groups, the time required to send and receive communications will drop below that required for other media, such as telephone or face-to-face meetings.

4.   The user's short-term memory may be a factor in conditioning frequency of interaction with the system.

Users will tend to become conditioned to sign on the system so that on the average they have about seven items to send or receive per interaction.

5.   In accordance with social exchange theory, no participant will continue to use a conferencing system unless his/her "rewards" are greater than his/her "costs." Among the factors that increase reward for users are

(a) Ratio of items received to items sent. This increases with (i) size of active group; (ii) throughput rate of the system.

(b) Observable increase in skill and speed in using the system. This improvement is related to the richness of the design in terms of advanced features available to users once they have mastered the basic mechanics.

(c) Importance of communication with system members in comparison with communication with persons not on the system; relative cost in time and money of other modes for communicating with people on the system.

On the level of the social dynamics of group interaction through this communications medium, our hypotheses (inductions from preliminary data) include:

(a) There is a strong tendency toward more equal participation in synchronous discussions, as compared to face-to-face groups.

(b) More opinions tend to be asked for and offered.

(c) There is less explicit agreement or disagreement with the opinions and suggestions of others.

(d) There is a great deal of explicit sociability of an informal sort on these systems.

ISBN 0-201-03140-X, 0-201-03141-8 (pbk.)

(e) There are no significant differences in overall satisfaction of participants in face-to-face, audio, video, or computerized conferences.

(f) There tends to be a great deal of "electronic migration" among user groups on a CC system.

For Discussion

1. Can you hypothesize a specific example or situation in which each of the following is an advantage, and another where it is a disadvantage, for a group communication situation?
    (a) Use of written communication
    (b) Lack of time coincidence
    (c) Presence of memory
    (d) Nonverbal communication
    (e) System delays
    (f) Structure

2. How many group communication structures can you hypothesize other than those mentioned in this chapter? Or, consider some of the possible variations or combinations of those mentioned.

3. What is the likely psychological atmosphere in a video conference?

4. In what group communication situations is the voting capability provided by the computer likely to be useful?

5. When is the use of anonymity likely to be helpful or harmful?

6. What situations now exist for which an increase in size of existing groups engaged in a project would be a likely benefit if effective communications could be maintained?

7. What could be the beneficial or harmful effects of the increased throughput offered by CC systems?

8. In what circumstances and how are various emotions conveyed with the written word? Would you consider the strength of the emotion to be greater than, less than, or of the same intensity as that conveyed by phone or face to face?

9. In what situations is a person's ability to reflect or to take his/her time desirable? When is it useful to pressure individuals to respond immediately to a particular topic?

10. What types of training, job skills, or occupations might lead people to be considerably more skillful than the average person in a CC medium? Consider, for example, a poet. What types of individuals are likely to be considerably less skillful than average? Consider a high-pressure salesman who carries that style over into this environment.

ISBN 0-201-03140-X, 0-201-03141-8 (pbk.)

11.　What would you hypothesize to be the characteristics of a person likely to emerge as a leader in a CC discussion? Pick several specific types of discussion. Then try to develop a general theorem.

12.　What might happen if a schizophrenic were given a different ID for each "self"? Might the selves communicate with each other?

ISBN 0-201-03140-X, 0-201-03141-8 (pbk.)

Appendix

ADDITIONAL EXAMPLES OF INTERPERSONAL COMMUNICATION IN COMPUTERIZED CONFERENCING

This chapter illustrated generalizations about communication processes in this medium with examples from Group 72 on EIES. In order to demonstrate the generalizability of the phenomena we have described, we have culled a few examples from a very different conferencing system. This is a proprietary system and a conference whose participants cannot be asked for permission to reprint their remarks; thus all names have been changed.

Cue Emitting and Cue Searching

NUMBER 200 BY CHARLIE AT 1419 ON 3/12/74
Gary. Not sure I really understand your statement, but I'll try to respond. I believe that as I better understand me, I have a better chance of understanding you but I have to keep confirming with you that I am perceiving you with some accord with your perceptions of yourself. This is a spiral process in which we keep checking perceptions with each other. I do not seek to control perceptions of "me"—I need candid feedback with specific illustrations to compare your perceptions with mine. Then I (we) can explore possibilities of enhanced development. I have to understand to be understood. Are we on the same wavelength?

NUMBER 16 FROM FRANK 14 MAR 74 14:14:06
Peter: This is fine and answers all the questions raised in addition to some that were never mentioned: in fact, maybe we could title those fifteen messages everything you always wanted to know about incasting but were afraid to ask. (hahaha)

Candor

NUMBER 9269 BY CHARLES AT 1618 ON 11/02/75
well I for one am particularly concerned . . . . Come on Iris if you want another conference. Open it up. Don't fuck around with supposedly private ones . . . . We have enough problems without adding to them. Remember I did not sign that promise not to get pissed off at you.

ISBN 0-201-03140-X, 0-201-03141-8 (pbk.)

NUMBER 11 FROM DON 14 NOV 73 17:52:46 (EDITED BY DON)
Ya I guess that you are right Mark and I apologize to both you and Bill for my outburst. Please forgive my rudeness. I just seems that some strange things have been happening the last few days and I guess that I'm under a bit of pressure with the board of management coming up and at the same time I'm trying to coordinate the efforts of planning, network planning and all the new efforts of comprehensive planning . . . . Seems that I'm caught in the middle of another political game. So once again please accept my humble apologies

NUMBER 2 FROM DAVID 12 NOV 73 10:12:10
The next few steps have to be very, very carefully mapped out. I just cannot understand how real people like you and I and one hell of a lot of others can have the same basic objectives and yet can disagree on so many fundamental premises. You keep insisting that you do not want to get involved in the politics of this project, and if you still maintain this position, then I suggest to you that you are destined to nothing . . . .

Ability of the Medium to Support Very Expressive and "Personal" Content

NUMBER 5324 BY IRIS AT 1454 ON 2/12/74
Hello, Murray! I've been sick as a puppy dog, with food poisoning. Not my own cooking, but chinese food. Don't you dare suggest a chinese dinner when you come down here on thursday: You will be strung up on a willow tree.

NUMBER 3346 BY LAURA AT 1828 ON 8/11/74
Your extravagant electronic comments send me off to my French lesson with an extravagant electronic smile. A bientot!

NUMBER 284 BY LAURA AT 1721 ON 4/11/74
Yes, it is a coincidence. Do you come here often???? I'm new here myself. Confirmation?? Really, Murray! I think that's a religious thing, isn't it? Like transubstantiation?

The Influence of (Randomly Assigned) Pen Names or Anonymity on Communications

NUMBER 181 BY J. CAESAR CAE AT 1134 ON 3/12/74
I am getting paranoid worrying about the ides of march—where are you Brutus, Cassius (lean and hungry), Anthony, and the rest of you noble Roamers—oops—Romans?

NUMBER 182 BY M.A. LEPIDUS LEP AT 1410 ON 3/12/74
We are all hiding behind the gentle cloak of night, in the darkness of our thought, under the guise of true friendship, between cosmic doubt and eternal despair. Roman in the gloaman?

NUMBER 150 BY L. AT 1425 ON 10/01/75
I am showing our secretary how to use the system so that I may concentrate on substance rather than on form. This is a demonstration.

ISBN 0-201-03140-X, 0-201-03141-8 (pbk.)

NUMBER 151 BY ANONYMOUS AND AT 1430 ON 10/01/75
This is the dumb secretary practicing on the machine.

NUMBER 206 BY DR. YAKOMOTO YAK AT 1645 ON 2/11/74
A lot of negative things have been thrown around in this conference—mostly by anon. I have trouble sorting out reaction from overreaction and serious concern from peripheral concern. I also am confused about the great discrepancy in the comments, particularly when they are contrasted to the frequent positive feedback which I and other members of the organization receive.

NUMBER 135 BY J. CAESAR CAE AT 1113 ON 3/12/74
While I try to digest all of the statements about roles and responsibilities—tell me this—how do we as a directorate stack up as evaluated by program management criteria?

NUMBER 142 TRE AT 1543 ON 6/11/74
Well I don't think that success has the capability of being defined on one level. What notes success to a funding agency may run contrary to the hopes of the project, and different again from the expectations of the projects clientele. Even so if the program does not meet the goals of any of these it may still be a . . . success as defined by some other group.

NUMBER 145 BY ANONYMOUS AT 1800 ON 6/11/74
This success rather smacks of an inability on the part of the participants to render comprehensible their own fantasies, dreams, aspirations, fears, and foibles. Instead of coming to grips with demonstrably autistic behaviour, the participants compensate by muttering in their collective muffs about the difficulty in tying down universal criteria.

NUMBER 151 BY ROGER AT 0904 ON 5/11/74
**********SUGGESTION**********
Since statements (as opposed to messages) are a matter of "public" record, it would be desirable if the more high-spirited participants edited the more colorful phrases before entering their statements for posterity. This would avoid possible embarrassment to the directorate and perhaps would assist us all in stating our views more clearly and exactly.

NUMBER 153 ANONYMOUS AT 1339 ON 5/11/74
Do definitions of "high sprited," "colorful phrases" and "possible embarrassment to the directorate" exist?

Perceptions of the Medium

NUMBER 304 BY LAURA AT 1740 ON 4/11/74
Use of the concept of space to talk . . . I'm going to have to work on that one. what I really like—is this the same thing?—is the way you snatched me out of the air. Feeling about the process so far is great sense of liberation in this free-form mode. I love it, but a feeling of constraints re heavy conferences, esp. Since it is now clear how easy it is to blow your cover.

ISBN 0-201-03140-X, 0-201-03141-8 (pbk.)

Need for Strong Leadership in Conferences

NUMBER 123 BY LAURA AT 1107 ON 3/12/74

One of the things that we fussed about when the world and the conferences were young was the need for an active facilitator (not just a chairman) in each conference . . . .

The events . . . or lack of them . . . of the last couple of weeks in some of the conferences were likely inevitable, although I think that the activities of a cheerleader might have kept things simmering gently . . . . It seems too bad to let conferences go completely cold. Anyway, what do the rest of you think. Should future conferencing have more active stimulation when it seems necessary, and if so, how would it be done?.

NUMBER 153 BY ANONYMOUS AT 1658 ON 7/11/74

I suggest that we should take a poll as soon as possible in order to crystallize things a bit.

How about discussing that if you feel it is needful . . . . Otherwise, please submit (via statement) issues that can be voted on a rating scale.

Example:

Issue . . . . Implementation of a matrix organization that would ensure continuing contact and exchange between bureaus.

Rating   1.   very important
                     2.   important
                     3.   slightly important
                     4.   not important
                     5.   no opinion

Statement:

I concur with 153. Let's get a vote on the major criteria and other issues so we can see where the group of us stand. I would suggest a second scale to describe whether the statement is specific enough or too general. My intuition says there is more agreement among this group than implied by the discussion.

ISBN 0-201-03140-X, 0-201-03141-8 (pbk.)

PART II

Potential Applications and Impacts of Computerized Conferencing

THE BOSWASH TIMES

The Computerized News Summary Service of the Megalopolis

September 17, 1993

ISBN 0-201-03140-X, 0-201-03141-8 (pbk.)

Albany, N. Y. The state legislature voted almost unanimously to reduce the corporate taxation rate by 50%. It is hoped this will deter the sizable transfer of corporate headquarters over the past three years to North Dakota. With the reduction of headquarters staffs to the president, a few clerks, and a terminal, many companies have found it very easy to shift their corporate headquarters to the location having the lowest tax rate. The current state legislature's action was hurriedly undertaken when the state law locating the corporate headquarters in that state where the most employees live was overturned in the federal courts. Of course, the undesirability of living in North Dakota has reduced most corporate presidential positions to figurehead positions with the real power in the hands of the vice presidents and the chairmen of the board. In fact, last year the number of female corporate presidents rose to over 30%.

Washington, D.C. Owners of home terminals, go to sleep happy tonight! In two government actions today you have noticeably benefited. First, the Federal Citizen Review Office has overturned regulation 15,384, which stated that the costs of a home terminal may only be deducted for business expenses according to the amount of time it was utilized for business. Since most home terminals are utilized for family and recreational use 30–60% of the time, this placed a severe limit on tax deductions in this area. FCRO has ruled that as long as the terminal is required for business use its expenses are 100% deductible and other use is to be considered incidental in nature.

In a separate action the FCC ruled that phone companies cannot require owners of home terminals utilizing them for business purposes to install business phones. The FCC has stated that it is only the host computer's phone that can be required to be a business exchange. Only if the terminal is equipped to receive business calls may the phone companies require business exchanges. As an amendment they also concluded that home computers taking incoming calls and not charging for computing use cannot be construed as a business use by the phone companies.

Corporate Wizard Unmasked

Wichita, Kansas. A six months' investigation into the whereabouts of the mysterious president of the Emerald Communications Corp. has uncovered a bizarre chapter in American business history. Former President Ozzie Smith, it seems, has actually been dead for six months.

133

His daughter Dorothy and three of her friends have been acting in the name of the President, utilizing the corporate computerized conferencing system.

"Actually, it was easy," explained Dorothy, "since Oz never appeared personally at meetings. We just had him cremated, then followed the advice of the Vice Presidents, which came in over the system."

Suspicions were raised, however, when the normally reclusive former President did not make his annual video appearance at the stockholder's meeting.

"We did not mean any harm," explained Dorothy. "We were just trying to prevent a corporate takeover by certain Eastern interests, who would use the company for their own nefarious purposes."

ISBN 0-201-03140-X, 0-201-03141-8 (pbk.)

CHAPTER FOUR

Potential Impacts of Computerized Conferencing on Managerial and Staff Functions*

A Scenario

A Morning in the Computerized Communications World of
Elizabeth Armstrong Smith,
President of World Oil Company

7:00 A.M.   Upon rising, Ms. Smith logs onto her computerized communications system and requests the first line (showing sender, time, and subject) of all private messages that have arrived during the night. After dressing, she glances over them. Most seem fairly routine, but one begins, "Mediterranean hijackers ask $10 million ransom for our LNG supertanker and crew."

President Smith asks for the full text of this message, plus her schedule for the day, which is always delivered to her as a private message from her administrative assistant the first thing each morning. There is a meeting scheduled with the chairman of the board at 10:30. Good. She sends messages for the company's vice president for Middle Eastern affairs and vice president for governmental relations to join them, to consider what action shall be taken to respond to the situation. Copies of the situation report and the meeting announcement go to the Department of Defense, the Department of State, and the Executive Offices of the President. She sets the meeting to last through noon, shifting her scheduled 11:30 meeting to 2:30 and sending a message to the three prospective attendees.

*Portions of this chapter appeared in articles published in *Spectrum* and as "The Wired World of the Office of the Future," *Proceedings* of the ICCC. The authors thank Morton Darrow of the Prudential Insurance Company of America, Richard Wilcox of the U.S. Arms Control and Disarmament Agency, and John Burns of Princeton University for their comments on an earlier draft of this chapter; and Michael Wolf for his helpful editorial comments and revisions.

Starr Roxanne Hiltz and Murray Turoff, The Network Nation: Human Communication via Computer

ISBN 0-201-03140-X, 0-201-03141-8 (pbk.)

7:30 A.M.   During breakfast, the terminal prints out the company "Daily Bulletin," which she takes to read on the train.

9:00 A.M.   Once at the office, President Smith requests her assistant to retrieve the full text of all private messages waiting for her. She notices five of the six persons to whom she has sent messages about the 10:30 meeting have received the messages. The Department of Defense notes that it has also informed the Israeli army and that it will send advice by 11 A.M. She telephones the sixth person, then begins entering an agenda for the 10:30 meeting, listing the possible options. Copies are sent to the legal and treasurer's department, with requests for comments by 10:15 A.M.

9:30 A.M.   There is still an hour left, so President Smith enters her Government Regulations conference, where she has been discussing a proposal for a natural gas rationing plan with six other oil company presidents, representatives of the Congress and the President's office, and public interest groups. She enters her vote on several propositions that have been made, adding associated comments on two of the votes. Then she enters a two-page position paper that had been developed with assistants and stored in her notebook, and now seems apropos.

10:15 A.M.   Before the 10:30 meeting, President Smith does a quick search of all files on "hijacking" and "kidnapping," receiving the text of information relating to case histories and policy statements on the subject that have been entered in the Industry President's Policy Conference. A private message arrives from the corporate treasurer, informing her that a $1 million payment could be provided immediately in dollars, Swiss francs, or gold, should that seem necessary, but that $10 million would take at least one day to gather together.

10:30–12:00 A.M.   During the meeting, further responses from the Department of State and Defense are brought in by the administrative assistant. It becomes clear that there are really only two viable options.

12:15 P.M.   A message arrives from the hijackers, who do not believe that a woman controls World Oil Company, and refuse to negotiate with her. "We'll let that idiot vice president for public relations act as the front man, maybe he'll earn some of what we pay him," President Smith mutters to herself. Meanwhile she enters a message to all other parties that "The vice president for public relations will handle the negotiations. He can fly out to the ship by helicopter to meet with them."

But it's going to be another long day. Before leaving for lunch, she receives a message from her husband on the terminal, reminding her of the dinner party tonight. She enters a message for her secretary: "Tell him no, but say it diplomatically."

ISBN 0-201-03140-X, 0-201-03141-8 (pbk.)

Changes in Organizational Styles and Effectiveness

One image of the work of the top corporate executive is that of the modern mandarin, sitting behind his gleaming, absolutely empty desk, encased in a cocoon of secretaries and assistants who create a specially protected environment, not unlike the courts of the feudal empires of old. Into the corporate cocoon of the executive suite, no uninvited and unscreened communication may penetrate; no physical work other than the ceremonial handshake, signing of contracts with golden pens, or lighting of expensive cigars is visible. From this isolated, unworldly world that pierces the clouds at the top of those modern palaces, the corporate skyscrapers, issue the decisions and orders that shape and control the world-wide centralized corporate empire.

The opposite corporate image or style is that of the harried, overworked executive, his "in" basket jammed, every available surface stacked high and overflowing with half-digested memos and reports, his appointment book brimming with so much travel and so many conferences that he never quite recovers from a chronic case of jet lag and coffee nerves. Centralization trends seem to keep pushing smaller and smaller decisions higher up the corporate ladder, and a multitude of new concerns, such as governmental regulations and environmental impacts, chew voraciously at all of his available time, until he collapses from a heart attack in his fifties.

Both of these executive styles may soon seem as outmoded and inefficient as the mandarin's four-inch fingernails, and one of the forces that may reshape the executive role is modern telecommunication. In this chapter, we will review some of the potential applications of computerized conferencing to managerial functions in an organizational environment, whether corporate, government, or nonprofit. If used as a major channel of communication among managers, it can facilitate new styles of integrating and overseeing both the information and the interaction that is needed to guide a large organization. It can facilitate the emergence of new styles of managing (as well as of new kinds of managers, such as Ms. Armstrong Smith) by helping to provide the communications structure necessary for a form of decentralization or delegation of authority that nevertheless provides for the accountability of the decision makers.

The thought of sitting at a terminal to conduct a business conference instead of around a restaurant or boardroom table is probably as foreign to the corporate or government executive of today as spaghetti was to Marco Polo when he visited the Chinese court. However, we predict that computerized conferencing will soon become as much a staple of their existence as the strange new noodle eventually became for the Italians. To illustrate that our subsequent exploration of the potential applications and impacts of computerized conferencing on large-scale organizations is not simple "pie in

ISBN 0-201-03140-X, 0-201-03141-8 (pbk.)

the sky," we will begin with a review of the results of one current use of a computer-based communication system to support managerial functions.

Experiences of Managers in the ARPANET

In addition to the long history of the use of EMISARI recounted in Chapter 2, a computerized message system that has acquired most of the characteristics of a conferencing system has been successfully used in managerial applications for several years in the U.S. Army Materiel Command. Uhlig et al. (1975) report on the results of an initial experiment, which used the medium for informal communication among five of their managers at headquarters and two managers at locations 200 and 700 miles away.

Though five of the managers had been co-located, they were so constantly tied up with travel and meetings that "Sometimes even getting two of them together can be difficult." The experience with this experiment bore out the kinds of potential advantages that have been described for managerial applications of CC systems:

> We have found a marked improvement in our communication with one another and a substantial increase in our group awareness of the actions going on within our organization. There are several reasons why use of the message handling facility has improved our communication. First, it is easy to leave an urgent message for someone and to be confident that he will get it as quickly as he is able to receive it. When one sits down to read messages via a network mailbox, one is generally in a frame of mind to receive what the sender of the message has written . . . . Other factors we have found include the willingness to be completely candid . . . . With regard to increase in group awareness, we found that people are willing to send messages to all people involved in an action because it doesn't require anything more than typing multiple names into the address list [Uhlig et al., 1975, pp. 117–118].

As the authors point out, "Many face-to-face meetings are still essential," but the experiment was successful enough that the use of the system was subsequently expanded to include a network of about 200 managers in 20 more locations around the United States.

Reporting on the results of two and a half years of use, Uhlig (1977, pp. 120–126) finds the following advantages and limitations:

> The ability to interact much more rapidly, in writing, than with any previous medium has generated many time savings within our organization. The head of one of our field activities estimates that he has reduced the turn-around time in getting decisions on urgent action to a matter of 2 to 3 days. To get the same decision would have taken 1 to 2 weeks prior to the availability of the Computer Based Message System . . . .
>
> We found a decrease in tension on the part of the managers using the system to be a real benefit. The potential to be "connected" to the office, no matter

ISBN 0-201-03140-X, 0-201-03141-8 (pbk.)

where you are, can take away the concern that something major may be happening which you need to know about . . . .

[But]—we have had some monumental misunderstandings arise when individuals tried to do everything via messages. The ease with which a reply can be sent is sometimes the opposite of what is needed. An individual may quickly read a message, and, because of his "personal mental filters," answer something quite different than what was asked . . . .

Most people were able to adapt to the [user] interface much more rapidly than we thought they would. However, many people have a built in aversion to anything that is related to computers, and a very good interface is needed to make them feel comfortable . . . .

The biggest single lesson we have learned from our experimental use of CBMS is that this is a major new medium of human communication and interaction, with a very positive impact on the way we do business.

Potential Advantages of Remote Meetings by Computerized Conference

Computer-mediated communication systems are not meant to totally replace all other communication forms. Executives and managers will still use the telephone, personal meetings, and letters for some purposes. However, the system, if wisely used, should cut down markedly on the amount of travel and personal meetings that are necessary, and help make those face-to-face meetings that are held more productive by establishing a common information base ahead of time. In addition, the computerized conference may be more efficient than face-to-face meetings for sharing information, eliciting points of view, and facilitating decision making. Many dysfunctional things occur in face-to-face decision-making meetings, as described in more detail in Chapter 3. First, one or two people tend to dominate the discussion. This dominance is aided by several factors:

1. Only one person can "speak" at once; others are reduced to the passive condition of being an audience.
2. Persons with high rank are deferred to; persons with low-status but irrelevant characteristics such as being black, female, or strange looking have their contributions devalued.
3. The quieter or less active participants tend to "tune out" by thinking about or doing other things. These participants stop participating, in fact, unless called upon, and sit there fidgeting while waiting for the meeting to end.
4. It has been observed in laboratory experiments that verbally dominant individuals tend to reinforce their dominance with aggressive nonverbal mannerisms which intimidate the others (pointing, leaning toward the person or group, etc.).
5. "Jet lag" means that persons who traveled a long distance to the conference are not able to function well.

ISBN 0-201-03140-X, 0-201-03141-8 (pbk.)

Other factors that can negatively affect the range of options explored and the quality of decisions realized include

6. The press of schedules, which means that a meeting cannot go on for more than three hours without a break for meals. Sometimes important matters get little consideration because they are brought up late in the meeting.
7. Managers are sometimes inhibited in bringing up risky options for fear of losing face if rejected.
8. Groups tend to get "hung up" on a topic, going over the same ideas rather than turning to new approaches. Often considerable time is lost at the beginning of a meeting in simply finding out what the issues are, and "where everybody is at."
9. Face-to-face groups seem, according to experimental evidence on the risky shift, to be prone to arriving at riskier decisions than would individuals who were not subject to social pressure.

These factors tend not to be operative in computerized conferencing.

On the other hand, there are disadvantages, which were also explored in detail in Chapter 3. For instance, some people find it easier to refuse to answer messages sent over these systems than to ignore a telephone call. Since there is generally no particular hour that must be set aside for participating in asynchronous conferences (those in which the participants are not all on line simultaneously), some people never get around to participating much at all. Without the richness of nonverbal cues, there is a greater chance that the meaning of a communication will be misinterpreted. While the discussion in a computerized conference may generate more ideas and better decisions than a face-to-face meeting, it is possible that the social cohesion necessary for effective implementation of those ideas by the group may not develop without face-to-face contact.

It is also possible to structure an autocratic and tightly controlled discussion if that seems desirable. This might be particularly helpful in delicate international or labor–management negotiations. For example, the computer could be programmed to enforce Robert's rules of order; or there could be limits placed on the length or number of entries allowed to each party before voting; or there could be a vote count and decision process that incorporated veto powers, as at a U.N. Security Council meeting.

Implications for Alternative Organizational Structures

The structure of a computerized conferencing system can be adapted to fit whatever form or type of organization exists. One typology of organizational forms, for instance, has been summarized by Dodd Bogart (1973) as "feudal,"

ISBN 0-201-03140-X, 0-201-03141-8 (pbk.)

"bureaucratic," "human," and "system." Let us examine the central characteristics of these four types of organizational structures and the implications that computerized conferencing might have for each of them.

1. *Feudal* (or authoritarian) organizations place an emphasis on the individual leader and on hierarchy, with influence and ideas flowing downward from the leader at the top; the loyal followers are expected to obey without question. Most procedures follow tradition; the organization acts in ways that have proven to work well in the past until the leader acquires a new insight or ambition. The "feudal" leader could use the CC system simply as a cheaper and faster way of gathering information, options, and advice from trusted followers, and then of disseminating his decisions and orders.

Feudal systems function and succeed because of the charisma of the leader and loyalty toward this leader. Coordination in organizations that are feudal and hierarchical in nature (with the leaders at the second and/or third levels also having a degree of charisma and their own followers) is always difficult and time consuming.

Universities and churches are frequent examples of multiple-level feudalistic structures, with each department head essentially like a baron, having his or her own following. Many fruitless committee meetings stem from the frequent personality conflicts among competing department heads, especially if the top-level leader is absent or in fact has no charisma.

A CC system might decrease the amount of time necessary for department heads to agree among themselves on common actions. It can also help to spread the "presence" and influence of the leader. A feudal system can actually be quite flexible and adaptive if this is the nature of the leader. However, to maintain wise and flexible leadership, the leader needs good information and needs to be able to be in constant communication with the organization at all levels and in all corners of the "empire." His or her inspirational presence must be felt down among the toiling workers if motivation and loyalty to the organization are to be maintained. It is very difficult for a leader (such as a university president or the governor of a state) to be present at all meetings that would benefit from his or her presence. However, in a CC environment, the leader's assistants could retrieve the discussions, and occasionally drop in an encouraging message to exhibit the needed "presence."

2. *Bureaucratic* organizations, first analyzed by the 19th-century sociologist Max Weber, emphasize the rational distribution of responsibilities among positions or offices, with formal rules and procedures governing all actions. In principle, any qualified person can fill any position, as symbolized by the organizational chart and job description manuals. The rules and procedures are oriented toward achieving reliability of performance; reducing uncertainty; and maximizing productivity and efficiency as measured by fairly short-term profits.

ISBN 0-201-03140-X, 0-201-03141-8 (pbk.)

One of the chief problems in modern bureaucracies is pinpointing the responsibility for a decision. This is particularly true if actions taken (or decisions not to take an action) are based on telephone calls and undocumented committee meetings. Somehow, there always seem to be later discrepancies about what was said and who took what view and who agreed to do what. We feel that CC systems will facilitate the documenting of responsibility for decisions, providing accurate records of who took what positions and actions at what time. For the dedicated bureaucrat who does wish to make the organization work to meet its objectives, computerized conferencing can provide the ability to establish for the record her or his actions and positions, and the reasons for them, should the decision be reconsidered at a later date. Since one characteristic of bureaucracy is frequent rotation in office, this fuller documentation is potentially very valuable for providing continuity and accountability.

In the bureaucratic organization, a "legitimate" application might be for interdepartmental discussion of issues that do not clearly fall within the domain of any one department or within existing rules and procedures. However, computerized conferencing is more likely to be used to subvert the bureaucracy. Bureaucratic organizations are like a stale block of Swiss cheese: a structure of rigid rules, but full of holes in the rules and regulations. Crafty bureaucrats can learn to find these holes and work around the formal structure to accomplish specific objectives. Often this is the only way significant change comes about in bureaucratic environments. With computerized conferencing, the potential would be increased for a group of motivated individuals to form and work in concert through private messages to find the appropriate way to circumvent the usual regulations and procedures; this of course would have a devastating effect on the organization—the cheese would begin to crumble—but perhaps a desirable one in terms of accomplishment.

3. The *human* organization; most fully described by management theorist Rensis Likert, places emphasis on small groups, rather than on the individual or position, as the locus of communication, influence, and decision making. The manager should direct his attention toward facilitating group cohesion and group participation and toward acting as a communication link to other groups, in order to maximize both group performance and individual self-actualization through group contributions.

The human organization model specifically defines the proper flow of information to be up, down, and across. There is no better mechanism than computerized conferencing for allowing the group leader to stay abreast of all the processes and projects within his own group and to simultaneously keep in frequent contact with other group leaders at other locations in the organization. However, the availability of computerized conferencing might en-

ISBN 0-201-03140-X, 0-201-03141-8 (pbk.)

courage the humanistic approach to be carried to the extreme, where group task accomplishment is replaced by group therapy. The heightened sense of "real" communication with others can become addictive in nature and seem an end in itself, causing the group to lose contact with the rest of the world.

4. The organizational *system*, is symbolized by mathematical models and the flow diagram featuring a feedback loop (which, unfortunately, unlike spaghetti or cheese, is not necessarily nourishing). Information, influence, individuals, and resources (money, production facilities, etc.) are seen as dynamic variables that may be allocated and reallocated to various functions and problems in order to maximize the organization's ability to deal with a changing environment (the macrosystem of which the organization is a part) and with changing priorities over a long term.

A computer-mediated communication system makes the system approach much more feasible for a large organization. It allows the organizational leaders to draw on large amounts of data and on models that are stored in any computer, and to structure decision-making and task groups on an ever-changing basis in order to respond to changing conditions.

In "Beyond Bureaucracy," Warren Bennis (1964) predicts that such a system-oriented organizational form will become necessary in the future, based on "adaptive," problem-solving, temporary systems of diverse specialists, linked together by coordinating and task-evaluating specialists in an organic flux. But how can this type of organization be reconciled with the increasing resistance of executives and their families to endless transfers from one geographic location to another? We believe that computerized conferencing would allow complex team projects to exist where the members are geographically dispersed.

Project management conferences would provide a complete written record of all specifications, changes, clarifications, and suggestions that took place, in addition to providing complete accountability if ever needed. It would also have a significant impact on the manager's ability to regulate his own time, and to be fully involved in more than one task force at once. On EIES, it is up to the user to choose when he or she wishes to review the new items and make contributions, rather than being a slave to the ringing telephone and the meeting scheduled at someone else's convenience.

Further Decentralization of Corporate Structures

Overall, what do the capabilities and features of computerized conferencing as applied to different organizational forms have to do with changes in corporate structures and functioning? If you change the communications structure of an organization, you inevitably change also the nature of the decision-making process within it and the kinds of decisions that are likely to result. Ultimately, you also change the form of the organization itself. Let us

ISBN 0-201-03140-X, 0-201-03141-8 (pbk.)

switch historical analogies and think for a while about the outcome of the introduction of the telegraph and the telephone. The growth of today's dominant business organizational form, the multinational corporation, would not have been possible without the world-wide linkage of corporate outposts to the Home Office by fast and pervasive communications lines.

By contrast, computerized conferencing can facilitate the decentralization of information exchange and decision making. The Home Office might become simply a supplier of services to relatively autonomous units of the organization. For example, Bell Northern Research found that as a result of a recent field trial of computerized conferencing among businessmen in an organization, the pattern of communicating with other branch offices had shifted from funneling requests through the central office to direct communication among branch office managers. With the introduction of the conferencing (or "Computer-Mediated Interaction") system, branch managers were able to solve their problems among themselves, rather than involving the central office.

What is the logical consequence of this for corporate hierarchies and functioning? If decisions are being made autonomously, at the local level, they might be made much more quickly and with a better understanding of the nature of the problem. For the corporate executive himself, his real power may be usurped by the local managers, and he may become reduced to serving as nothing more than a figurehead, like modern monarchs.

On the other hand, executives who adapt to the new communications tool might find that they can become much better informed and much more able to try out controversial ideas than ever before. Computerized conferencing allows the lateral coordination necessary for decentralization of authority with a speed and efficiency not possible with other communication systems. Ongoing transcripts of all conferences among middle managers permit monitoring and/or intervention if an unwise decision seems imminent.

It is well known that one of the characteristic tendencies of current bureaucratic or hierarchical communications structures is that "good news" travels with great speed, whereas "bad news" is filtered out as much as possible, in order not to arouse the ire of the recipient. Like the biblical king of old who donned common clothes in order to go among the people and find out what they really thought, the pen name and anonymous features of computerized conferencing may enable the executive to find out what middle-level managers are really thinking and doing. Although the chief executive might be reluctant to introduce a very new or different idea into a face-to-face conference, for fear of either looking like an ass (or an emperor with no clothes?) or of swaying the decision by virtue of his lofty rank, no such inhibiting factors need be present in the computerized conference. It should be noted, of course, that if an executive wants to exert authority based

ISBN 0-201-03140-X, 0-201-03141-8 (pbk.)

on greater knowledge of the overall situation, then he or she would not choose to use anonymity or a pen name when making a suggestion.

Crisis Management

Your general manager and three other executives suddenly quit . . . a wildcat strike of assemblers jeopardizes delivery of vital parts . . . an important customer has just rejected your first shipment . . . your university's tuition deposits are 25% less than expected. Such major and unexpected events in the routine of an organization are crises. Robert Kupperman, Richard Wilcox, and Harvey Smith, who encouraged the development of EMISARI for dealing with crises at OEP, have emphasized that such crises are distinguished by extreme unpredictability with respect to the amount and importance of readily available information. Consequently, most conventional information systems are inadequate for the task.

As they state (1977, p. 404)

> It is useful to note the characteristics common to most crisis management. Perhaps the most frustrating is the uncertainty concerning what has happened or is likely to happen, coupled with the strong feeling of the necessity to take some action anyway "before it is too late." This leads to an emphasis on gathering information . . . . Unfortunately, few conventional information systems are equal to the task of covering unconventional situations . . . . Thus, with limited information and resources the manager may find it difficult just to keep up with rapid developments, let alone improve the overall picture of the situation . . . .

As we pointed out in Chapter 2, the very first operational CC system, EMISARI, was designed to handle such crises by facilitating the collection and dissemination of data, news, policy statements, etc., to all of the persons throughout an organization who were concerned with resolving the situation. Crisis management often requires the following sorts of things:

1. Factual information from many sources;
2. Value judgments from key individuals;
3. Discussion on proposed alternatives and their ramifications;
4. Anonymity in voting or discussing such issues, so that middle managers do not feel inhibited about criticizing the ideas of their superiors;
5. A way to determine a general consensus within a short period of time, when a decision must be reached.

EMISARI demonstrated that computerized conferencing, with its ability to quickly structure and categorize the information flow, can clearly fill these needs. Richard Wilcox, who used a descendant of EMISARI when he had to report twice daily to the White House during the 1974 truckers strike, points

ISBN 0-201-03140-X, 0-201-03141-8 (pbk.)

out that the typical management information system (MIS) is highly structured and designed essentially for one-way communication from the field to headquarters. In a crisis the advantage of computerized conferencing is its flexibility, he says.

> Headquarters can communicate quickly back to the field, saying, for instance, the guidelines have changed, forget the Table 3 we asked you for yesterday, today we need the following information instead. This request can, in turn, be questioned and perhaps modified through rapid communication with many people simultaneously.

The Budgeting Process

If solutions occurred as a function of the words that have been written on a problem, then there would be no need to say anything about the budget allocation process. The accountant, the economist, the operations researcher, the management scientist, the systems analyst; all offer methodologies, approaches, and "solutions" on how to deal with this common decision process. Thus, we have seen the emergence of concepts such as cost-effectiveness, cost–benefit analysis, program planning and budgeting process, and zero-based budgeting.

The formal budgeting systems incorporate numerous objective and subjective measures dealing with concepts such as probability of success, timeliness, mission contribution, and problem significance. By the time all the considerations relating to any one budget item in relation to others have been crunched through a budget-assessment-oriented computer model, frequently nobody understands exactly why various priority and funding level choices were made. It is not clear that such mathematically derived budgetary solutions are any better than such time-honored seat-of-the-pants approaches as the following.

The Squeaking Wheel: Propose a cut in the budget and see who squeaks. If the anguish seems justified, readjust the budget.

Follow the Leader: Continue the budget allocation as in the past, merely making across-the-board adjustments for factors such as inflation.

Nothing Succeeds Like Success: If the people doing a job have been very successful in the past, increase their budget in the hope of more of the same.

White Charger: If the individual has a well-laid-out plan that is aggressively and convincingly presented, give him some funds so he can charge ahead.

The Group Think: Give the requests to a committee to review and recommend changes.

None of the foregoing are fundamentally more or less scientific or objective than any of the more elaborate approaches. The real problems in the budget

ISBN 0-201-03140-X, 0-201-03141-8 (pbk.)

allocation process cannot be solved by a particular "magic" technique; rather, solutions can be found only by optimizing the necessary communication processes.

The closest analogy to the communication elements of the budgeting process as it now takes place in most organizations is the polling process. In effect, the line managers each develop their requirements in isolation from the other units. These "polls" of requests with associated supporting material are passed up the hierarchy and may go through one or more levels of review and revision before a decision point is reached.

The first major problem that this method produces is a sense of conflict and competition among managers at the same level in an organization. This is particularly true when the overall funds are much less than the total budget being sought, or when severe cutbacks are being made.

Our argument is that an increase in lateral communication would improve both understanding of the budgetary priorities that emerge, and the realistic nature of the assertions made (since a manager's counterparts in other parts of the organization would be able to see and unmask "inflated demands" made for bargaining purposes).

It is a reasonable hypothesis that the translation of the budget process from a poll to a group communication process would bring about a more cooperative analysis of resource allocation decisions and heighten the potential for managers at the same level to place their individual tasks into a cooperative framework. It would remove the impediments to cooperation that exist when one manager is unaware and suspicious of what another is trying to "steal" from him in the budget process. This shift in the nature of the communications usually associated with the budget process is probably the key element to allow managers to mutually support one another and to establish cooperative efforts and objectives toward attainment of organizational goals. In the long run it would foster both the decentralization of decisions and the accountability for them.

Text Processing and the Automated Office

The industrial revolution and the rising status of blacks has given the working-class female better paying and more interesting alternatives to domestic service. Among the signs of this structural change in the socio-economic stratification system of Western societies are the castles of the rich, abandoned or turned into museums now that they can no longer be adequately staffed by an army of servants; and the complaint of the middle-class matron: "You can't get good household help anymore . . . ."

The movement toward equality of opportunity for women is likely to have the same sort of consequences for middle-class working women and for the corporate castles that have relied on armies of them to do all the clerical

ISBN 0-201-03140-X, 0-201-03141-8 (pbk.)

work. Able and ambitious women will no longer settle for being low-level secretaries or clerks.

There will still be good secretaries, or administrative assistants; but these will generally be well-paid highly skilled college graduates, both male and female, whose duties, rank, and pay will mean that only top executives will have a "private secretary."

Faced with a crisis in the quantity, quality, and cost of skilled office support work, management has opted for the piecemeal solution of buying whatever machine seems able to automate specific aspects of the process. There is little thought about modularity or compatibility among the gadgets purchased; about how they fit into the present and future organization, seen as a total system; or about their effects on overall organizational morale or productivity. Thus, we have, as an answer to the difficulty of getting mail delivered, a computerized robot that roams the halls beeping and bleeping and stopping at each office door for 20 seconds to allow the occupants to put mail on or take it off the cart (Blau, 1978). This helps to speed up the internal transmission of mail, but does nothing about the total mail-based communication system, such as mailroom delays within office buildings, processing and delivery delays in the U.S. Postal System, and the cost of typing and postage for letters with multiple addressees. The expensive machines ($10,000–$12,000 each, and at least one is necessary for each floor of an office building) will become totally useless if the organization attempts to attack the total process by moving toward computer-based mail or message systems.

The emerging technology of the "office of the future" can be used to implement very different concepts in terms of the effects on the organization as a social system. That office automation can creep in from the bottom, replacing typewriters with word processors and keypunches with various data-entry units, means that it can evolve without the necessary consideration taking place at appropriate management levels.

Too many people are viewing word processing as a logical extension of the factory assembly line into the office. They tend to assume that the way to use computer-assisted technology for text composition and modification is to centralize and specialize. The word-processing center and centralized dictation and copying operations are aspects of this approach. This strategy separates the producers of the service from the consumers and does not provide adequate communication links. By following the logic of specialization and dividing the diversified job of the secretary or middle manager into several distinct, specialized full-time tasks, morale and ultimately efficiency are destroyed. However, modern information–communication systems such as NLS (on Line System) can allow each individual to increase the diversity, responsibility, and communication richness of her/his job.

ISBN 0-201-03140-X, 0-201-03141-8 (pbk.)

If the effect of word-processing systems were assessed on a total-system, life-cycle cost–benefit basis, the systems might not seem quite the paragons of efficiency and savings that short-term studies of time and motion and cost suggest. In addition, a high degree of specialization and segmentation of office work means that the whole process can grind to a halt whenever one of the specialized kinds of machines or personnel fails to function. Flexibility and the ability to adapt to change and to unforeseen conditions (such as the two-week absence of both operators of the word-processing equipment, if they get sick at the same time) are benefits and costs that must also be considered.

Office automation is fraught with fundamental management issues such as decentralization, accountability of management roles and functions, major capital expenditures, and restructuring of business functions. Computerized conferencing, when viewed as an integral part of an information–communication system, offers the potential for maintaining the office worker and the manager as generalists, able to adapt to changing situations. With the existence of the central computer in the loop there is no economic need to create functional pools dedicated to one very narrow piece of the process. The functions of creating, editing, filing, and retrieving material can all be handled via a central computer, probably with less training effort per manager and secretary than is now required to learn to use many of the current generation of word-processing systems. The trend is largely the result of a transitory phase in the office environment where paper is still considered necessary for filing documents and records. The shift to complete electronic forms of records for even free text information will occur when the cost of electronic storage and retrieval drops below that of paper and file cabinets. This is likely to occur within the next five years and could be said to be already here, when the human costs are properly factored in. Incremental decisions to purchase this and that piece of equipment add up to major capital investments, which the organization is then reluctant to abandon or replace, since doing so would require that someone take the blame for a bad purchase decision.

Other Possible Applications

Boards of directors could become more heavily utilized as a resource that would contribute real expertise to the formation of corporate policies, especially with the diversity of background and outlook common to the membership of a typical board. Rather than a perfunctory monthly meeting in which participation is limited to superficial debate and unstructured comments, the board member would have time to think about the matters to be discussed and feed back comments by computerized conferencing beforehand. Some

ISBN 0-201-03140-X, 0-201-03141-8 (pbk.)

might argue that a corporation likes to have a rubber-stamp set of outside board members, but in this era of "corporate social responsibility" and stockholder's suits, it is wiser in the long run to assure that a broad range of viewpoints be brought to bear on all policy decisions. Public scrutiny of corporation policy and of the actions of individual board members would seem to favor some means of improving the manner in which corporate boards transact their business, such as computerized conferencing.

The working of interindustry, intercompany, and professional society working committees or special-interest groups could also be greatly enhanced by the use of computerized conferencing. It is virtually impossible for a geographically scattered group to reach conclusions, agree on the wording, and publish a polished report without the effort either being concentrated in one or two people who dominate the proceedings or going by default to the one with the best means of clerical support. Public hearings for comments on draft reports would be much more efficient in a computerized conferencing mode; there would not be the concern that the comments would not reach the committee in time; comments could be acknowledged and clarifications requested in an extremely efficient manner. In the case of private sector interindustry or intercompany dealings, it is extremely important for a great deal of attention to be paid to openness, avoiding any hint of collusion or conspiracy to monopolize or act counter to public interest. The record of proceedings that could be taken from the backup files of a computerized conference would satisfy any investigating group as to the reasons why certain decisions were made by such committees and trade organizations.

Still another imaginable application to managerial functions would be to replace the face-to-face job interview with the computerized conferencing interview. Since pseudonyms (or just last names) may be used, there is no way in which the prospective employer can know the applicant's age, sex, or racial or ethnic group, or whether he/she has a physical disability; thus there is no possible way in which discrimination based on such factors could unconsciously creep into the employer's evaluation of the applicant from the interview.

Communicating and working on these systems also facilitates flexible hours and flexible work location. There is no reason, for instance, why most of a person's work could not be done from home, and at times other than normal office hours. (See Elbing et al., 1975, for a discussion of the inevitability and advantages of a flexible working hours system.) This is especially important for working mothers, who could work at home on days when a child is sick and home from school. In addition, people who are traveling can be kept in touch with events at the office through their portable terminal plugged into a hotel or telephone booth phone.

ISBN 0-201-03140-X, 0-201-03141-8 (pbk.)

Some Possible Negative Consequences

It is to be expected that computerized conferencing, just like computers before it, will not only help to solve some old problems but will create some new ones, too, for organizations that adopt it.

For one thing, the spur to decentralization provided by such a system can only accelerate the abandonment by corporations of the central cities and their further loss of tax base and employment opportunities.

A category of employee whose job would be threatened by such systems is the person who acts largely as an "information controller." Every bureaucracy seems to have several of these senior clerks, usually with the company longer than its President, who jealously guards his numbers, dribbling them out one by one upon request. Such employees would no longer enjoy the pride and security of being the only one able to retrieve such data, once it was on line and receivable by a whole set of authorized persons through the conferencing system. It would seem only just for an organization to provide a generous retraining and/or early retirement plan for them and for other employees whose information- and communication-related jobs would be replaced by the computerized system. Since companies have not always acted in such a humane manner, however, the potential displacement of employees by the system is a matter for concern.

Another question is whether interaction via computerized conferencing can substitute for daily face-to-face conferences with co-workers, in terms of satisfying the social–emotional needs of the participants, as well as their task-completion needs, and of supporting the degree of cohesion necessary for motivation and coordination of work efforts.

A fully developed management communication–information system would have as its goal the ability for managers to obtain the information and analysis that they need, when they need it, without having to go through intermediaries. However, this also poses the danger of inexpert and uncritical use of the output. We constantly witness the misuse of computer output in the sense that the great volume of numbers generated tends to numb the recipient and cause him/her to flip to the solution or conclusion and accept it on faith. That is, the computer legitimizes information that may in fact be incorrect or inappropriate, but the recipients believe that "the numbers must be correct because it took a tremendous computer effort to generate them."

Of course, there is also the privacy and security issue. Existing systems have features to try to prevent unauthorized use of the system or receipt of entries. For example, a private message on EIES may be completely destroyed after delivery, so that there is no record of it in the computer. Each user has a unique number and password and is notified, each time he or she gets on the

ISBN 0-201-03140-X, 0-201-03141-8 (pbk.)

system, of the last time and date a user with such a password was active. Thus, unauthorized use may be quickly detected. There is always the possibility of abuse of the system, however—an unscrupulous executive might secretly have a system programmed to deliver him a copy of a private message, for instance, or to notify him of the identity of the "anonymous" contributors. On the other hand such a person is also likely to engage in illegal wiretapping of telephone calls or opening of correspondence. From a technological point of view the security of such systems presents the same problems as corporate data bases, files of internal memos or correspondence, and staff reports. Although a reasonable amount of physical and hardware security can be bought, the ultimate security is provided by the people involved as users and suppliers of the services.

A NEW MODE OF OVERWORK?

The flexibility of computerized conferencing means that since employees can continue to work while traveling or even in the hospital, they may be *expected* to do so. It becomes impossible to have a real vacation or to "get away from the office." For example, the users of one computer-based system report that formerly when the boss went on a business trip to Europe, they had time to catch up on unfinished tasks and reports. Now, he sends a constant stream of demands and questions, around the clock, and expects a same-day answer. While away at professional meetings or conferences, the coauthors of this book have found that since it is possible to answer communications by using the usual lunch or dinner or after-dinner hours provided as free time at meetings, there is a kind of pressure actually to do so. Thus, the workday at a conference now stretches from early morning to late at night, with the CC system making it possible—but somewhat exhausting—to simultaneously fully participate in the face-to-face meeting and fully participate in the work groups that are communicating via the CC system. For hard-driving bosses and workaholics, in other words, computerized conferencing can represent a dangerous reinforcement of their bad habits.

Building a Management Communication–Information System

In all of the potential kinds of applications to managerial functions described thus far, we have been assuming that what a CC system can do is build a more effective kind of communication and management information network; and that this will have very fundamental impacts on the kinds of social interactions and organizational outputs that will occur as a result. As Mears (1974) defines a communication network, it is "the interaction required

ISBN 0-201-03140-X, 0-201-03141-8 (pbk.)

by a group to accomplish a task." He states,

> An organization's effectiveness depends upon the performance of numerous small groups which function and interact within the overall organizational system. Since the activity of a small group depends to a great extent upon its information flow, the communications act has been studied as a means of influencing efficiency [p. 71].

What kinds of communications structures and capabilities would we want in a CC system for managerial applications in order to maximize its usefulness for different kinds of communication tasks and its overall efficiency in terms of the kinds of decision-making processes and outcomes that we would like to support? In this section, we are going to take an excursion into the best of all possible communication–information worlds. We will begin with the "bare-bones" kinds of features available in current computer-mediated communication systems, and then think through the kinds of additional capabilities and features that ought to be added to support this kind of application, and that are possible with current technology.

In other words, looking at the impact of computerized conferencing on the organization as a whole is the view from the outside. Another form of understanding can be gained by looking from the inside out. In the next several pages, join us as we think of ourselves as a manager within an organization contemplating the capabilities we would like to have in order to accomplish the tasks set out for us. Our understanding of a computer may be nil, but our understanding of what it is we have to do certainly is not. We will begin to develop our requirements in an order that is possible as a function of our growing appreciation of what the computer can do for us through actual use of it.

Obviously we send many messages or much news to our associates, and we also try to get others on the phone. Depending on where our associates are, the mail room or the post office often leaves us wondering when or even if the item is received. Phone calls often turn into a minor comedy of calls, return calls, and calling again. If we put this straightforward process on a computer, we would immediately gain the advantage of allowing delivery to be as fast as the person on the other end wishes. Also it is quite easy to have the computer inform us when a person receives a particular memo we send. We now have for our use an electronic mail system; however, a little experience with it would enable us to realize that the computer can add to the message process some automated features that in some instances replicate what we usually do and in other cases provide new capabilities.

Many of our memos go to more than one person and there are some definite groups of people to whom we are always "copying" things. We can let the computer remember these groups so there is only one group name we have to type in when we send a memo to all the members of that group. In addition, very often we send a memorandum to certain individuals for their approval before it is sent to others. This routine introduces a new type of message for the computer to handle, one that requests an approval of the message from certain addressees before it is allowed to proceed to others, with, of course, notification to us as the writer when approval or nonapproval has occurred. Another of many variations to this message could be a request to a group, each of whom could answer whether he/she were able to take care of the request. The first "Yes" reply would terminate sending the message to additional individuals.

It occurs to us that there are times when we remind ourselves that there is a memo we should send, but do not want sent yet, such as reminding everyone on next Tuesday morning about a meeting that afternoon or a deadline that will occur. Therefore we instruct the computer to hold the message we are now writing until some date and time when it may begin the delivery process. Thus we take advantage of the computer's memory to allow us to do things as we think about them and have the computer carry out the actions when appropriate. In fact, we can send these delayed messages to ourselves as reminders. Hmm! Reminders are an interesting concept. Let's create a new type of item—our personal list of reminders, which we can create, retrieve, edit, and delete to keep abreast of the tasks we have to accomplish. We can place alarms on these reminders as well.

Most certainly our system has some text editing, so that we can correct our typing errors. However, we soon realize that many of our memos contain data, which raises a set of text-editing requirements for creating columns and tables of numbers or items. But it soon occurs to us that we often seek to collect information from others in the organization—both qualitative and quantitative information. This means that we would like the ability to design and send some standard forms as a "message." When others receive these "messages," they are really being asked to fill in these forms, and the computer will automatically generate back to us a new message which represents the completed form.

It would seem that sets of these messages are going to be about a particular topic—a project that is under way, a problem being dealt with, a plan being laid. Much of the material must be retained for review and updating. Let us then organize something, called a conference or discussion, that has a specific topic, perhaps an agenda, and certainly a chairman who is responsible for seeing that the material needed is provided, the outdated items deleted, and the current items properly organized. We can call these items comments, so they do not get confused with messages. We will give each conference its own name and a list of conferees, so that the computer can keep each conferee up to date on what is new in a particular conference. It would also be nice to take further advantage of the computer by building in ways to categorize or

ISBN 0-201-03140-X, 0-201-03141-8 (pbk.)

organize the items by such things as key words or phrases, titles, associations among items, separate sequences of discussion threads, and a record of edits or changes to older comments.

Over time in a discussion we begin to realize we might cut out much superfluous commentary if we had a mechanism for quickly determining the consensus or nonconsensus of the groups about a particular topic. Therefore we put in a voting capability. The computer, we are pretty sure, is capable of counting votes for us and displaying the results. We can have the computer provide a variety of scales: the desirability of and/or confidence in an observation; the probability and/or degree of impact of a possible event; etc. Also we had better put in the ability to get a numeric estimate and have the computer exhibit the distribution of the estimates, such as the potential cost of a product. If we make careful use of the voting, it could be a real time-saver.

Naturally all this does not occur to us at once. As we utilize the system, we continuously discover ways it could be improved to aid us and we are fortunate in having a computer staff that is either one step ahead of us or very responsive to user needs. Although we may not have known much about computers when we started sending messages, our experience begins to give us an appreciation of what it can possibly do for us.

It occurs to us that with all this power to handle text we wish to write things that are bigger, like staff reports of one sort or another. Unlike a message or comment, we do not do this in one sitting at a terminal. We should have a personal composition space where we can compose items over a period of time and store them for later use. It might be a good idea to call this a notebook. Maybe we should have the option of allowing particular individuals to read or write in sections of this notebook, in case we want a co-worker to review or help draft a report. We will need a place to make these reports available to their readership and we might just call this set of reports a bulletin, journal, or some other suitable name. This place might also have some special sections to deal with news, notices, and other categories common to organizational communication. Maybe a corporate newspaper is in order. Also, there are certain offices preparing rather specialized memos that are sent around and perhaps these offices should have public notebooks in which only they can write but that others can read. Some examples might be the abstracts of news stories or press releases done by the public relations department, the usual stuff the legal department offers, analyses of competitors' products, and administrative announcements. These are very much a function of our particular company, but they are all serviced by the same public notebook capability, each having its appropriate title.

Now we have our different communication spaces (private, common, personal, and public) but we realize there are different ways of handling basically the same item of text. Of course, the nice thing about text is that it

ISBN 0-201-03140-X, 0-201-03141-8 (pbk.)

can also include numbers when we need it, but those old-fashioned systems on computers never let us handle text. What we must have is an easy way of moving these text items around, like making a comment out of a message. In fact, we should be able to bury text items into others so we can include the contents of a message in the middle of a comment by just referencing it with a very few key strokes at the terminal and not actually having to retype or copy it. When we have done this, the system is now completely fluid or adaptable to anything we might decide we wish to do with the textual material available.

The system has now built up a considerable library of textual material. It is not unusual that many of us begin to become impatient with the size of some of the text items. Certainly we need to have the complete story in a report, but in referring back to it we prefer to deal with summaries. Therefore, we have the computer people introduce a method whereby when we write a report, we can specify that the text itself can ask questions of the reader like: "Do you want more detail on this point rather than moving ahead to another point?" This is an ability to write "adaptive text" that can adjust its content to the desires of the reader, not dissimilar to what is done in computer-assisted instructional systems (CAI). It is as if we could write a novel that asks the reader if in the next chapter the villain should succeed in committing another murder, and thereby choose the appropriate version of that chapter. Now we can deliver reports that are as short or as long, or as concise or detailed, as a particular reader may wish. We will probably have to institute some composition course for managers to get this new writing style across, including some guidelines on report organization, if the full benefits of this feature are to be realized. After all, this type of thing is not really doable with paper and pencil, and takes some getting used to.

By now our simple message-sending system has evolved considerably into a rich collection of alternative communication structures and facilities, as well as some data processing capabilities. There are pieces of the system we make little or no use of and there are certain things we do almost every time we get on the terminal. What we need now is some way to collect all the operations we regularly do into some macro operations, each with a unique name or label. With this we can just call on that one label and have the system carry out all those operations we do regularly, such as review a particular set of conferences for anything new or search a group of report files for anything new having certain key words in it. In essence we are developing our own set of personal commands or abbreviated codes for tailoring the system to our personal needs, as every other user will be doing with his or her personal commands.

It is interesting to observe that nothing we have designed for the software of the computer system has any content that reflects our particular company

ISBN 0-201-03140-X, 0-201-03141-8 (pbk.)

or organization and its special concerns. A communication structure knows nothing about the problems being dealt with or the content of the text. The particular forms have been defined to relate to our objectives, but another organization or another division of our company can tailor the same system to its needs. This is confusing because for decades now people have been trying to tell us that extensive analyses are needed for the design of a management information system, and that the design is then tailored around particular data items and the resulting software is only appropriate to our particular operation. It would seem, on the other hand, that this communication system is coming very close to fulfilling our needs for an MIS capability. Maybe a few additional capabilities will allow us to cross that fine line and begin to call it a management information system.

The first added capability would appear to be a need for a permanent data base of numbers that are reported from different segments of the organization. We extend the concept of a form so that we can set up files of permanent forms, perhaps called report tables or estimates. These permanent forms have the property that each is "wired" to a particular individual, who is the only one able to supply the data to fill out the form. If we retrieve it, the computer indicates the date and time the data were updated and who is responsible for keeping those data current. In other words, we have a data base that is subdivided in one dimension by the human source of the data. If we see anything questionable about the data reported, we always know whom to message to inquire about what we observed. This is very different in nature from the anonymous data bases we are all familiar with.

Now, we should allow a new type of "message"—call it a footnote—that can be appended to a particular data form, so that the permanent data form can build up an associated list of signed footnotes. Therefore, the person reporting the data can make qualitative statements regarding anything unusual, or a person responsible for analyzing or interpreting the data can footnote the data with his or her observations. Any time we go to get some data we are given the opportunity to review the footnotes. What we now have is the ability to deal with an associated mix of qualitative and quantitative information. Since, once again, we have maintained communications "purity" and have not frozen the software to deal with particular data items, we can adapt these permanent data reporting forms to the particular management needs of the moment.

Remember that our text items can have indirect references to other items, which are then incorporated in the text item whenever it is printed. We can also allow indirect specification of references to these permanent data forms, which means that we can write a report that incorporates any of them. Whenever someone asks to print out that report, the computer fills in the individual data forms and utilizes the most recently reported data. This

ISBN 0-201-03140-X, 0-201-03141-8 (pbk.)

makes it very easy to have a standard weekly report for which we may only have to alter a few paragraphs of text every week and the computer pieces in the data that are needed on a regular basis. When that report becomes outdated because the problem has been taken care of, we find it very easy to eliminate the report and the obsolete data forms because they are not fixed in the software, but totally user defined.

Our organization undoubtedly has a number of large-scale data base systems to deal with the normal required flow of data through the system for such purposes as scheduling production, accounting and taxes, and control of inventories. Suppose some very large-scale planning, econometric, and marketing models have also been set up by various talented groups to serve particular needs? Usually these groups think that we should become experts in every one of these blankety-blank systems, as if we had nothing else to do in this organization! However, there are times when we do need some of these data, and it is useful to make runs of various models in slightly differing circumstances. We could use these outputs as part of the proceedings in individual conferences or in particular reports.

Let's have a special conferee—really a very small computer (a microprocessor)—that can be taught to phone up these other data bases and extract particular data or run particular models. We will have some of the experts work with us to design forms we can fill out to send to our computer conferee to tell it what we need in any instance. The microprocessor will obtain the results we need by dialing the other computer, requesting a run of the model or a search of the data base, receiving the results, and putting them back into the communication system according to our instructions. We will also let some of the experts participate, so they can act as human transponders to their particular data and model systems when we are having problems with the forms or in interpreting the output of some of those systems.

At this point, we have tied together through our CC system all of the data sources in our organization, whether human or computer based. There is really no longer any need for one central computer system. We leave the models and data base systems with those segments of the organization having primary responsibility for development and use of those systems. Our CC system now becomes the primary mechanism for coordinating all of these resources and utilizing them across separate segments of our operation.

We soon discover that one thing we are doing all the time is tracking actions of particular types. A particular case, project, piece of outside correspondence, report, etc. must move through a series of steps or processes in the organization. What we need is a special form that carries with it the ability to process the data it contains. On this form we can define each of the steps an item must move through, and who is responsible for indicating when a

ISBN 0-201-03140-X, 0-201-03141-8 (pbk.)

particular step has been completed. When a particular item has to be tracked, a copy of the form suitable for that project is titled with the name of the project and made active. All those responsible for various steps are notified automatically by the computer of the existence of the new form. Any time a particular step in the process is completed, the person responsible updates that step and the person having primary responsibility is notified. The form might have time limits on various steps, so an alarm can be automatically triggered when the time limit is exceeded without the steps being completed. Furthermore, the computer will allow individuals to retrieve items by type, how long they are taking, which steps they are in, etc.

It may also be that data are gathered and sent to a big PERT* model somewhere that integrates the various projects making up some effort, such as a particular new product development, for continous scanning of what appears to be the bottleneck at the moment. This would make use of our computerized conferee, of course.

Certain tables of data being reported might represent cases for which a column of data requires some degree of numerical processing, such as additions for totals, computing averages, doing a regression analysis, or making a projection. All of these can be set up as special data types for which the computer performs these specified calculations any time data is updated. We would also desire adding some display graphics so numerical data can be plotted on the terminal for those who prefer graphical output. We can also choose to extend the text processing with automatic spelling correction and such things as automatic hyphenation. More subtle but feasible is to do semiautomatic abstracting and the automatic analysis of what should be used as key words. In addition, there are fascinating potentials for what may be termed indirect communication channels. For example, if those responsible for maintaining and publishing a public notebook can retrieve the key words others are using to find information, and in particular what words do not produce any relevant information, then they may be able to schedule some of their efforts to fill in gaps. More futuristic, and perhaps having some dangerous aspects, is the analysis of user preferences as exhibited by retrieving, voting, and associative behavior, to determine if very different interpretations of the same text items are being made by different management groups. This could be useful in clearing up fundamental differences in values or goals of which the groups and individuals are not even conscious.

We have now evolved a true management information system which can be viewed as a pure communication system with extensive data processing embedded in the communication structure. Almost all the capabilities mentioned in the preceding paragraphs have been exhibited in existing systems

ISBN 0-201-03140-X, 0-201-03141-8 (pbk.)

---

*PERT is one example of a critical path analysis system for tracking project tasks.

(EMISARI, EIES, CAI systems, data management systems, and other CC and electronic mail systems). It is not as if we have evolved anything beyond the capabilities of existing hardware and software technology, nor are the costs such that a proper cost–benefits study would not show a positive benefit, when compared to current noncomputerized communication approaches in the organization. The system we have evolved here can serve numerous functions and purposes:

> Message routing, meeting scheduling, personal calendars, time scheduling, report preparation, correspondence tracking, project coordination, budget coordination, formal MIS reporting, notification for overdue tasks, filing, word processing, retrieval, personal note handling, browsing, forms implementation, display graphics, indexing, editing and updating, confirmations of delivery, centralized filing, annotating, footnoting, postscripting, delayed transmission, gathering approvals, data maintenance, interface to other systems, and status summaries.

In each case, the computational and logical capabilities of the computer are used to relieve the human of the burden of carrying out these communications bookkeeping functions manually. The potential labor savings in any organization are tremendous, and will, we hope, result in the manager's having more time to deal with the contents of problems rather than wasting effort on the mechanics involved in the solution process (that is, the communication process).

No doubt we have had a marvelous time over the years throwing our weight behind the implementation of the system we have been designing. First, there was all that money that some idiot spent on word-processing equipment that could not also be used as terminals for the system. As we know, it was very difficult to get the organization to admit to lack of foresight and wasted investment. Trying to sell flexibility and adaptability as objectives for investing money has never been easy. Getting management to realize that the rules about centralization of computer services for efficiency purposes were the result of an era when hardware costs were dominant over people costs was a very trying experience. For management to unlearn current management guidelines is like pulling teeth. This took a tremendous effort in cost analyses repeated too many times. Sensitizing the computer people to the need for good user interfaces was like asking the blind to describe the colors in a sunset. We finally made home course materials for typing available in the library when it was realized that none of the management types were really going to go to a classroom with their peers for such a course.

There were enough casualties in terms of those who could not adapt to the new environment so that there was a sizable countermovement to defeat the system. Most of these individuals have now been moved to dead-end jobs commensurate with their talents. The boss still gets a little upset with wives

ISBN 0-201-03140-X, 0-201-03141-8 (pbk.)

ISBN 0-201-03140-X, 0-201-03141-8 (pbk.)

and kids using the system via the home terminals for family communication, and at some of the anonymous comments and graffiti that come across, but he seems to have reached a state where the benefits are such he does not bring up these issues any longer. Currently management people are working at home an average of two days a week. This seems to have occurred gradually, so that by the time the personnel department realized what was happening, it was a *fait accompli*. However, the union has now requested the same privileges and provisions of home terminals for all secretaries. This promises to be the major battle this year. Personnel has taken the position that they want to tie pay to the actual amount of work done through the terminal, if homework is allowed for nonmanagement staffs. It is strange that "homework" is now a word with new connotations, more desirable than before.

When it is all summed up, those data-base-oriented management information systems always seemed to put us in the position of reacting to the input demands of some computer for data and trying to make sense out of reams of computer output. There was very little room for our intuition or our informal judgment. One got the feeling of being a cog in some machine. Perhaps some managers think that mode of operation will lead to better company profits. However, we are not about to spend 12-hour days for the benefit of the organization, unless it also benefits our own image and our own goals. MIS as a communication process puts the control of what is taking place back in the hands of us manager–user types and provides the opportunity for exhibiting what we have to offer the organization. In the long run, the ability to gratify our own needs and ambitions is what will pay off for the company, and no model has yet been able to incorporate that concept in its equations.

Why Not?

Regardless of the potential benefits ascribed to computerized conferencing, we do not expect it to immediately saturate the corporate environment. There are a number of significant factors that will inhibit its growth. The most important of these are the following.

1. Most of the people who currently design, "sell," and utilize management information systems are committed to the data base/model approach. There is no recognition, even in our university courses on such subjects, that the management information process is primarily a *communication* process. To compound the problem, there is a reluctance in any organization to view communications as a problem at all or to perceive that one can do something about it that does not revolve around the psychology of specific individuals. This is further complicated by the

fact that it is difficult for individuals to actually understand the possibilities offered by a new form of communication until they can experience it.

2. There is a natural and probably justifiable aversion to computers on the part of those managers who are guided by "intuitive" approaches to management action. It will take some education for them to realize that this is one of the few computer systems that can lend support to intuitive and judgmental approaches by widening the range of intuition that can be brought to bear in a coherent manner on a given issue.

3. The availability of terminals, particularly for use in the privacy of one's office and home, is limited. This item represents the major capital investment and would be hard to justify on the basis of this application alone, in today's cost-conscious environment.

Despite such problems with gaining initial acceptance, the use of these systems for managerial and staff functions will grow. Sometime in the late 1980s, we hypothesize, computerized conferencing and associated message systems will represent the major use of computers, when measured by how many individuals make use of any available computer service. Once the price of terminals drops to $500 (projected to occur by 1980), the economy and convenience of this form of communication will make its growth inevitable.

Summary

Operational systems have demonstrated that managers in dispersed locations in an organization can communicate more fully and efficiently with CC systems than without them. Many potential applications appear promising in the managerial area, including task forces, crisis management, budget allocation processes, and the replacement of single-purpose word processing systems with multipurpose conferencing systems.

We stand today at a significant decision point in the evolution of business and organizational use of computers. During the last 20 years, these machines have brought about an increase in the efficiency of operations for most organizations.

We can continue down that path and it is the easiest of all the paths that face us. The computer can continue to foster the "assembly lining" of the white collar worker from secretary to manager, and the continued specialization of those workers until there is no accountability for the system as a whole. We can choose another path where "quality" is as highly regarded as "productivity"; where the goal is not blind optimization but the resiliency of the organization in responding to the uncertainty of its environment. We believe that the industrialized nations are entering an era in which very little

ISBN 0-201-03140-X, 0-201-03141-8 (pbk.)

is certain and organizations are bound to be faced with all the unexpected problems characteristic of a limited resource environment. The directions in which computers have taken us to date may result in short-term savings but if continued are likely to create rigidity and a lack of adaptability to change. We have tried in this chapter to exhibit the potential for a new direction based on using computer technology in what we believe is a flexible manner to deal with organizational communications.

We also believe that computerized conferencing within an organizational context is a system for job enrichment and raising the level of all workers as well as the pride they will have in the work they do. It is in a real sense a humanistic use of computers where the computer is no longer the overseer but the servant of the humans involved.

For Discussion

1. In what kinds of situations is very unequal participation in decision making desirable for corporate applications? How might managerial computerized conferences be structured to produce unequal participation or weighing of influence in decisions?

2. Think of an organization with which you are familiar, perhaps a college or university. Do any of the four models of organizational types seem to fit it? What applications of computerized conferencing can you think of for this organization? How might they affect management styles and processes?

3. Take an organization you are familiar with and itemize those jobs in the organization that will become more important or less important with the wide-scale introduction of computerized conferencing.

4. What type of computerized conferences can you hypothesize would take place between individuals in different corporations or institutions, pertaining to company or institutional functions?

5. In what situations in a company would role-playing conferences be useful? (This would be where one employee assumes the role of another during discussion.)

6. For what specific corporate situations is voting desirable? When are anonymity or Delphi-like processes desirable?

7. In what ways can a person take control of a CC discussion and unduly sway the outcome, as often goes on in face-to-face meetings? Could you hypothesize ways to design the features of a CC system to make this easy to do?

8. What kinds of advantages might computerized conferencing have for

multinational corporations? What kinds of unique features might be desirable for this application?

9. In what ways would accountability for decisions, actions, or responsibility be improved in a CC environment?

10. What would be the elements to consider from a life-cycle cost view in evaluating the introduction of computerized conferencing? How would these differ from a short-term cost study?

THE BOSWASH TIMES

The Computerized News Summary Service of the Megalopolis

ISBN 0-201-03140-X, 0-201-03141-8 (pbk.)

Feds Handicapped by Regulations

San Francisco. Team Spirit Incorporated is a small and unusual business that is currently suing the U.S. Government for discriminatory hiring practices against the handicapped. It was financed under an HEW grant to train small groups of handicapped people to work as teams and seek employment as teams. Recently one of their work-at-home Information Facilitator Teams made up of a blind person and a wheelchair-bound individual applied for a single Civil Service job and was turned down on the grounds that Civil Service can only hire a single individual for a single job, even if the applicants are willing to split the pay.

Mr. Wallenstern, president of Team Spirit, said the left hand of government does not know what the right hand is doing. "We were financed to create a new form of team employee, which we are successfully placing in jobs in the private sector and yet they refuse to hire these teams," he states. This reporter was unable to reach the Director of Civil Service for comment; however, the Deputy Assistant Vice-Director for Administration of Employment of the Handicapped stated that in their analysis they found that over 300 legislative acts, 450 rules and regulations, and the reprogramming of a number of accounting systems would be required to allow Civil Service to hire a team of people for one job slot. It seems their hands are tied in legislative knots.

Grandmother of the Year

Columbus, Ohio. Mrs. Bessie Ferguson, an 85-year-old lady who is an ex-schoolteacher, was named today Electronic Grandmother of the Year. Mrs. Ferguson is credited with active counseling and communication with over 30 children around the country via the computerized conferencing system run by Friends of Children, Incorporated, which provides aid for emotionally disturbed children. Mrs. Ferguson is one of the 30,000 elderly active in this program serving hundreds of thousands of youngsters.

Mrs. Ferguson will receive one of the new wall-size CRT units and was quite excited. She stated her eyes are not what they used to be and those four-centimeter-size letters are really what she needs to ease the strain on her eyes. In addition, the award carries with it a gathering of all her children to visit her this summer. She seemed both excited and hesitant about this. "At 85, bound to a wheelchair, and rather scrawny, I'm not physically too nice to look at. It might be a bit of a shock to a few of my children who picture me as an authority symbol."

ISBN 0-201-03140-X, 0-201-03141-8 (pbk.)

CHAPTER 5

Computer-Mediated Communications and the Disadvantaged*

Success is counted sweetest
By those who ne'er succeed.
To comprehend a nectar
Requires sorest need.

Emily Dickinson

Whenever a useful new technology is developed, one policy question that should be vigorously pursued is how to make it available to those who cannot afford to buy it themselves.

Both the "Other America" and the "underdeveloped" or third world countries have been with us throughout the postwar decades. At first glance, the advent of such computer-mediated communication systems as computerized conferencing, computer-assisted instruction, and home terminals from which white collar work can be done would seem to be just one more technological factor that will be used only by the well educated and well-to-do, who have the skills and the money to take advantage of the opportunities they offer. However, like "Sesame Street" on educational television (another high-cost, high-technology medium), an enlightened and purposeful public program might be designed to make the new communication medium serve the disadvantaged rather than compound their disadvantages.

In this chapter, we will review the categories of persons who have tended to be poor and unemployed in America during the last two decades; the nature of and opportunities offered by computerized conferencing and associated communication developments; and the way in which these developments might be used to overcome the lack of mobility and lack of reading and writing skills, which contribute to the invisible barriers imprisoning the disadvantaged.

---

*We thank Elaine Kerr and Howard Gage for their collaborative efforts related to this chapter.

Starr Roxanne Hiltz and Murray Turoff, The Network Nation: Human Communication via Computer

ISBN 0-201-03140-X, 0-201-03141-8 (pbk.)

Since the inequities within our society are dwarfed by the gap between the "have" and "have-not" nations, we will also take a look at how this technology might be made available to village people in the poorest and least developed areas in order to solve their communication problems.

Who Are the Disadvantaged?

In 1973, there were 23 million Americans with incomes below the official poverty line-($4,540 for a nonfarm family of four, which is indeed poverty). They were exactly the same categories of people that Harrington had found in *The Other America* 15 years before, despite the aborted "war on poverty" and the civil rights and women's rights movements. Fifteen percent were old (over 65); 33% were very young (under 14); 32% were black, and 31% of all blacks had incomes below the poverty line, as opposed to only 8% of all whites. Another 10% were Spanish speaking. Some 36% were in families with a female head (a category that had increased greatly since 1959, when only 18% of the poor were found in female-headed families). Looking at this the other way, we see that 45% of all families headed by females had incomes below the poverty line (summarized from *Current Population Reports*, Bureau of the Census, 1975). The majority of the poor in all of the foregoing categories had low levels of education and no skills that could command a steady job at decent wages in our increasingly service-oriented economy. Also occurring disproportionately among the poor are the handicapped, and those who are or have been institutionalized in prisons, mental hospitals, or similar places where they receive little attention or other assistance that could make them able to get and keep a well-paying job upon release.

It is the thesis of this chapter that computer-mediated communication systems, through portable terminals that can be used in the home or even in a prison cell, can be used to bring education, counseling, peer support, and job opportunities to all of these categories of the poor. We will take as our examples the handicapped, the aged, and prisoners, but similar applications could also be devised for other categories of disadvantaged persons.

Computerized conferencing offers a unique opportunity to return mobility-limited persons to the mainstream of society by increasing their participation in educational opportunities, counseling programs, friendship networks, community activities, special-interest groups, and occupational endeavors. As mobility restrictions are lifted with the aid of computerized conferencing, friendships may be created or resurrected, community activities joined, professional advice received, and occupational or professional skills learned or reutilized.

A key part of our argument is that the mobility-limited disadvantaged can be trained for gainful home-based employment using the same home terminal

ISBN 0-201-03140-X, 0-201-03141-8 (pbk.)

that serves as an educational and social network. This is based on the fact that during the next decade, most paper records and monetary transactions will be replaced by electronic transfers of information among computers. For example, we will move toward a "cashless society," in which instead of writing checks and sending bills, there will be made a point-of-sale entry that transfers the amount of the sale from the computerized account of the buyer to that of the seller. In other words, the technology of computers and of data communication is improving so fast that the trend toward the creation of new kinds of white collar jobs will continue.

So far, this trend has merely accelerated the disadvantages of the poor. It has enabled large companies to decentralize and move most of their operations out of the big cities, since the decentralized operation can act much like a centralized one by being tied to the same big computer center. As the companies leave the cities, the central cities, where the poor are trapped, lose jobs and tax revenue.

The new jobs are located in the suburbs and in exurbia, without mass transit, where the poor cannot afford to live and to which they are unable to commute. However, there is no reason why this economic moat cannot be spanned by the very kind of computer communication technology that helped to create it in the first place, with central city dwellers doing much of their work from terminals in their apartments or neighborhood work centers. (A neighborhood work center would have terminals, telephones, desks in cubicles or offices, clerical support, and other equipment and services needed to support white collar work. A person would work at the closest center, along with workers from many other organizations. Conversely, the workers for any one company or organization might be spread among 50 to 100 such centers, linked together by telecommunications.)

Bringing the Interaction Location to the Handicapped

The biggest advantage of computer-mediated communication is that it spans space and time barriers, allowing a person to work, learn, and communicate from those places and at those times that are most convenient for him or her. Thus the mobility limitations of the physically handicapped make them a disadvantaged group that can benefit greatly from this technology.

Following Hamilton (1950, p. 17), we define a disability as "a condition of impairment, physical or mental, having an objective aspect that can usually be described by a physician . . . . A handicap is the cumulative result of the obstacles that disability interposes between the individual and his maximum functional level." A person may be physically impaired without being socially disadvantaged. Although the disability may hinder the performance of everyday tasks, it is the handicap that impedes the attainment of goals by

ISBN 0-201-03140-X, 0-201-03141-8 (pbk.)

interfering with normal social relationships. The handicap is made up of social barriers that must be surmounted if the person is to be considered fully rehabilitated. Three kinds of disabilities are ordinarily distinguished: physical, emotional, and mental. We shall concentrate on physical disabilities.

Computer-Based Education for Handicapped Children

Currently, education for homebound handicapped children is both expensive and inadequate, and job and social interaction opportunities for adult handicapped individuals are very scarce. The results are an economic drain on society, since almost all handicapped persons are obliged to assume the status of unemployed, dependent individuals characterized by poverty, isolation, and low self-esteem. One application of the home terminal and computer-mediated communication systems would be in home tutoring. Computer-assisted instruction could be provided for each student, in the areas corresponding to a regular elementary, high school, or college curriculum.

The widest application of CAI for handicapped children was done by Sandals (1974), who has successfully used individual terminals to aid in several diverse phases of learning, including drill and practice, tutorials, and simulation. Thompson and Johnson (1971) report that CAI can help to increase a child's span of attention and heighten the enthusiasm shown toward his teacher.

However, in existing applications, the learners and the teacher are all brought together in the same spot at the same time, with the teacher physically overseeing the students. With the marriage of computerized conferencing to CAI, the terminal could be in the learner's own home, with the teacher's supervision and review of progress and sending of comments done remotely. This is important to anyone for whom travel is difficult or for whom long periods of exertion, rather than many short learning periods, are physically taxing, or whose condition necessitates activities (such as periodic medical regimens) that prevent full-time day attendance at a school.

Many special terminal interface devices now exist to enable almost any type of physically handicapped person to write and read from a computer terminal. There are Braille typewriters, special types of units for deaf persons, terminals and coding devices that can produce typing through the pushing of a button with an elbow or foot, or through utterance of any sound. Although these specially designed interface devices now cost $1000 to $8000 each, the cost would be much cheaper if a large-scale program in the design and use of the devices introduced the economies of large-scale production. (For more details about interface devices, see Kafafian, 1970, 1973.)

How would a computerized conferencing and instructional system for the homebound handicapped differ from current home tutoring practices? Currently, a child receives only a few hours a week of individual instruction in

ISBN 0-201-03140-X, 0-201-03141-8 (pbk.)

the home. There is no equivalent to the peer-group interaction that occurs in the classroom. For the tutor, many hours are spent traveling, rather than in actually giving instruction. Any one tutor must deal with children who have a wide variety of levels and types of achievement, needs, and disabilities. In addition, a single tutor is supposed to be responsible for teaching all subjects, whereas the usual secondary level curriculum has become quite specialized by subject area.

With a computer-based system, a group of 8 to 20 children at a similar ability level could have a tutor. Through computerized conferencing, the tutor would be apprised of the daily progress of each student and would be sent questions, problems, or perhaps English compositions. At his or her convenience, the tutor would answer academic or personal questions, advise on what modules the student should review or use next, correct compositions or tests sent to the student over the conference, etc.

The youngsters could also have formal or informal group therapy sessions over the conference system. Perhaps this could be scheduled as a "synchronous" conference (simultaneous users on line) at a certain hour each day, with participants able to have prepared responses or comments on the "topic of the day" in their computerized scratchpads or notebooks, if they preferred. The group leader and each of the handicapped persons would engage in this session from his or her own home. The children might have their own peer-group conference. There might also be "big brothers" or "big sisters" with whom they could communicate over the system for advice and social support, or surrogate "grandparents" (mobility-limited older persons who would welcome the opportunity to be of help without the risk and effort of having to travel to do so).

A computer-based system would allow teams of teachers and other helpers to interact with each individual, thus offering the handicapped student access to expert tutors in specialized subject areas. It is not meant to eliminate all home visits or personal contact by the primary tutor for each individual student; but these sessions could be greatly decreased, and the time freed from travel could be devoted to actually giving instruction and guidance to the students. We will expand on such potential gains in the following section.

Economics

How can our society afford all this? Current systems for providing education for the handicapped are not cheap, and are so inadequate that they never enable the recipient of the services to attain independence. For example, in New Jersey, the laws provide for each homebound handicapped child to receive five hours of private tutoring a week, through a home visit from a trained teacher. These five hours are of course not equal to the 25 or so hours of classroom time received by children who attend the regular schools, so the

ISBN 0-201-03140-X, 0-201-03141-8 (pbk.)

child is liable to fall further and further behind. Nor is it cheap, especially with costs at about \$7.00 per hour for the tutor's time, since the tutor must add travel time. In addition, the long, intensive sessions are not likely to be particularly effective for many homebound handicapped youngsters, for whom frequent half-hour to hour-long learning sessions would be optimal. In addition, if the child or the tutor is sick on the appointed day, another week is lost under the current system.

With computerized conferencing and CAI modules, a tutor could handle a "class" of 8 to 20 students, as a full time job at a smaller cost per student. This class could be assembled from all children over a statewide area, if necessary, since actual home visits need not be made except to start the child on the system, and perhaps every three months or so to keep the feeling of personal contact between tutor and pupil. The student could work for as many hours a week as he is able, and thus perhaps receive more education than those attending regular classrooms do during their 25-hour school week. Hence she/he might actually catch up with age peers rather than constantly falling further and further behind. (For more details on educational applications to the handicapped, see Turoff, 1975b, Turoff and Gage, 1976.)

The Deaf and Deaf-Blind

The deaf are perhaps the group that could benefit most from computer-mediated communication. A large give-and-take discussion among a group of persons using sign language is very difficult to coordinate; moreover, few of the nondeaf are able to communicate in sign language. Many deaf children seem to lack the motivation to learn to read and write, when all of the important persons around them communicate in sign language. This greatly decreases their capability for employment and social interaction with the nondeaf. A computerized conferencing system should provide a strong motivation for deaf children to learn to type, express themselves in written language, and read (see Turoff, 1975b).

Adult Employment at Home

Turning attention to adults, we have previously reviewed the trends whereby there will be more and more written records created and stored electronically in computers in the future, rather than handwritten or type-written and stored in physical files. A handicapped person could thus do useful work at home, on a piecework basis, if necessary, working at those times and for those lengths of time that will fit in with their other routines. It would be very possible for two or more handicapped persons to contract to share a job, or two or more jobs. Such work teams could allocate tasks according to the abilities and time availability of their members while guaranteeing to the employer that the entire job would be completed.

ISBN 0-201-03140-X, 0-201-03141-8 (pbk.)

Heterogeneous Groups

Whether a conferencing system is being used for peer discussions, therapeutic discussion groups moderated by a professional, or work-related communication, the various participants need not be aware of the disability that any of them suffer unless a person wishes to volunteer the information. This can greatly facilitate unbiased interaction among a group that includes the "temporarily able" or the "nonsocially handicapped"* as well as those who are handicapped. Even if the participants are aware that a particular person is blind or deaf, the social salience of the characteristic is much less, because it is not visible. Moreover, if it takes a handicapped person longer to read and/or write into a system, this does not slow down or inhibit the speed or ease of participation of the other members.

The Mentally Handicapped

The mentally handicapped could also benefit from computer telecommunication technology. For those who are only moderately retarded, the constant repetition facilitated by CAI can be a useful way to review and reinforce classroom lessons. With the adult mentally retarded, for instance, job training can be provided so that there is adequate repetition at a speed suitable to the individual. In addition, modules could be produced that explain and drill on simple tasks such as how to take a bus from the person's home to a shopping area.

However, it is the communication facilitated by computerized conferencing that most intrigues us, especially when it is used to create a support system for the parents of the profoundly and severely retarded. One of the most serious problems that educators of the severely and profoundly retarded child must deal with is that the lessons so painstakingly taught in the classroom are not reinforced in the home. They complain that especially on Monday mornings and after vacations, the "loss" is terribly apparent. Meanwhile, the parents are not informed of the details of what and how their children are being taught, and frequently feel helpless and "unqualified" to try to teach their children in the home.

---

*These terms are used because almost everyone is physically disabled or socially handicapped in some way, or will be for a considerable period of their life. The only differences are in the severity of the disability and whether or not technology has found an effective prosthesis; and in the degree of social stigma attached to various conditions. For instance, how many academics could continue to function normally in social or professional roles if it were not for eyeglasses? Many of us would be homebound and unemployed, groping and squinting through life. Not only has technology provided us with eyeglasses that almost completely obviate the effect of our disability; but society has made the wearing of these devices not only acceptable, but—with the emergence of "designer" frames—positively chic!

ISBN 0-201-03140-X, 0-201-03141-8 (pbk.)

We envision a computerized conferencing network that would tie together the teacher in the classroom, the parents, and the psychologists and medical experts who specialize in the problems of retarded children and their parents. The teachers and the experts would have their own portable terminals, so that they may easily enter materials on nights and weekends. The parents might borrow terminals or have daily access to public terminals located in a library or other public facility. The following are the kinds of activities that might take place.

- Each week the teacher could enter a set of lesson plans and suggestions on how the parents might augment in-class work with work at home. This would be a "group message."
- The teacher could send frequent private messages to each parent on the progress and particular problems of his/her child.
- The parents could send questions and problems to the experts or teacher for advice (private messages). These could be sent and replied to using anonymity.
- The psychologist could lead group discussions among the parents on such topics as how to deal with tantrums or with older or younger siblings of the retarded child (group conference).
- The parents could form a supportive peer group among themselves, sharing daily problems and giving advice to one another.
- For parents of institutionalized children who are mentally retarded or emotionally disturbed, the conferencing facility could greatly increase contact among parent, child, and staff. Currently, contacts are almost entirely limited to occasional visits, since telephone calls would be too disruptive of the institutional routine, and letters are too time-consuming to exchange on a daily basis. Short daily exchanges among parent, child, and caretaker would be much more feasible to build into a routine, and would greatly increase the morale of parent and child.

Penetrating Prison Walls

That our prison system dehumanizes and neither educates nor rehabilitates is one of our national shames. One problem is security. Many guards and citizens in surrounding communities feel safe only when prisoners are kept in their cells, alone. Visitors are severely restricted, and many persons, such as students who would like to tutor and/or counsel a prisoner, are dissuaded by prison authorities, who do not want to be responsible for the visitors' physical safety.

Computerized conferencing and CAI could be used by prisoners during the long hours they spend in their cells to learn saleable skills such as program-

ISBN 0-201-03140-X, 0-201-03141-8 (pbk.)

ming, as well as basic reading and writing and mathematical skills needed for passing high school equivalency or CLEP exams. Meanwhile, the tutors/counselors could interact from the safety of their own homes, and at their own convenience. The only modification to the usual computerized conferencing capabilities would probably be that in high-security institutions the prisoners would not be allowed to send messages to each other over the system, but could interact only with outsiders who were cleared by prison authorities.

At least one experiment with this type of application is under way. Computer-assisted instruction through terminals in Caldwell Penitentiary and Essex County Youth House in Newark, New Jersey, is being funded by a grant from the Victoria Foundation. Several grade levels of skills in mathematics and reading are now stored and ready for use, and there are plans to phase in courses in history, biology, chemistry, electronics, foreign languages, and computer programming. A local official is quoted as saying that the program is an example of how modern technology can provide "a giant step forward in helping the inmates to improve their skills with a resultant benefit to society on their release" (*Sunday Star-Ledger*, Newark, N.J., August 15, 1976, p. 71).

The main goal of a computer-mediated communication system for prisoners would be to change attitudes, motivations, knowledge, skills, and credentials needed to prepare the prisoner for a successful employment experience, and thus to offer a viable alternative to deviant or criminal careers. Glaser (1964), for instance, found that most of the variation in recidivism of parolees and released prisoners is explained by whether or not a released prisoner can find a job. Given that computer-assisted instruction and systems designed for communication and counseling can enable a prisoner to be prepared for employment, something must also be done to match the newly qualified prospective employee with job openings.

It would be possible to have an on-line Bulletin listing job openings and qualifications solicited from employers who indicate that they are willing to hire an ex-convict. Any prisoner, within three months of release, could apply for the job by sending a message over the terminal. It is even possible that preliminary job interviews and/or screening tests for the prospective employee could be carried out over the CC system.

Furthermore, some jobs in a training mode could be made available through the terminal. An important social aspect is that co-workers could get to "know" the prisoner through the system, without feeling threatened. When the ex-convict actually met co-workers face-to-face for the first time on the job, it would be as a "real person" for whom they had developed some feelings as a result of interaction over the CC system, rather than as a stigmatized, unknown "criminal."

ISBN 0-201-03140-X, 0-201-03141-8 (pbk.)

A Support Network for the Aged

Contrary to the public image of great numbers of old people crammed into understaffed firetraps called nursing homes, only about 5% of people over 65 live in such institutions. However, a very large proportion of older people live alone or with another older person, and suffer from fear, isolation, and inadequate services because their physical condition limits their mobility. For many older people, any public transportation system involving stairs is too difficult to utilize.

Computerized conferencing could be used to tie together a network of older persons and community support personnel. For security, there could be a "daily sign on" conference, with moderators designated from within the group, for the exchange of personal news and information about the well-being of each of the members. If a person has not signed in before 3 P.M. or so on any given day and has not stored a message on her or his whereabouts, telephone calls and then a home visit could be made to make sure that they are okay.

Many older persons suffering from chronic problems become anxious about changes in symptoms, but hesitate to "bother" a doctor with constant questions. A doctor specializing in gerontology could be on the conference and would check in no less than twice a day to receive private messages and questions about symptoms or other health problems of the group members, and to advise what to do or whether an office visit seems necessary

The Bulletin, edited by one of the members, could contain announcements of senior citizen events in the vicinity (perhaps the menu for the lunch at the Senior Citizens Center that day, for instance, and the activity schedule). It could also carry news of interest about Social Security or food stamp or Medicare changes, for instance, and instructions on whom to write to about impending legislation of interest. If the network were tied into a senior citizens' minibus transportation system, so much the better. Having reviewed possible activities for the day, the person could send a message about the time and destination for desired transportation or services for the day.

Perhaps most important, the conference system could tie these older members to other persons on the system whom they might be able to help by communicating over the system, such as a handicapped youngster or a prisoner whom they have "adopted." Many older people have training and skills as educators, social workers, or managers. Whereas their physical mobility may be limited, they have not lost the ability to make useful contributions to society, if they can do most of their work without extensive traveling. Margaret Mead, among other social scientists, has commented on the importance of "grandparenting" and of emotional ties between the very young and the very old. With geographic mobility, the replacement of extended family households by two-generation nuclear families, and the

ISBN 0-201-03140-X, 0-201-03141-8 (pbk.)

emergence of age-segregated "retirement communities,"'there seems to be less and less of such close contact. Perhaps CC systems can provide the communications network to reestablish ties between the over-65 and under-21 age groups.

The home terminal might also enable those with the time and energy to do certain kinds of volunteer or paid work from their home, such as making corrections or entries to computerized files. In addition, it could facilitate the life-long learning that is an ideal for many persons, but that becomes difficult when travel to classrooms must be accomplished. There might be CAI modules on the system about nutrition, for instance, or foreign languages, or poetry. The modules could be designed to meet the requests of the users, and could be changed from time to time.

A Social Service Referral System

The various special-purpose local, county, state, national, public, and private social service agencies that exist in metropolitan areas such as Newark and New York City do not at present form anything that could be considered an integrated system. There are some county-level directories that have been made up by the United Community Fund for some areas, but these tend to get out of date almost as soon as they are published.

A CC system offers a unique opportunity to provide an information and referral system to better inform all of the agencies and programs in an area' about changes that take place in programs and regulations and personnel. Each agency would need to have at least one terminal available. The conferencing system could then be used in the following kinds of ways:

1. An on-line directory of agencies, contacts, and programs could be constantly updated. It should include a description of each kind of service or program offered by the agency, who is eligible, waiting times or costs, and the people in charge. It would be keyed by the type of population and/or problem it is designed to serve. For example, the nursing homes in an area could keep a frequently updated record of the availability of beds and rooms and costs, by category of patient (private paying patients versus Medicaid-dependent patients, etc.). When a hospital social worker or other counselor has an aged client who needs to be placed, she/he would immediately obtain a complete list of the current possibilities, and could report this back to the client and family at once rather than having to make dozens of telephone calls.

2. The messaging system could be used to smooth and speed up referrals. When a client is to be referred to other agencies, the caseworker could

ISBN 0-201-03140-X, 0-201-03141-8 (pbk.)

first send a summary of the client's situation and a request for an appointment to the person to whom the referral is being made. Before actually referring the client, the caseworker would receive a confirmation that the client seems eligible and able to be helped by the second agency, and a date, time, and person given for an initial appointment. This would prevent the all-too-frequent situation in which a client is sent to another agency, only to be turned away as inappropriate for the services offered by that second agency.

3. The conferencing system could be used by groups, such as persons working with aged, abortion counselors, or agency directors, to hold short-term conferences on topics related to their own professional concerns. The "group message" associated with the conference group might also be utilized by professionals who need a specific kind of information or advice. Such conferences would facilitate a constant exchange of experiences and ideas among social service professionals engaged in similar kinds of work, without taking them away from that work for whole days at a time.

For example, those working with the aged might start a conference on social/recreational activities. They could describe the activities available in their organization, and the degree of success and problems encountered. The conference would be a mechanism for pooling the experiences of many workers in many different agencies.

Integrating Social Services via Computer: The Case of "Riverville"

Though there has been no field trial of a computer-mediated system that would allow communication as well as information about social services to be shared among agencies, there has been a large-scale system financed by HEW in California that was designed to provide for automated welfare-client tracking and information and service integration. Called UMIS, it has been studied through interviews with staff and other participants. Since over a dozen similar automated systems are under development, and since the kinds of information it collected and disseminated would be building blocks in any complete computer-mediated communication system for social services dissemination and tracking, we will review the situation in which it was implemented, its design capabilities, and its reported impacts. This review will be based on an unpublished project summary by Rob Kling (1977).

"Riverville" is a medium-sized city of 170,000 that also serves as a regional center for surrounding rural counties. Approximately 26% of its families are black and 20% live below the poverty level. As late as 1966, there was little

ISBN 0-201-03140-X, 0-201-03141-8 (pbk.)

administrative coordination among the approximately 150 private, city, county, state, and federal agencies intended to provide health and welfare services. When a new mayor was elected in 1972, he instituted a program to try to obtain needed information and integration among all the services serving similar clients. A "technical strategy" utilized an automated information system (UMIS) that would track the path of particular clients through the maze of public and private agencies. An "administrative strategy" consolidated all but one of the 36 city-supported agencies into a common administrative unit. The design goals of the UMIS project included the following (Kling, 1977, pp. 7–8):

1. Provide baseline information about the needs of people.
2. Provide for and monitor the sequencing or scheduling of (social) services on an orderly basis.
3. Track individuals and families through the service system to ensure they received services as planned.
4. Provide information for management decisions about the amount of services individuals and families have received and their progress in breaking out of the cycle of poverty.
5. Eliminate duplicate records.
6. Increase the control over welfare funds.
7. Automate follow-up to keep people from "getting lost."
8. Evaluate the social service programs.
9. Eliminate duplicate services.

This was to occur by having each client who came to an agency fill out a standard information form on all of the household members, including income and health data, etc. They were then to be checked to see which programs any member of the household might be eligible for; advised which agencies could provide these services; and given a written plan describing all the agencies and services that were set up to provide these services. Each agency to which a client was referred was to complete an "outreach form" that would be entered into the record, describing in a standard brief manner the services provided and progress of the case.

However, the program has failed to provide all of the potential for information and integration that it was designed to produce. Most agencies refuse to take part because there seems to be no value in it for them. The various agencies within the city Human Services Department and a few others are "fully on" UMIS, following the initial form-entry procedure. Another 25 cooperate in filling out the outreach forms for their clients. The rest of the 150, including the state welfare agency, refuse to participate at all. Among the reasons are that they have their own automated system with different reporting conventions; or for small agencies, a small case load and a

ISBN 0-201-03140-X, 0-201-03141-8 (pbk.)

fear of computers among the staff makes them feel that the "paperwork" is not worth any summarized statistical information that they would receive in return. Finally, it does not make caseworkers feel that it is of any help to them in discussing and tracking the progress of their clients through multiple agencies. Among the reasons for this are that the cryptic entries of types of services received by a client are no substitute for the kind of rich detail contained in traditional case histories and retrievable from other agencies only by a telephone call or letter to the person giving the help.

The kinds of capabilities that are present in a CC system, we argue, are exactly those that were missing from the Riverville statistics-only attempt to integrate and track services. They would have enabled the caseworkers to exchange the kind of information they would actually find helpful in improving the services given to a client who is being served by a multitude of different agencies, either simultaneously or sequentially.

Among the serious problems that would have to be overcome in instituting such a system, of course, are guarantees for the protection of the client's records from access by unauthorized persons. One method of approach would be that the administrator in charge of each agency might be the only person empowered to authorize access to the materials (on-line continuous case history) that were built up about each client. In addition, it would be best if no name or address or other directly identifying information about the client were used to label the record; rather, an ID could be kept in a separate file in which it is matched with name and Social Security number.

Some Problems and Opportunities

The key barrier to implementing the kinds of suggestions that have been presented here is the cost and availability of terminals and initial resistance to any computerized system by social service professionals.

The aged and the handicapped are by no means homogeneous groups; indeed, they overlap considerably. Many variables may impinge on their receptivity to computer conferencing: sex, race, age, socioeconomic status, education, ethnicity, age when disabled, the severity, duration, prognosis, and visibility of the disability, current or previous job level, etc.* An extensive set of exploratory field trials will be necessary in order to discover the particular

---

*See Elaine B. Kerr, "The Vocational Experiences of Physically Handicapped—Poorly Educated Workers after Job Placement," unpublished dissertation, Columbia University, 1971, for a description of the impact of these variables on the vocational achievements of the handicapped.

ISBN 0-201-03140-X, 0-201-03141-8 (pbk.)

kinds of training, special software aids, and uses that are appropriate for various subgroups of users among the disadvantaged.

One way to start this application area, even with a shortage of home terminals, is the introduction of terminals into community gathering places such as public libraries, various community centers, schools, and clinics for general public use (see Turoff and Spector, 1976).

Another problem is that most of our delivery programs to the disadvantaged and the governmental offices that sponsor them are divided along clinical categories: the aged, the orthopedically handicapped, the deaf, the blind, the poor, prisoners, etc. There is no institutionalized support for system approaches that cut across these categories. The nature of a CC system allows it to cut across these divisions and facilitates integration of service delivery, such as educational, vocational, and rehabilitation programs. In terms of economies of scale, a lot more can be done with a large user population base than by attempting to build separate systems for each separate disadvantaged segment. However, this would threaten vested interests in the multitude of specialized, separate programs.

Terminals and computerized conferencing made available to the public might help support new programs or social movements among the disadvantaged. Computerized conferencing is unique as a communication system because the content of a message can serve as an address. This is not possible with the telephone or lettters, which require an address independent of content. A person could go to a computer system and choose to enter a discussion by picking a topic. She/he would not know who the other people might be, but a group can be formed with those who have a common concern. For disadvantaged persons, such as the mobility-limited aged, prisoners, and those who are handicapped, this provides an opportunity to easily form groups with common concerns and interests. One should realize that it is not an easy process for people to find one another or to get to know one another with the current pace and complexity of our society. This is true for the typical citizen, let alone for those without money or mobility. In most city and suburban areas we are largely an automobile-dependent society, especially during the evenings and weekends, but the cost of ownership is rising far faster than the salaries of those in lower income brackets. The conferencing system can help concerned people to "find" one another and to generate a set of priorities and proposed actions to meet their needs.

However, this brings us to the most serious problem of all—the fact that the middle and upper classes, which control spending and policies for social welfare programs in the United States, are not inclined to implement something that could help the disadvantaged and the "misfits" to articulate their needs. Most of the money for existing computer systems in social welfare has thus been directed at the administrative goal of programs to minimize

ISBN 0-201-03140-X, 0-201-03141-8 (pbk.)

"abuses," in terms of fraud. Although these abuses need to be corrected, the number not receiving the aid they are entitled to probably far exceeds those who are abusing the systems. We seem to put far more emphasis on minimizing delivery costs than on trying to improve the quality and equity of our social services. This indicates a very difficult atmosphere for innovations such as the computerized conferencing concept (see Laudon, 1976).

A False Problem

Many educators and researchers working with the problems of the handicapped have been trying to maximize their mobility and to advocate a policy of "mainstreaming," or putting handicapped persons in normal school and work environments and treating them as much as possible like everyone else. For those with mild disabilities, this seems to us to be optimal. For those with severe limitations on their mobility or ability to communicate through normal channels, however, this seems nonsensical. It is true that the handicapped can easily become isolated from the rest of the world. Because of this, the initial reaction of some people to the idea of computerized conferencing for the handicapped is a fear that it would lead to their greater isolation.

We believe that this fear is not warranted. Computer-mediated systems for communication should be used in such a way as to supplement regular face-to-face interaction and to expand the total interaction and participation of handicapped persons in all types of activities as much as possible. Such a system can enable handicapped persons to associate with a much larger number and variety of persons in all types of activities than is now possible. Increasing the number of communication ties will eventually lead to more friendships, and to face-to-face meetings with new friends and co-workers, in most cases.

Even those handicapped persons who have achieved considerable mobility, such as by the use of wheelchairs, often face periods of immobility. For example, during winter months in northern cities inadequate snow removal often prevents the maneuvering of wheelchairs. In at least one incident in a government agency in Ottawa the introduction of a computerized conferencing system has allowed a wheelchaired person to remain active on his job during winter months via a home terminal.

Computerized conferencing is also very useful for periods of temporary disability or hospitalization. By participant observation, it has been discovered that a person can engage in communication on these systems while receiving an intravenous feeding or other types of routine hospital treatments that would make "normal" social interaction impossible, without disrupting or even being noticed by hospital personnel.

ISBN 0-201-03140-X, 0-201-03141-8 (pbk.)

ISBN 0-201-03140-X, 0-201-03141-8 (pbk.)

Communication for Developing Areas*

It is possible, using a small generator and a communications satellite, to bring a combination system of telephone and computer-assisted store-and-forward message systems and computerized information systems to even the most remote of areas. For example, satellites are now being used to bring telecommunications to native peoples in the Canadian northwest, in an area where there was no electricity and where weather conditions also blocked out radio signals for much of the year.

The most immediately practicable of the uses of computerized conferencing in order to alleviate the disadvantages of developing areas is to use them for international scientific meetings and technology transfer, as will be described in Chapter 7. However, with a great deal more imagination and much field-testing, it would also be possible to use the new technologies to directly serve ordinary citizens.

By adding a computer terminal and access to a CC system to the communications available in a central place in each of the villages in a developing area, here are some of the things that can occur:

- A "daily newspaper" of national and world events can be sent in each morning and printed out for dissemination to local leaders.
- People can come and enter a complaint or suggestion or question for an official in the national or regional government. They can then come a day or two later to receive the answer, after there has been plenty of time for the correct official to receive the inquiry and to prepare a reply.
- Local professionals such as doctors and teachers can be permitted to spend an hour or so a week "in conference" with their peers around the world about problems and issues of concern to them. This would help them to feel less cut off from professional developments and peers, and might lessen the brain drain.
- "Bid-and-barter" conferences: If someone in one village has something he wishes to exchange, sell, or rent, he can put in a notice. Likewise, a person can advertise for a desired item and the price he/she is willing to pay (perhaps even for a spouse, with requirements fully listed!). In this way, the market area for exchange of goods and services might be greatly expanded. Such a conference might cover borrowing or rental of equipment such as trucks or agricultural machines in addition to sales and trades, thus facilitating resource sharing.

---

*Developing "areas" rather than "nations" is used advisedly. There are developing areas within industrialized nations, such as Indian reservations in the United States or the Northwest Territories in Canada.

- Regional planning conferences: Leaders from each of several villages could set up a series of conferences to discuss issues that affect all of them, such as a new road or school or irrigation system. By discussing all of the issues, they might arrive at a common program of priority projects to present to the central government, without having to spend weeks traveling away from their own villages.
- Special problems or crisis conferences: An expert assigned to a region could be the moderator of a region-wide conference to share problems, advice, and reports on a problem such as drought, insects, pests of a particular type, or livestock diseases. In time of an acute emergency such as flood or other natural disaster, outside experts, government officials, and local officials could all stay in touch with the total situation through a temporary emergency conference.
- Personal messages (births, deaths, marriages, illnesses, etc.) might be sent immediately and simultaneously to all interested persons in each of the villages on the network. Besides the convenience to the users of such message services, it is important to allow such use in order to familiarize villagers with the use of the system, so that they might be more likely to use it for other purposes.

It may seem absurd to begin with such a high-technology system to provide for the communication needs of the people in an underdeveloped area. However, the cost per user is much lower than the costs of trying to provide individual telephones, radios, automobiles, etc., on the model of the developed nations. Current costs are about $200,000 for the computer system and $1000 to $2000 per terminal. Dividing this over, say, 100 villages, the cost would be about $4000 dollars per village (plus the satellite, which could also be used for other purposes). In addition, of course, the system would need a full-time "operator" in each village to help users formulate their messages or entries, type in the entries for those who were unable, explain the use of the system to the better educated, and generally act as an intermediary and facilitator. (See Turoff, 1974, for additional details.) This might be a good role for a Peace Corps or national service volunteer to play. In any case, in terms of a "per message" cost, any such system is much cheaper than individual mail or telephone, and offers many capabilities not available on these more traditional media.

More important than technological or economic barriers to the introduction of computer-based communications systems in underdeveloped areas is the attitude of the prospective recipients of this technology transfer. It is likely that the reactions and resistances will be similar to those that have been experienced in attempting international Delphi studies. There is a strong

ISBN 0-201-03140-X, 0-201-03141-8 (pbk.)

impression in developing areas that the computer and computer-related technology are an American imperialist tool, and somehow will disadvantageously affect the autonomy of the recipient nation.

Conclusion

The tone of this chapter has been speculative, dwelling on uses that could be made of some of the developments in telecommunications by the disadvantaged within and among nations. Unless planning for and experimentation with such systems is financed by government or foundation sources, however, they are likely to be used only by the largest and richest of corporate groups, which can afford to purchase such a system on their own. Experience with applications of the information-processing capabilities of computers has thus far not been very helpful to the poor and disadvantaged, however. As Laudon (1976) points out, they have been maximized for efficient regulation of the poor, rather than for equity or justice.

The new applications of human communication via computers described in this chapter offer those concerned with lessening inequality of well-being in America and the world an opportunity to help to "equip the [disadvantaged] individual to become a participating member of a participant society" (Lee, 1973, p. 27).*

Many of our current approaches in this area, by emphasizing short-term tradeoffs, have tended to evolve human service systems for the disadvantaged that perpetuate their dependency on the society and lead to a far greater long-term cost to society than would appear necessary. Computerized conferencing in this area requires a long-term perspective and a general system perspective to justify its introduction. However, because of unique design characteristics in this area, it is very necessary that experimentation get under way now if we are to have the right systems available when the costs permit wide-scale implementation.

---

*It should be noted that Lee does not presently see computers as potentially part of a humanistically oriented remaking of society. As he puts it rather explicity, "The great inventor of the geodesic dome, Buckminster Fuller, has such an oversimplified notion of human society and such a faith in mechanical gadgets that he sees mankind's salvation in the computer" (p. 22). We are inclined to share Fuller's vision that the computer can be used to enable humans to more fully realize their potential and actively participate in the decisions that affect their lives.

ISBN 0-201-03140-X, 0-201-03141-8 (pbk.)

For Discussion

1. One very large group of the poor in America is the female-headed family, many of which include young children. How might computer-mediated communication systems be utilized to increase the employment opportunities and social and supportive contacts of this segment of the population?

2. Another large population segment that was not treated explicitly in this chapter are blacks and Spanish-speaking persons with incomes below the poverty level. What particular modifications might be necessary to CC systems to make them acceptable to these groups? In what applications might such systems be particularly useful to undereducated unemployed members of minority groups? How would human translators function in such systems, and what design requirements might this introduce?

3. What kinds of specialized training and motivational techniques might be necessary to get the aged to try computerized conferencing? What uses of computer-mediated communication systems for the aged can you think of besides those mentioned in this chapter?

4. How feasible do you think it actually is to tie rural villages in developing areas with one another and the rest of the world through computer networks?

5. How could the disadvantaged be exploited through the use of computerized conferencing? (Consider, for example, the modern equivalent of an indentured servant utilizing this technology.)

6. In what specific situations would it be of benefit not to know that the individual you are communicating with is handicapped?

7. What are the phases a therapy group goes through and how would these be modified by computerized conferencing?

8. For a welfare family whose youngster has been charged with a crime, how many different people or agencies and institutions might get involved in this one situation? Itemize them and include the processing of the paper work as an involvement. How could CC systems be applied to this process?

9. In rural areas consumerism is unheard of and if one does not sample the salt it might turn out to be 50% sand. How would you tailor a rural bid-and-barter marketing system to inhibit this sort of abuse?

10. Itemize specific design requirements for communication structures or special features desirable for various disadvantaged applications.

ISBN 0-201-03140-X, 0-201-03141-8 (pbk.)

THE BOSWASH TIMES

The Computerized News Summary Service of the Megalopolis

Home Section

June 1, 1993

All the News
Fit to Display

Overheard:

George, I don't think we can afford to write your mother this week. At a dollar a letter we have used up our postage budget already.

Jimmy, you had better catch up with your lessons on the terminal tonight. Tomorrow we need both terminals for a game of bridge with that couple in San Francisco.

Andrea, I am trying to explain to you we cannot afford a trip to Los Angeles to meet your computer pal.

Dear, shall we join a conference tonight on Speculative Stock Investments or Sexual Excursions? What's your mood?

Really, John, two terminals are not enough. We are always fighting with the children over their use. You are just going to have to buy a third. Anyway, I saw the cutest terminal that matches the drapes in our living room, with nice little flowers all over it.

If we route our trip through Philadelphia we can visit that couple from our therapy conference. I am dying to meet her.

Social News

Capulet–Montague

Juliet Capulet and Romeo Montague announce with joy their forthcoming marriage. In order that both families may attend without undue risk of disturbing the ceremony, it will be conducted by computerized conference.

DeBergerac, Incorporated

Love letters written to suit any situation. Each request given confidential and custom service. Our experts are drawn from the finest English departments in the country. We have specialists in sincerity, teasing, provoking, stimulating, amusing letters; "Dear Johns"; "yes,", "no," and "maybe" responses. Foreign language and slang experts available to match the ethnic background of the recipient. Your money refunded if the response to any of our letters indicates that it did not effectively express the feelings you wished to convey.

Excerpts from the Boswash Times

Modern Living Section

CC Education Proposed

Washington, D.C., December 10, 1992. The Consumer League pressed today for universal Computer Conferencing Education at the high school level. "Without training

ISBN 0-201-03140-X, 0-201-03141-8 (pbk.)

187

in how to make use of the multiple services and options available on these systems," they argued, "a person has less chance of successfully navigating through the tempting array of loans, services, and recreational activities available on the computer networks than an untrained driver used to have on the highway." Furthermore, they proposed that driver education teachers be retrained for this new function. Since people began to conserve on fuel by doing most of their working, socializing, learning, and transacting from their home terminals, automobile traffic and insurance rates have declined to the point where driver education classes are severely underenrolled.

Couple Unintentionally
Reunited

Spokane, Washington, August 12, 1991. Ms. Mildred Gable entered a suit today in district court against AJAX Conferencing Systems. Mrs. Gable claims that one of the anonymous members of her computerized conferencing therapy sessions offered by AJAX is her runaway husband. AJAX has refused to divulge to Mrs. Gable the current location of the gentleman in question and claims his identity is protected under the agreement to which all members of a therapy session are subject. Dr. Proctor, moderator of the session, believes the violation of this protection would be an infringement of the doctor–client relationship. However, the attorneys for Ms. Gable feel that these privacy rights

cannot be used to continue a violation of the law.

The Association of Conferencing Networks has filed as friends of the court a brief describing what would be involved if service suppliers had to have an investigative staff to handle a possible multitude of such cases, if Ms. Gable were successful in her plea.

In talking with Ms. Gable, this reporter was told by her that no other man but her husband could have related some of the episodes described by him to the group as a whole. She merely wants the child support to which she is entitled for their seven children.

SEC Opposes High Risk

Tulsa, Oklahoma, July 5, 1991. The SEC has filed a court injunction in this city against High Risk Incorporated. High Risk is a small firm that Mr. Claude Stevens runs out of his basement on his own Minicomputer. Mr. Steven's one-man company offers a game to the public over the local digital network that appears in the opinion of the SEC to be an unregistered commodity market. The SEC claims Mr. Stevens must join one of the standard commodity exchanges in order to continue his operation. In talking with Mr. Stevens, it was explained to this reporter that his High Risk Game involves a virtual commodity the value of which is tied to the rainfall in this area on a day-to-day basis. Winners receive additional credits for charges on any system offered over the Tulsa Digi Net. He claims it

188

ISBN 0-201-03140-X, 0-201-03141-8 (pbk.)

is pure entertainment and no more than a contest-type operation.

Mr. Harold Winwart, assistant deputy administrator for the SEC, claims the SEC cannot let this small operation go unchecked because it exhibits all the properties of a commodity exchange and the fact that money is not the final transaction item is merely an attempt to disguise the issue. Mr. Stevens has pointed out that the cost of joining an exchange is in excess of his yearly budget for operating the system and his profits.

Overheard:

Alice, let's propose a conference on how to deal with 13-year-old girls. These books are for the birds. Surely there are some other parents having our problems.

Dad, I've decided it's more attractive to take my sophomore year of college over the terminals. The conference professors received a higher rating than those at Princeton for these courses. Anyway, with what I save I can get my own terminal.

ISBN 0-201-03140-X, 0-201-03141-8 (pbk.)

CHAPTER 6

Public Use

The area of general public use will probably be one of the last to develop on a large-scale basis, but we can expect to see its emergence soon among hobbyists, much as ham radio operators preceded the mass use of citizens band (CB) radio. It is likely that this application will first emerge in the form of innovative demonstration projects before industry is convinced that there is a mass market for computer systems that incorporate conferencing components.

Wide-scale use in the home or by the general public awaits the development of (1) familiarity with use of such systems, gained at work or in school; and (2) access to terminals. It is also conditional upon the development of policies and regulations that do not place economic or legal barriers in the path of public access to and use of computer-based communication systems, as discussed further in the chapter on policy and regulation.

Access to Terminals

One way to begin to give the public access to specific conferences or message-based services is to make computer terminals available in

- Public libraries (where they can also be used to search computerized bibliographic data bases and abstract services);
- "Town halls" or community centers (where messages requesting information or aid from public officials can be filtered through an intermediary to the proper recipient and the answers left under an ID number or printed and mailed out);
- Public schools, where terminals would be used for educational purposes.

In other words, conferencing services could be made available to the public through publicly located terminals that were also used for other purposes. As has been pointed out by Turoff and Spector (1976, p. 702),

> The library is an institution that is available to the public, its personnel are familiar with serving the public, and it is relatively neutral with respect to

Starr Roxanne Hiltz and Murray Turoff, The Network Nation: Human Communication via Computer

ISBN 0-201-03140-X, 0-201-03141-8 (pbk.)

political, social, ethnic and organizational polarizations. From the point of view of a person who is interested in delivering computer technology to the public, the library is a convenient place to do it.

In addition, many middle-class employees may be given a terminal to use at home. Ultimately, however, mass use will depend on the availability of inexpensive, high-quality terminals for the home market. This is just a matter of time, and we believe that the calculator is the model of the kinds of decreases in cost and increases in quality that are probable in the next decade. Five years ago, huge noisy elephants of terminals sold for thousands of dollars. By 1976, a cathode-ray tube (CRT) was available for about $1000 and a lightweight portable hard-copy terminal for about $2000. In the spring of 1978, Teletype Corporation made available for about $1400 a 30-character-per-second typewriter quality hard-copy terminal that uses regular paper (the TY 43); it can also be rented for about $45 a month. By 1985, we project, there will be CRT terminals available for the home market for about $300, with a printer attachment available for another $200 to $300. Given the apparent willingness of Americans to purchase such expensive hobby equipment as TV games and CB radio equipment, the availability of computer terminals at this price (comparable to that of a color TV set) should make them attractive for the home market.

In this chapter, we will explore the kinds of applications of computer-based communications systems that will be possible, as soon as terminals and education in their use are available, in education, politics, economics and professional services, and general household and recreational use.

Education: The Trend to Life-Long Learning

In post industrial societies, larger and larger proportions of the population have been spending ever greater parts of their lives in educational activities. In addition to initial formal education, housewives, employees, and retirees have been returning for "continuing education," whether to improve their chances of promotion in their career; to change careers, or simply to enrich their lives. Melvin Webber (1973, p. 293) has commented on this trend and on its probable relationship in the future to improved educational technologies:

Already somewhat over 40% of American youths are going to college. (In California, it is well over 80%.) With mass distribution of books, magazines, music, painting, radio and television programming, with ready access to schools of extraordinary variety, and with increasing ease and lowering cost of travel, America is becoming a knowledge-hungry society. Learning has become the prime occupation for many young adults, and it's now a major avocation for the millions of older adults who attend night schools or follow personal programs of

ISBN 0-201-03140-X, 0-201-03141-8 (pbk.)

study. In the middle-range future, learning might become the dominant activity for the mass of Americans. In fact, this eventuality would follow as a direct consequence of the revolution now building in various communications technologies—in two-way cable television, video-cassettes, facsimile transmission, electronic access to libraries, and so on.

What this forecast does not take into account are the facts that travel might become increasingly expensive and difficult, rather than easy; and the specific emergence of computerized conferencing as one of the communication technologies that might feed the "knowledge-hungry society."

Enthusiasm for an "electronic Aristotle" in the form of CAI was expressed by Patrick Suppres in *Scientific American* in 1966:

> One can predict that in a few more years millions of school children will have access to what Phillip of Macedon's son Alexander enjoyed as a royal prerogative: the personal services of a tutor as well-informed and responsive as Aristotle [Quoted in Peters, 1976, p. 42].

However, this prediction proved much too optimistic. One reason was cost: the PLATO system, for instance, will cost $5 million for software and $6000 per terminal in the Control Data Corporation's commercial version (Peters, 1976). A second was the lack of "human interaction."

> We were all a little optimistic about how CAI [computer-assisted instruction] would solve all the problems of education . . . .
> We were forgetting the social aspects of the educational environment, the need to combine personal and social contact with reinforcement from the machine. It's a question of balance that we are just now beginning to understand [Ernest Pope, quoted in Peters, 1976, p. 42].

The self-paced aspect of computer-based educational systems is especially promising. In a self-paced (but not computer-based) political science survey course taught at Southwest Texas State College, for instance, it was found by the professors who evaluated the method that not only did self-paced students score higher on tests of mastery of the material than their conventionally taught counterparts, but the Spanish-American students in self-paced sections managed to perform as well as the Anglo majority, despite their linguistic handicap, which was not true with traditional teaching methods (Miller, 1977). Similar results were found at Harvard. In an economics course taught by self-paced instruction, final exam grades were 10 to 15% better on the average, and the performance of women and minority students was improved even more.

Add to self-pacing the individualized feedback and stored information equivalent to an endless library shelf, and the computer-mediated educational system becomes an educational tool with enormous potential. It can be adapted to learners of all ages and levels of ability, as demonstrated by Omar

ISBN 0-201-03140-X, 0-201-03141-8 (pbk.)

K. Moore's computer-supported learning laboratories called "responsive environments." (See Moore, 1977, for an overview of this work.) For example, three-year-olds have learned to read at a second grade level and to write and type; and inner city Chicano children made substantial progress in closing the measured-IQ gap between themselves and middle-class white children in just one year (Moore, 1977, pp. 6–7).

Computerized conferencing offers the opportunity to combine a unique set of educational resources or capabilities that build on the strengths of CAI and self-paced learning:

- Administration of CAI modules with the result regularly reported to a supervising teacher.
- Conferences devoted to class discussions. For example, imagine how exciting it would be for first-year students in, say, French in the United States and English in France to have a joint conference in which the participants all wrote in their own language for the first half of the year; then had to switch to writing in the other language for the second half. A "real" interchange with foreign acquaintances would make mastery of the language much more interesting.
- A seven-day-a-week private communication line for exchange of questions and answers between teacher and students.
- On-line access to abstract services and computer-stored data bases and educational materials, such as journals.
- Access to on-line simulations or games related to the subject matter.

Computer-based resources could be combined with periodic face-to-face or video-taped lecture sessions, the usual books and other printed instructional materials, and audio cassettes (particularly for language or poetry courses). This combination makes for a powerful educational tool, facilitating a highly participatory style of learning at one's own pace. We believe that such uses of computer systems might best be introduced at the elementary school level. In order to familiarize children with the use of computer terminals and interactive routines for retrieving information, they could do calculations, compose and correct compositions that are then ready to be put into a "class newspaper," and carry on learning-oriented discussions on various topics. Producing "computer literacy" at an early age should be seen as a fundamental goal of education for the generation who will live and work in the twenty-first century.

Computerized Conferencing at the Secondary Level

In an educational system, the teacher–student relationship is the first communication channel that comes to mind in thinking about applications of computerized conferencing. However, there are many existing and possible

ISBN 0-201-03140-X, 0-201-03141-8 (pbk.)

role relationships within the formal educational process that could be facilitated and mediated by a computerized conference.

Classes of students might join together on a town or regional basis to try to discuss "real" problems and apply their classroom learning to them. For example, questions of pollution or local political issues might be discussed in this way, with outside "experts" or government officials occasionally participating. A local environmental problem might be treated as a multiclass project by students enrolled in biology and civics classes.

The role of the guidance department in counseling students about course selection, study habits, career or college planning, and personal problems is another educational function that could be aided by a conferencing system. Many questions are fairly straightforward; rather than have to make an appointment (which sometimes requires weeks to obtain), a student could send the question and receive an answer through the system. For sensitive personal problems a pen name could be used by the student in sending and receiving a reply. This could encourage many students to obtain professional guidance about problems that they might otherwise be too embarrassed to discuss.

Another interesting possibility along these lines is to have topical conferences on problems, such as drugs or alcohol, which involve representatives of parents, students, and teachers, all using suitable pen names that identify the point of view, but not the individual. This could provide a relatively nonthreatening medium for the thorough presentation and discussion of opposing points of view on issues that divide the generations.

Among the other kinds of possibilities for educational uses at the secondary level suggested by a graduate class of teachers at the New Jersey Institute of Technology, as a result of their experimentation with EIES, were

- Conferences among teachers in a region about the way in which they conduct their classes and the new materials they integrate into them. In many high schools, a teacher of chemistry or biology or other specialized subjects may be the only representative of that discipline. Computerized conferencing could provide a forum for peer-group discussion that is not now present.
- Communications among board members, administrators, teachers, parents, and students on policy issues related to the school.
- Extracurricular planning committees.
- Teachers' union contract negotiations.
- A continuous link among the State Department of Education and the superintendent of schools in each community, to provide input into the decision-making process at the state level and to allow local school districts to obtain interpretations and explanations for various policies.

ISBN 0-201-03140-X, 0-201-03141-8 (pbk.)

ISBN 0-201-03140-X, 0-201-03141-8 (pbk.)

College and Continuing Education

The continual rise in the cost of a college education has led to a growing educational gap that is likely to be filled by the concept of the college without walls and buildings. It is easy to imagine a complete college education being made available via this communication method. The cost of a college education at $5000 a year averages out to $10 per class hour per student. This is already more expensive than the EIES system on a nationwide basis, and double the costs on a local basis, not counting the costs of commuting or the inconvenience of having to be physically present for classes at certain hours each week. One could afford to have the most talented teachers prepare the CAI modules and other educational materials that would be made available on a mass basis over such systems. A rich variety of courses could be made available in specialized areas not offered on any one campus, drawing from a prospectively national or international student body.

One educational field trial that has produced some additional data was conducted in the Spring of 1977 at New Jersey Institute of Technology. Turoff's graduate computer science class for high school teachers was supplemented by a free-discussion computer conference. Near the end of the semester, the (edited) exchange shown in Fig. 6-1 took place. Note that the only argument seems to be the extent to which computer-based instruction should be combined with other teaching tools and methods.

There is no reason why college level courses could not also be taken by college graduates or nonmatriculated students, in addition to students enrolled for credit, to broaden or update their knowledge. In addition, participation in a conference or on-line Delphi on technical or scientific issues in various fields is in itself a form of continuing professional education. Take, for instance, the EIES panel of leading experts in hepatitis convened to act as a mechanism to review current research and make it available to the general practitioner. In the course of the argument, debates, and information exchange involved in this process, the experts themselves will undoubtedly learn things of value from one another, relating to different points of view and the evidence and reasoning that supports them.

Another interesting idea is to use interactive computers for training of advanced specialists. For example, Heiser et al. (1977) have described a project in which first- through fifth-year psychiatric residents were exposed to communication from a paranoid patient and from a computer programmed to communicate like a paranoid person. Five judges misjudged half of the time whether the typed responses were from a computer or the real patient (in other words, they guessed correctly only as often as they might have by chance; so the program passed the "Turing test" of not enabling the users to tell if they were communicating with a person or a machine). The point is that

Figure 6-1 Attitudes of students toward a computerized class.

C 90 CC30 MURRAY TUROFF (MURRAY, 103) 4/11/77 6:37 PM
I want each of you to specifically answer (you can use your pen name) what your reaction would have been to taking this course from a terminal located at your school, with the class carried out entirely through the terminal. Could the course have covered the same material? Would more of the teachers at your school have participated? Benefits and disadvantages please.

C 90 CC31 DONNA (913) 4/11/77 7:20 PM
Ref 24 c—I would have reacted very negatively to taking this course completely over the terminal primarily because I find the concept of the "mechanical teacher" so distasteful. I am of the opinion that face-to-face contact is very important between student and teacher and that a full understanding of the course material occurs more completely. In this type of a situation visual contact on the part of the teacher is very important in determining if a concept is understood by the students.
RE: CC 30 To answer the third question first. Yes, I think more teachers would have taken the course because of the convenience of using the terminal at any time. Time is an important commodity to all of us and if the commuting time could be saved I think that time could be used at more profitable tasks. Teachers could use parts of their TPH's [teacher preparation hours] in learning programming and in overcoming their general fears and misgivings about computers. At my high school we are thinking of instituting in-service programs for the teachers for credit and I can envision the utilization of the EIES system in this manner. In fact, this is something I want to seriously pursue because I am a member of the committee that will be responsible for these in-service programs and this idea intrigues me.

C 90 CC32 JOHN 4/11/77 7:43 PM
If I had found out that the course was being taught by the use of terminals at my school, I am certain that my first reaction would have been negative. The major objection would be the same as was mentioned in an earlier comment. That is, the personal contact with the teacher would be lost.
However, after having finished a majority of the course, I think that the use of terminals at the high schools would have been an excellent idea. The major problem I've found in completing the work has been that in order to check our programs we had to be here. Since we have class only once a week it was very difficult to get onto the computer. If we each had our own terminal, at which we could send our programs to the computer, we could have saved a huge amount of time.
I'm sure that this advantage is far more important than the personal contact in a course of this nature. This course requires that the student do a large amount of practice work. Any questions are usually answered by finding out if something "works" on the computer, or by consulting the text. Questions to the instructor are generally brief and could be answered easily through the conference system.
I don't think that all courses could be taught as easily on this system. Most science courses, like biology, chemistry, and physics require a large amount of explanation from the teacher. Without that explanation I think many students would be lost. For those courses a more complex system would be needed than could be used in teaching this course.

C 90 CC33 ALVIN (979) 4/12/77 2:12 PM
I think that the EIES computer system is a good idea for a teaching aid. However, when a technology such as this is developed, the idea of using the system effectively

196

ISBN 0-201-03140-X, 0-201-03141-8 (pbk.)

Figure 6-1 (Continued)

must be considered. A question, how might this communications system be used to overcome the disadvantages of the lack of the reading and writing skills that may be encountered by some urban students, which will contribute to the "un-seen" barrier that holds back the disadvantaged?

C 90 CC36 (ANONYMOUS) 4/18/77 3:38 PM
   Would the availability of these types of systems for fun and games in the home as well as teaching change the outlook of students toward learning the English language?

C 90 CC37 KENNETH (918) 4/18/77 7:07 PM
   I would not object to this course being taught completely by a computer system as long as it had the ability to conference as displayed in this system. This would allow us to ask questions concerning topics we do not understand. It would supply student–teacher interaction. The advantage of a computerized program is increased time available for computer use. A disadvantage would be the fact that not all high schools have computer terminals.

90 CC 43 GARETH (911) 4/18/77 7:09 PM

RE C36
   I have seen a decline in English language proficiency over the past 16 years of teaching. I firmly believe that the influence of TV has been one part of the problem. Less reading and writing is being done as people sit in front of the "tube." To what extent educational trends have been a factor is difficult to say—I am certain less emphasis is being placed on skill development and rote learning than there once was. Perhaps old fashioned, I believe English usage, spoken and written, is a skill and as stated above, they are developed with practice. Here again, computerized teaching could possibly be used to great advantage.

---

such interactive communication-based programs might be profitably used in the training of psychiatrists and others for whom learning on real patients is a dangerous and not optimal educational method:

> We claim this experiment validates the method or model of simulating paranoia. . . . There is no empirical evidence that "being nice" to paranoids is therapeutic. There are unquestionable advantages to having a model to practice on. . . . Students in the mental health field are expected to learn by practicing on suffering human beings, with the methodological goal of never making a mistake. In such a setting the potential for superstition and habit to supplant hypotheses and tests is obvious. (Heiser et al., 1977)

Participatory Democracy through Conference Groups

The most exciting and potentially revolutionary political application of a CC system is the facilitation of the direct participation and voting of citizens on important state or national issues. Etzioni and his colleagues have extensively explored and field-tested such an electronic direct democracy, based on

ISBN 0-201-03140-X, 0-201-03141-8 (pbk.)

the use of conference telephone calls; but the possibilities of asynchronous participation and a written record of all discussions and votes makes the computerized conference form a much more attractive and practical possibility, once computer terminals are available in homes and offices.

As they explain the idea (Eztioni et al., 1975, p. 64):

> The form of democracy found in the ancient Greek city-state, the kibbutz, and the New England town meeting, which gave every citizen the opportunity to directly participate in the political process, has become impractical in America's mass society. But this need not be the case. The technological means exist through which millions of people can enter into dialogue with one another and with their representatives, and can form the authentic consensus essential for democracy.

The base for such a democracy would be groups of approximately 30 citizens. At this level, all participants could fully discuss and examine an issue before voting and reaching a consensus. Each of these base-level groups would then elect one representative to a higher-level group of approximately 30; they in turn would elect one representative to present their views to a yet higher-level group. In this manner, it is possible to reach 24 million participants in five steps and 198 million in six (Etzioni et al., 1975, p. 65). In Etzioni's version of this process, "The consensus arising from this communication tree can then be put, for final ratification, to a vote of all participants."

Another model for participatory democracy via computerized conferencing is offered by Coates, who foresees the possibility of a mass referendum on crisis issues (Coates, 1977a, p. 197):

> Telecommunications is likely to influence collective participation in voting and decision processes. Technological developments are coming together, for example, to make the following possible. A major hurricane is about to hit the coastline of the Carolinas. The launching of airplanes or small rockets loaded with suitable chemicals could seed the hurricane and reduce its energy intensity. But there would be a risk of the storm turning north and hitting Virginia. Assuming a suitable network of telecommunications, a fair if not total sample of the population at risk from the hurricane and from its potential seeding could in real time interact, be informed, discuss and be surveyed to inform the decision maker on whether or not he should launch the countermeasures.

He then expands upon other possible participatory processes that can be made possible by computerized conferencing and other interactive communications media:

> direct interrogation systems to permit any user to obtain any desired information; face-to-face audio-visual conferencing with indeterminate numbers of people over indeterminate distances in real time or *ad lib; ad hoc* national conferences; interregional conferences, contests and tournaments; the building

ISBN 0-201-03140-X, 0-201-03141-8 (pbk.)

of unique constituencies irrespective of geographic areas; the formation of special interest groups; and bureaucratic and legislative direct interaction with constituents [Coates, 1977a, p. 203].

Congress now has an on-line system to inform its members of the status of pending legislation. This is just one example of a public policy-making data base that could be opened up to direct interrogation by citizens. There is no reason why the *Congressional Record* could not be produced and distributed through such a data base, to provide same-day access to anyone with a terminal.

To sum up, there is a great deal of promise in using a computer-based system both for the timely and detailed dissemination of information on ongoing debates, and for providing an input channel to the political decision-making process for the general public.

Hawaii's Con-Con Project

Perhaps the first operational use of CC systems to facilitate "participatory democracy" will be J. W. Huston's Constitutional Convention ("Con-Con") project in Hawaii. Funded by grants from local and mainland foundations, it is being designed to establish 21 community centers throughout the state to allow public participation in the 1978 Hawaii Constitutional Convention, either through computerized conferencing or videotapes, and to incorporate three rounds of polling on each issue to provide public opinion inputs to the convention (J. W. Huston, 1977, private communication). The effort is explained as follows:

The Constitutional Network is an attempt to broaden, educate, and intensify public participation without expanding the size of the delegation. Through the Constitutional Network, the dynamic segments of Hawaii's communities (our business and labor organizations, our neighborhood boards and social clubs, our community organizations and schools, our huis and our kumiaia) will be able to see and hear much more of their delegates' work, will be able to interact with each other a great deal, and will be able to tell Con-Con what they like and what they want. At each center there will be a computer terminal. These will be connected by telephone lines to Constitutional Network's communications center, probably located in metropolitan Honolulu. Local opinion will be entered at the local terminals and stored for retrieval at the communications center. Issue-oriented polling, statements from the public, group presentations, trial balloons, survey techniques are all easy when they are computerized. Groups which have traditionally made inputs to the legislative process will find their voices enhanced by the Network. The newly-enfranchised and the seldom heard will learn and discover the benefits of participation in democracy. Citizens will feel closer to their government, government will grow closer to its citizens.

ISBN 0-201-03140-X, 0-201-03141-8 (pbk.)

Lobbying

Costs to put together an effective lobby are tremendous and consequently there are many groups that do not participate as effectively in the political process as other groups that are smaller but better funded. In principle, a CC system would make it very easy for individuals interested in the same objective to find one another and to organize as a group. This could well lead to a much more fluid situation with respect to political alignments. Needless to say, we do not understand all the implications for the political process of this potential for uniting like-minded people.

Another interesting concept is computer-linked community councils consisting of representatives of neighborhood, ethnic, or other groups. In many urban communities the leadership of various community groups usually have no real opportunity to meet with their counterparts except at public meetings where images must be maintained. Because of this it is usually "city hall" that negotiates, acts as intermediary, or plays off one group against another. A group of community leaders able to engage in peer-group communication could well form more united fronts, eliminate small problems before they become big ones, and perhaps develop a more cohesive urban community working in cooperation rather than competitively.

Citizen Advisory Groups

Some people think that citizen advisory groups to local and state governments are ineffectual because they do not meet often enough or have sufficient influence. Computerized conferencing would allow such groups to hold a continuous meeting between face-to-face sessions. It would also allow government officials to more easily participate selectively in those portions of the discussion where their information, advice, or encouragement is needed; and to retrieve the advice or poll the members of these groups when a decision is at hand.

Citizen Participation and Aid

Often citizens feel at a disadvantage when participating in the political process because they do not have access to the same knowledge and expertise that government officials and industrial or business groups have. Although there appear to be scientists, engineers, lawyers, architects, and other technical and professional people who are willing to give knowledge and advice to such groups, often they are not located within the given community and travel would require more time and expense than these people wish to give. Computerized conferencing would make it possible for citizen groups in different areas to pool the technical and professional talent available to them. This pooling would at least provide the opportunity for citizen groups to get

ISBN 0-201-03140-X, 0-201-03141-8 (pbk.)

better handles on facts available and the opportunity to take well-informed positions on complex topics. This would lessen the likelihood that well-meaning citizen groups would take unreasonable positions because of a lack of knowledge.

Use by Elected Officials

There are many potential benefits stemming from the use of computerized conferencing by legislators and other elected officials. One area is the improved ability to stay in touch with advisors from the official's constituency who may be located hundreds of miles away from the official. The second is the potential impact on the caucusing or political negotiation process, which is one of the most time-consuming communication activities a legislator engages in. A group wishing to coordinate their activities in support of a particular bill could do so much more efficiently with a CC system than with the current limitations of phone and face-to-face meetings. We have no real understanding of the potential long-term impacts of this communication medium on this form of political negotiation. However, it is of course possible for computerized conferencing to be used to simply make traditional political processes more efficient. Jimmy Carter, for instance, used a message system to coordinate his presidential campaign. Ultimately, one could imagine legislative sessions being held through a computerized conference, particularly special sessions or emergency sessions called to deal with a specific issue.

Professional and Other Services

There are a wide variety of services that might be provided more quickly, more cheaply, and more conveniently if they were delivered via a computer communication system. Some of these (such as professional advice or counseling) consist largely of communication to begin with, whereas others (such as EFTs—electronic funds transfers) may simply deliver or perform the service electronically.

Psychiatric or marriage counseling, either on an individual or on a couple or group basis, would not only be more convenient through this medium, but might also have some intrinsic advantages. It could be that people will be more open and candid when it is possible for them to discuss their problems without their real identities being known, or without having to actually look at the other members of a therapy group or the counselor.

Likewise, imagine the advantages of having legal or accounting advice available on a one- or two-day turnaround, without having to make an appointment and take time off from work to go to see the professional.

ISBN 0-201-03140-X, 0-201-03141-8 (pbk.)

Complex problems would undoubtedly require a visit, but most preliminary inquiries ought to be able to be handled through a standard set of questions prepared by the professional and answered by the client, who would then subsequently receive the advice, simple contract, or will, or perhaps completed tax return. Specialized consultants of all types might be available at an hourly charge in this manner—say, a plant professional to answer questions about ailing formerly green things, a statistician to advise on the correct statistical test for a data analysis problem, or a personnel expert to advise on human relations problems in your organization. Paid consulting of all types would be facilitated, since a computerized registry of consultants with on-line references could replace the word-of-mouth system that now prevails.

There are actually a host of volunteer professional services that might become more widely available because of computerized access, such as retired businessmen willing to help the small business person; aid for junior chamber groups, extensions to the scouting merit badge program, and Vista operations. Problems of distance and travel time, and the fact that those who need the help and those willing to provide it have difficulty in easily finding one another currently discourage many potential volunteers.

Telemedicine is another area in which, especially for those in remote areas without easy access to a doctor or to specialists, advice might be delivered electronically. One version of this is to have specialists on-line act as consultants to general practitioners who are not sure how to diagnose or treat a particular condition. Another is to invoke the power of the computer to store the collective wisdom of the medical profession about a particular condition. A computer program could then lead a paraprofessional or general practitioner through all of the suggested tests for a suspected condition, ask for the reading or reports on each of these parameters, then give a diagnosis and prescription for treatment.

An example of the use of the computer to provide "artificial intelligence in medicine" is the CASNET/Glaucoma system at Rutgers, which provides consultation in diagnosis and therapy of glaucoma; the MYCIN system at Stanford, which assists in the treatment of infectious diseases; and the INTERNIST system at the University of Pittsburgh, which provides clinical consultation in problems of internal medicine. These programs are at or near "expert" status now, are in experimental use, and are ready for operational trials. Several other artificial "expert" systems are being developed in medicine, biochemistry, genetics, psychology, business management, language and speech processing, and computer programming (Amarel et al., 1977, p. 1).

Variations of this are what might be termed "referential consulting networks" of professionals, and direct access to codified stored knowledge by the public.

ISBN 0-201-03140-X, 0-201-03141-8 (pbk.)

Referential Consulting Networks

Given the availability of computer terminals in public libraries for public access and use, and for on-line searches of computer-based bibliographic and abstract files, a CC system might be used by librarians themselves to form what Manfred Kochen calls a "referential consulting network" (Kochen, 1973). The argument is, now that "information" is located in many places other than the traditional books and journals that were under the librarian's care, it is time for the formal reference librarian to become a general community knowledge resource, an "information please" professional ready to find an answer to any question asked, in terms of where to get the information needed, if not the actual answer. For those that cannot be answered with resources stored in the local library, a large network of reference librarians and "on-call" experts might be used to share their knowledge resources. The idea is that every reference librarian is to some extent an expert consultant, to some degree, for some class of questions (Kochen, 1973, pp. 11–12). Through some kind of "bulletin board" conference, a reference librarian who does not know where to go to answer a particular request could describe the inquiry. The answers supplied by others could then be automatically appended to the original request. Ongoing conferences on the availability of and experiences with new kinds of information resources might also be conducted.

On the other hand, it may be that libraries and librarians as intermediaries between the public and stored information will be replaced to a large extent by the computer-based systems that give this access directly to the public.

The British Post Office's VIEWDATA

In-the-home computer-based interactive systems are not futuristic dreams, but rather a kind of service that is available now, and awaits only the capital investment and marketing needed to turn available technology into a mass market item. In Great Britain, the Post Office has taken steps to make such facilities available. The only information we have been able to obtain, without a field trip, is promotional literature put out by the Post Office itself. Given the limitations of such information, we nevertheless can conclude that the VIEWDATA service is a precursor of the kinds of in-home consumer services we have been projecting. According to Bright (1976, pp. 2–4), the following characterizes VIEWDATA, which will be offered as a public service at modest rates.

- Employing the telephone network as the communication medium, the Post Office commenced the pilot trial of its computer-based information system

ISBN 0-201-03140-X, 0-201-03141-8 (pbk.)

in January 1976. It is aimed at telephone users at home and in the office who will be able to call up information on a wide variety of subjects for display on their television screens.

- Users of Viewdata will be in direct two-way conversation with the computer system and able to enjoy a wide variety of [interactive] facilities in addition to information retrieval. These features all stem from the dialogue techniques embodied in the system which enable programs to be employed to assist in answering difficult and multifaceted questions, for example advice in completing income tax returns, by breaking down the complex problem into a series of simple structured yes/no answers from the user. Other significant advantages of the interactive mode are the eventual inclusion of message facilities, calculations, games, and quizzes.
- Some eighty organizations have now undertaken to participate in the pilot trial. They will be supplying contributions spanning over two hundred main and subsidiary topics, ranging from sports and news to house mortgages and holidays in the domestic sector, and from market reports to economic indices and financial analysis in the business sector.
- The use of computers for storage and search of this wealth of information will put it at consumers' fingertips; an extremely simple operation and search procedure, and response times better than three seconds, will make available detailed and extensively cross-referenced information on subjects of local as well as national interest.
- Up to 1000 terminals will be placed in London offices and homes in a market trial in 1978. The earliest likely date for the actual Public Service to commence on a national basis is envisioned to be during 1979.

EFT: Part of an All-Pervasive Computer Communications System?

Electronic funds transfer potentially represents a blending of computer power, communications services (including information services), and financial transaction services. Current attempts to demarcate and differentiate what is a computer system, a communication system, and an EFT system fall apart when it is considered that by the mid-eighties the computer terminals will probably be located in most homes. A number of banks already allow corporations to access their bank accounts via ordinary computer terminals— why not individuals? (The current regulatory issue of what is a "branch bank" then becomes somewhat absurd.)

The computer terminal in the home raises the possibility that those services that will be most competitive or desired by consumers will be those that make it easiest to carry out at least four kinds of functions in an integrated or compatible manner.

ISBN 0-201-03140-X, 0-201-03141-8 (pbk.)

Finding: the retrieval of information about what is ava
terms of goods or services. Various goods and services cou
priced in extensive on-line catalogues.

Negotiation: the user-to-user communication about th
or transaction; the prospective buyer could direct specific ques..
to the prospective seller.

Processing: the actual carrying out of the transaction and the generation of a record to interested parties.

Analysis: the summary of transactions by categories of interest to various users of the system. For instance, the computer could do the equivalent of "balancing one's checkbook" after each transaction, and preparing tax and accounting reports.

The tasks of finding and negotiating a transaction require human communication. Therefore we can expect that the future of electronic funds transfer involves the incorporation of computerized conferencing technology. However, the fallacy of current EFT concepts is the hidden assumption that the nature of the way transactions are conducted in the society will remain the same, and the nature of money and financial institutions will not change. In principle, however, the computer system and the associated communications capability can allow for the reintroduction of a bid-and-barter-oriented economy. Such a process could, in the long run, rejuvenate the applicability of some of the axioms that underlie the free enterprise system, and possibly do more to minimize inflation trends than the current tendency to manipulating factors such as interest rates. There are many startling aspects about reintroducing the bid-and-barter system. In terms of the involvement of the public, one of these is emergence of local public groupings for the exchange of goods and services, which will be discussed further in the final chapter of this book.

The Domesticated Computer: Uses in the Home

Family and Friendship Networks

Computerized conferencing can make it very easy to keep in touch with family and friends and colleagues who are located some distance away. A person could generate the equivalent of a "Dear Everyone" newsletter a few times a month, for instance, adding a few sentences at the beginning or end specifically directed to each person. In this manner, it would not be much of a time-consuming chore at all to keep in touch.

ISBN 0-201-03140-X, 0-201-03141-8 (pbk.)

Sociologist Elaine Kerr, in EIES Conference 72, has speculated as follows about this particular kind of application of a conferencing system:

C72 CC 10 ELAINE KERR (ELAINE, 114) 12/13/76 1:38 PM
My comments concern the impact of computer conferencing upon family and friendship networks in our society.

One consequence of our highly mobile and rapidly changing society is the more rapid disintegration of family and friendship ties than had been the case in simpler times. Most of us today consider ourselves very fortunate if we retain friendships over the years, recognizing at the same time the "inevitability" of problems imposed by distance (for all but the very wealthy, money constrains travel and extended phone visits; letter writing becomes tedious).

The plight of the extended family in our society, and the consequences of its disintegration, are well known.

Would the widespread use of computer conferencing impact upon these two social problems?

I hypothesize the following:

1. The mean duration of friendships will be longer in a "computer conference society" than at present.

2. Friendships terminated in a "computer conference society" are more likely to be a function of changed interests than distance.

3. Because friendship and family ties will endure, there will be a concomitant increase in:
   (a) Individual ability to cope, with friendships functioning as support mechanisms;
   (b) Individual "happiness" (satisfactions, adjustments);
   (c) Social cohesion, with spillover for efficiency, effectiveness, etc.

4. There will be a decrease in:
   (a) Individual loneliness, psychological problems, suicide rates, marital happiness, as traveling partners retain contact, etc.
   (b) The frequency and significance of social problems (juvenile delinquency, alcoholism, divorce, suicide, etc.).

Whether these optimistic hypotheses will be borne out we cannot say, but certainly computerized conferencing provides a convenient and low-cost channel of communication for staying in touch with friends and family who no longer live nearby, and who can enlarge the effective support network available to individuals.

People Finders and Recreational Conferences

One unique characteristic of communication via computer is that the title of a message or conference can be used as a kind of "To whom it may concern" address. This provides an efficient mechanism for people to form groups to conduct discussions on topics of common interest, even though they may never have met one another.

Some such conferences may be purely recreational, such as a gourmet recipe exchange, a discussion of specialty stores in the area and the goodies

ISBN 0-201-03140-X, 0-201-03141-8 (pbk.)

that they carry; a fishermen's or ornithologists' exchange on where to find the big ones or the rare ones.

Public "bulletin boards" might host a local bid-and-barter service for goods and services, indexed by type of thing being offered. Unlike the local newspaper (which often has everything jumbled together, so that you have to read through 500 ads only to discover that there are no antique china cabinets this week), the offerer can withdraw the ad as soon as the sale has been made or the wanted service obtained, thus saving many fruitless telephone calls on the part of too-late seekers. A variation of this might be the "labor exchange" in which people are given some sort of skill-equivalent rating, and then trade hours of their special skill (such as plumbing or lawn mowing) for hours of cake-baking or baby-sitting.

On a more serious level, people might have topic-oriented group discussions of problems that they share in common, and their solutions for dealing with these problems. Parents of teenage children in the community might discuss curfews or dress codes or how to handle experimentation with marijuana, with or without the group's sharing the cost of a professional leader to guide and advise their discussion. To give a few more examples, separated or divorced women who feel guilty about working, or women who feel guilty about not working, could likewise engage in such generally therapeutic peer-group discussions, using pen names to protect their privacy.

Games

Computer-based "hangman," or tic-tac-toe, chess, or "Star Trek" represents the "traditional" game structure on a computer, in which the individual plays against the computer. With the introduction of the communication feature to the memory and computing power of the computer, many truly interactive games involving two to many hundreds of persons may be imagined. For starters, the computer could facilitate remote "bridge," dealing the hands and making sure that only "legal" types of communications passed through the system. Based on their past record, players might be given "handicaps" or ability levels, so that when seeking partners for a game, they could be assured of a person at approximately their own level of ability.

Another interesting idea is large-scale battle games, or role-playing games in which the participants try to simulate a local or national problem and its solution by playing the roles of the actors in the situation.

Concepts such as "Boy's State" or "Model United Nations" or a mock Congress can now become a new form of adult game. For example, a model "U.S. Congress" could deal with the very same issues that the real Congress is debating at the time. In effect, such a set of mock Congresses across the United States may become a very interesting new form of polling. The real politicians might closely watch how an issue resolves itself in the different

ISBN 0-201-03140-X, 0-201-03141-8 (pbk.)

cities of the United States. We can imagine a real Senator observing with interest how those acting this role across the country vote and what they consider the most effective sort of speech to give. This would be a very interesting form of political psychodrama.

For battle games, all the major roles on both sides of the battle can now be filled by real individuals. What we then have is a scale of gaming involving many hundreds of people in one game. The computer provides the communication structure and protocol reflective of the particular game.

It is even conceivable that professional teams would emerge and people would pay to watch the action occurring. Over the long term it is conceivable that intellectual games would achieve a position as national pastime equal to that now occupied by football or baseball. (Might the electronic "pros" even be immortalized on bubble gum cards?)

For those who do need physical reality in a game, the field of robotics would make possible the remote manipulation of mechanical players in a modern unbloody replication of a Roman circus. There could be a "choice of weapons"; then the robots could be maneuvered via computer terminals, and the action watched on video screens, until one had completely demolished and incapacitated the other. A very expensive game, of course, which would probably make it an upper-class or spectator sport. Whether such mechanical murder would be unhealthy for society or represent a necessary emotional outlet for pent-up societal frustrations depends on our solutions to other societal problems and unverified theories. At least today, people seem to want to see individuals bang one another up in various forms of individual and group violent conflict (otherwise known as "contact sports"); perhaps this need could be more humanely met by computer-controlled robots. Besides, the act of causing walking computers to destroy one another might be a natural outlet for pent up hostility towards these machines if many of the current abuses continue.

The Life-Style Change Game

Families or other household forms often tend to get in a rut as far as their activities and interaction patterns are concerned. The same round of activities is carried out by the same persons and decided in the same way. It is hard to break out of a life pattern that has become dull and unsatisfying. What follows is speculation on a computer-based system for families to optimize and add variety to their life-style decisions.

In the "life-style change game," each member of the household unit would be given a certain number of points or votes. Suppose we are dealing with the choice of recreational activities. Each family member would list her/his desired activities, in order, with associated criteria, such as time requirement, cost, requirements for certain weather or other participants, and whether it is

ISBN 0-201-03140-X, 0-201-03141-8 (pbk.)

a passive, intellectual, social, or active sports activity. The total family recreational budget, in terms of hours per week and monetary resources, would be set as limits on the combinations of choices that could be made. The next step might be for the family to choose one of the four following life-style patterns:

cyclic: a regular, scheduled system of rotation among the same activities, with a month or a quarter being the length of a cycle.

random: no pattern at all; completely random choice generated each time.

balanced: random choice within limits specifying that a certain minimum and maximum number of choices of each type must occur within each cycle.

feast or famine: fluctuation between extremes is made a characteristic within each cycle.

The computer would then use the choices and their weights to generate a schedule for each cycle that met the requirements of the unit for "life-style" and overall minima and maxima for each of the tagged criteria. The computer would then print out the unit's schedule for the month or the quarter.

Similar procedures could be followed, with assignment of household chores and choice of menus. No one could resent "arbitrary" choices and requirements laid down by the parents; the computer would be the one making the decision for the unit!

The Willing Young and Information Power: A Plea

It is possible to design a CC system usable by anyone who can read and write. It does not have to be more complicated than using a telephone. However, what will ultimately evolve are systems that have the power to gather, organize, process, and circulate information as a part of the communication process. In order for individuals to be able to use such systems effectively, they will have to have some basic understanding of information technology and its application. This will determine the difference between those who merely utilize these systems and those who can master them to their benefit. It is probably true that the mass of the adult population today has neither the initiative nor the desire to obtain the necessary reeducation. Our educational system as it now exists has not been particularly effective in teaching people how to utilize information readily available in libraries and public records, so that we have a heavy reliance on information gatekeepers who obtain or provide information to the public.

The emergence of computerized conferencing and information technology presents society with a choice. One option is to perpetuate the use of intermediaries to provide the interface between the information technology and the public; the other option is to raise the educational level of the public

ISBN 0-201-03140-X, 0-201-03141-8 (pbk.)

so it can cope with the technology directly. It is our view that the latter is preferable for society as a whole, lest we raise a new priesthood—the information specialists—which has the potential for making our current difficulties with doctors and lawyers seem trivial by comparison.

Considering the time scale for development of major technological changes in this area (every five to seven years) it would seem imperative that curricula in computer and information science be introduced now into public school systems, beginning at the third to fifth grade level. For those who have had the opportunity to do it, there appears to be little difficulty in introducing eight- and nine-year-olds to computers. The emergence of "the little things that count"- (calculators) and their introduction into many grammar school math courses is an encouraging sign, but only a very small first step toward what is needed.

Information and its use will become a far more pervasive form of wealth for the generation now entering the public school system. With the current model of the educational process, only a minority of those emerging a decade or more hence will become masters of the use of information and its associated technology. Unless something is done, the mass use of this technology will emerge as something akin to CB radio and TV. Although there is nothing wrong with entertainment and fun as one of a number of public uses, it would represent a major loss of opportunity for society as a whole if that application emerged as the only general public use of computerized conferencing.

Summary

In this chapter, we have suggested a few of the ways in which "a computer terminal in every home" might be utilized to provide a wide range of communication services, ranging from the diversion of computer-mediated games to the life-and-death questions of the diagnosis of a serious medical condition. We have merely scratched the surface of the possibilities, suggesting only those kinds of applications that are currently under development and testing, or that could be achieved with existing technology.

Like all new technologies, there are both potential gains and possible dangers. The biggest danger for the public is that this technology will be available to only a few. If that is allowed to happen, we face the consequence of becoming, perhaps, a technocracy in which effective control of decisions in the society lies with those who have their hands on the valves regulating the flow of information.

The objective of making this technology available to the public as more than a new toy will require both foresight and planning by those who

ISBN 0-201-03140-X, 0-201-03141-8 (pbk.)

represent the public interests and who influence policies, laws, regulation, and most important, our educational systems.

For Discussion

1. What would be the advantages and disadvantages of having profession-ally conducted marriage counseling done through a computerized con-ferencing system, rather than through office visits to the counselor?

2. Can you imagine religious worship through a CC system? What form would it take? (How about confessions, for instance?)

3. What might a courtship look like if conducted through a CC system? Is it possible for people to "meet" and "fall in love" over the system?

4. What consumer-information-based services do you think are absolutely not suited to delivery through computerized conferencing? Why?

5. Widespread use of computerized conferencing is dependent on the acquisition of typing skills by males as well as females. Do you think that this is likely to occur? What steps could be taken to make sure that typing is a skill that every educated adult has, like the basic arithmetic and handwriting skills?

6. What might be the long-term impact of 50 or more mock Congresses going on simultaneously across the country, regularly reported on via the news media and intepreted by public opinion experts? What impact would this have on current polling practices?

7. How popular would a Roman circus be with mechanical gladiators? How about a mechanical football team?

8. If mass voting (a given feature of conference systems) becomes popular or available on current issues, what might be the impact on the political process?

9. Do you see a new professional emerging in terms of the "scribe" who writes letters for others? To what types of situations would this new form of "ghosting" be applicable?

10. What would be the impact of a major reduction in the amount of TV viewing by the general public?

11. What will happen to newspapers when classified advertisements are provided via information components of CC systems?

12. What sorts of information–communication services might represent a source of power for those who could control them?

ISBN 0-201-03140-X, 0-201-03141-8 (pbk.)

BULLETIN OF THE INTERNATIONAL SOCIETY FOR THE STUDY OF SOCIAL SYSTEMS

VOLUME XI, NO. 89 MAY 25, 1987
Preferred language: 1. French 2. English
3. German 4. Russian
Language choice ?   2
Get

Ads & Notices	(1)
News	(2)
Assessments	(3)
Papers	(4)
Bulletin choice?	2

EMERGENCY CONFERENCE!!!!!!!!!!!!
For the next three days, Conference 1000 is set aside for a discussion and voting on recommendations to the United Nations Special Force to Counteract Terrorism. They have requested our advice on how to handle the current incident in Euroafromania.

Bulletin choice?   3
A73   Resolved that universities shall be urged to dissolve traditional departments of sociology, anthropology, economics, psychology, and political science, and replace them with unified Departments of Social Systems.

Are you   (1) in favor   (2) opposed

Assessment?   1

Bulletin choice?   4
New articles since you were last on:

1. Subcultures without Territory or Tradition: The Ethnography of Emergent Electronic Communities, by Starr Roxanne Hiltz
2. World Public Opinion: A Review of the First Decade of the International Public Opinion Assessment System, by Robert Bezilla
3. On the Optimal Balance among Investments in Education, Food Production, Communication, and Solar System Migration Potential: A Mathematical Model with Derived Postulates, by Murray Turoff

ENTER THE NUMBER OF ANY OF THE ABOVE FOR WHICH YOU WISH TO SEE ABSTRACTS
Abstract #?+

Bulletin choice?   1
There are 4 Notices today—Notices choice?   All

N 17062 5/22/87 SOCIAL PSYCHOLOGIST
French social psychologist wishes to trade homes with Canadian or American for the summer of 1989. Study is equipped with microprocessor including floppy disk, two telephones, computer terminal. In the Bordeaux region, with kitchen and two bedrooms. Desires similar resources and introduction to potential bilingual research assistants.
Reply to 801-362.

ISBN 0-201-03140-X, 0-201-03141-8 (pbk.)

N 17063 5/22/87 RURAL SOCIOLOGIST
Have contract on environmental impact study of a new solar energy form In midwest area. Need an experienced rural sociologist.
EIES Box Number B1583

N 17064 5/22/87 ETHNOMETHODOLOGIST
A group of traditional ethnomethodologists working in the area of nuclear family norms seeking patron for setting up an ongoing conference—free consulting offered in return—complete vitas available. Reply to 801–400

N 17065 5/23/87 SEX SELECTION
We are putting together an interdisciplinary group for a research project on the impact of the new Select-a-Boy, Select-A-Girl kits for sex determination of children at the time of conception. Needed are a demographer, public opinion researcher, and an anthropologist. Send vita and brief statement of your proposed role in the project to Box B 1978.

ISBN 0-201-03140-X, 0-201-03141-8 (pbk.)

CHAPTER 7

Science and Technology*

Science and technology is the yeast that leavens the bread of our civilization. Even small amounts of new knowledge in these areas have tremendous impact on our society and its structure. However, each new finding requires tremendous investment of time and resources and the exploration of many dead ends or the production of much unleavened bread. Although a crucial endeavor to the maintenance of our society as we know it, research is a highly inefficient process when compared to other institutional functions. As might be expected, the process of the evolution of new knowledge in science and technology has been studied from almost every disciplinary aspect: history, philosophy, sociology, economics, psychology, management science, and information science. Explanations and theories have been formulated that seem to explain why some advances occur or do not occur and whether these are translated into realizable gains or even harmful outcomes for the society. However, there always seem to exist examples of scientific advances that defy or contradict any specific approach to the understanding of what brought them about.

In Chapter 1, we reviewed the existing nature of the communication system in science, and the way in which the EIES system was designed to supplement and improve it. To summarize, we adopted Diana Crane's image of "invisible colleges" of scientists engaged in research within the same specialty area, bound together by formal and informal channels of communication. Computerized conferencing appears to be ideally suited to increasing the speed and amount of communication and to supporting these geographically dispersed scientific research communities.

---

*Many of the ideas in this chapter were developed in cooperation with other members and consultants of the EIES evaluation team: Tom Featheringham, Ian Mitroff, Nicholas Mullins, Diana Crane, and Barry Barnes. We also thank Joshua Lederberg for his comments. Portions of this chapter appear in "The Impact of a Computerized Conferencing System on Scientific Research Specialties," *Journal of Research Communication Studies*.

Starr Roxanne Hiltz and Murray Turoff, The Network Nation: Human Communication via Computer

ISBN 0-201-03140-X, 0-201-03141-8 (pbk.)

In this chapter, we examine the expected outcomes of the use of CC systems by scientists and engineers. The amount and type of communication and the overall productivity of a scientific user group can be expected to be directly affected if the system is heavily used. In addition, some emergent or "second-order" impacts that are likely include changes in the social structure of the specialty group; the increased salience of scientific controversies; and changes in the overall stage of development of a scientific specialty. Other forms of scientific communication, such as the journal, the grant review process, and the scientific meeting, are likely to change if CC systems become widespread within scientific specialties. Among the unique opportunities that such systems offer are truly international participation in "meetings" and other exchanges of scientific and technical information, and increased access to scientific expertise by nonscientists. In addition to "pure communication," computer aids to a research project itself, such as models, analytic routines, data bases, and bibliographic services, may be made available and easily accessible to scientists through their conferencing system. Despite these potentials for application to scientific communication, there are certain characteristics of the reward structure of science that present incompatibilities or barriers for this area of application. Some of these obstacles are discussed at the end of the chapter. We begin, however, with a look at some of the circumstances that provide the matrix for the potential impacts of computerized conferencing on science and technology.

Problems and Bottlenecks in the Traditional Scientific Communication Process

For the United States, at least, this is a particularly crucial time to improve the size and efficiency of communication nets in science. The decrease in academic opportunities and the "academic crunch" (of stable or declining enrollments, while costs for heat, salaries, Social Security, etc. are rising), combined with the proliferation of scientific specialties and literature, have produced the following problems:

1. *Isolation.* Well-trained, potentially research-productive, recent Ph.D.'s find themselves unable to obtain employment at the top universities and end up at fairly small, isolated liberal arts colleges in the academic hinterland. It is very unlikely that they will find a single other researcher interested in their specialty at their teaching institution; indeed, a recruitment criterion at small colleges is to maximize the total teaching areas of competence within a department by having as little overlap in specialties between faculty members as possible.

This phenomenon is likely to continue if U.S. higher education evolves toward "teleducation," with most students clustered at small institutions or noncampuses, receiving national lecture courses with the guidance of local instructors.

ISBN 0-201-03140-X, 0-201-03141-8 (pbk.)

2. *Immobility*. Senior, tenured faculty find that they cannot move to another institution. The academic/geographic mobility of the past was undoubtedly one way in which new approaches spread from one university research center to another. More and more we must find better ways to move the ideas without moving the people.

3. *Travel*. The costs of travel are rising and the numbers of scientific meetings expanding. Meanwhile, another by-product of the academic financial crunch has been a cutback in such things as funds for travel to conventions, and a new "hard line" on many campuses on the use of the telephone for long-distance calls, and even the mails.

4. *Interdisciplinary Research*. A sizable portion of research efforts financed by the federal government is no longer oriented along traditional disciplinary lines. Researchers in new areas often lack communication facilities appropriate to the cross-disciplinary communication related to their work. Many such research communities have no common professional meeting or society.

5. *Information Overload*. The individual scientist must cope with an exponentially expanding number of scientific and technical publications. He or she must either spend more time on searching and reading and less on original research and writing; increasingly restrict the scope of his or her work to a narrower or more specialized field; and/or find more efficient ways to search and utilize the flood of information. As Rahmstorf and Penniman point out (1977, p. 2), "The capabilities of the individual to solve this overload problem are limited. He has a given time for scientific work and a given capacity to search and read literature. Spending more time on literature reduces the time left for creative and productive work."

The only alternative is to change the current form of dissemination of findings so that it is more concise and selectively disseminated. Nobel Laureate Joshua Lederberg, for instance, has suggested that the productivity of scientists might be significantly increased if there were *fewer* formally published articles and books, since the individual scientist would then not spend so much time fruitlessly wading through the journal literature.

6. *Joint Authorship and Joint Efforts*. Unless authors and/or research team members are in the same location, joint endeavors are rather difficult and time-consuming. Certain types of efforts, like the implementation of a computer model, are almost impossible to undertake unless the key members of the team are co-located. Furthermore, it is somewhat common today to find researchers who have discovered that the fellow

ISBN 0-201-03140-X, 0-201-03141-8 (pbk.)

researchers they most relate to are located elsewhere than their home institution.

7. *Refereeing.* The time delay in getting a paper reviewed and often re-reviewed prior to publication is well known. For many areas of professional activity, this can take a year or more. Likewise, the refereeing or review of a research proposal takes months at best.

8. *Evaluation.* This is perhaps the area that has received the least attention in current efforts at improving scientific information flow and transfer. How often have we retrieved an article based on an examination of title, abstract, and/or index keys, only to discover it was not what was expected? Where was the mechanism for the reader to update the system, indicating an appropriate change in the title, abstract, or keys, so that others would have a better chance of a more relevant search with respect to the particular item?

Ranking in importance with the original article are the later reviews of or reactions to it published elsewhere, or merely passed among the scientific group. Even if published, these are not well correlated with the original reference in most information retrieval systems.

9. *International Access.* Turning to the international level, we find a serious problem of brain drain. The badly needed scientists who are educated by developing nations do not stay there, partly because they cannot stay in contact with developments in their areas that are occurring abroad, or with their foreign colleagues.

In terms of the problems of scientific information, EIES or similar systems will allow a group of researchers to work together on a day-to-day basis regardless of geographical location and individual time constraints, since it does not require the time coincidence of phone conversations. The timely exchange of research findings or views and the resolution of differences can proceed as quickly as desired by the group. Joint authorship becomes a painless procedure with respect to the mechanics of the process. Actual projects can be undertaken by a dispersed team. Refereeing can now involve direct discussions between authors and referees with the latter identified by pen names. Reviews and critiques of published items can be rapidly disseminated. At least, these communications procedures seem feasible within the design of EIES or similar systems. What cannot be stated so firmly is that research groups will actually take advantage of these facilities. Assuming that they do, however, what impacts can we expect upon scientific research groups?

ISBN 0-201-03140-X, 0-201-03141-8 (pbk.)

Potential Impacts of Computerized Conferencing

We believe that there will be some very marked effects of the use of computerized conferencing on the scientific specialties that utilize it. Consider that existing communications structures are either very slow (printed journals), very fitful and expensive (yearly conferences or special meetings), or very exclusive (personal letter, personal visit, or telephone call). Computerized conferencing will enable the members of a user group to keep in constant communication with one another and to exchange ideas and findings on a daily-to-weekly basis, sending and receiving such materials at their own convenience.

Crane's recent review of "The Structure of Science: Implications for Scientific Communication" summarized some proposals for solving the communications problems of scientific specialties as follows:

> Suggested innovations for dealing with communication problems in science fall into four principal categories: (1) changes in some aspect of the formal communication system, such as the creation of a new type of communications outlet or information service, including replacement of the informal circulation of papers in advance of publication by a formal system that would accomplish the same purpose; (2) improvements in arrangements for oral (informal) communications; (3) replacement of the formal circulation of papers in "packages" in the form of journals by a system of selective dissemination tailored to the needs of the individual scientist; and (4) devices to aid the scientist's personal search of the literature or outright replacement of it by a computerized information retrieval system [Crane, 1972, p. 121].

The first three types of innovations are primarily designed to increase the "visibility" of materials in the scientist's own research areas. The fourth would aid the scientist in locating materials in other areas. Computerized conferencing systems can be used in all these ways.

The communications impact of EIES-type systems on scientific specialties is being conceptualized in terms of three dimensions:

1. The total *amount* of information flow among a group of researchers in a field (which may be subdivided into that which is considered potentially useful or valuable, and time-wasting "junk mail");
2. The *content* of this communication, in terms of such categories as
   (a) results of completed research,
   (b) drafts of preliminary results and conclusions,
   (c) descriptions of projects under way (their hypotheses and methods),
   (d) discussions related to the generation of research ideas or proposals,
   (e) discussions of broad methodological, theoretical, or ethical issues or directions in the field,

ISBN 0-201-03140-X, 0-201-03141-8 (pbk.)

(f) "professional gossip" on such things as available funding or positions, or who is going where or who is working on what,

(g) personal (social–emotional) communications among scientists;

3. The total scope or *size* of the communication net—how many different scientists in how many different research groups or geographic locations are sending and receiving information among themselves in each of the foregoing categories?

Our hypotheses are that CC systems can significantly increase the amount of communication among a larger network of scientists, particularly in categories (c) through (g); and that this increase will in turn improve the productivity of the researchers as a group.

Thus, the primary (desired) impact of computerized conferencing is on the productivity of user groups; however, there will undoubtedly be many other impacts on scientific specialties and the process of the production and dissemination of scientific knowledge should such systems indeed become heavily utilized.

The Productivity Problem

A recent review of attempts at developing operational measures of productivity provides a good overview of the various solutions that have been arrived at and their justifications (Blume and Sinclair, 1973, pp. 127–128).

> The only really valid assessment must come from within science; indeed it must come from within the specialty to which a given piece of work is pre-eminently directed, for only a specialist can judge the importance of a contribution to his speciality. In practice, a simple counting of published papers has often been used by sociologists (e.g., Crane, 1965; Gaston, 1969). Much better, and used quite frequently of late, has been the "Citation Index." It has been suggested that the extent to which a paper is quoted by succeeding authors is a measure of its impact on the field, and that this may be equated with its importance as a contribution to science. The use of such a procedure has been facilitated by the development of the Science Citation Index, in which succeeding references to any given paper are listed.
>
> Each paper published by a given individual could thus be "weighted" by its importance and his total "contribution" thus assessed.

It is also possible to use process data generated by scientists using CC systems to augment a simple count of publications and citations, which are partial and inadequate measures, but the only ones that have been available up until now. Using the system, it will be possible to capture in a permanent record the communication and production process itself, with the possibility of developing measures of peer ratings of the quality of contributions, actual

ISBN 0-201-03140-X, 0-201-03141-8 (pbk.)

counts of suggestions and ideas that are adopted by others, measures of contributions as a reviewer, etc.

It has been argued that publication is not an adequate measure of scientific achievement. However, it is difficult to determine the quality of performance of other roles necessary to the functioning of a scientific research community. Observation of scientific communications on EIES may open the door to better understanding of these other activities, and may ultimately provide a basis for their recognition in a documentable manner, as is the case with publications.

A Field Experiment with Energy Researchers

The Institute for the Future has had a field experiment under way that has provided computerized conferencing to scientists in two organizations and a single specialty area. Their approach to measuring productivity impacts has been a process-oriented one and their initial findings do indicate a positive impact on productivity.

The two-year trial of the impact of computerized conferencing on the work styles of energy researchers began in 1975. In a quasi-experimental design, groups within the Energy Research and Development Administration (ERDA) and the Electric Power Research Institute (EPRI) were given the use of conventional and CC media in alternating periods (Vallee and Johansen, 1975). Rather than try to assess scientific productivity directly, the strategy was to ascertain from the scientists themselves those communication and work pattern variables that they felt were related to productivity.

In an initial questionnaire, respondents indicated that they felt the following work pattern variables affected their productivity (Johansen, Vallee, and Palmer, 1977):

1. Communication with researchers in their own organization;
2. Communication with researchers in other locations in their locality;
3. Communication with researchers in different regions in the United States;
4. Communication with researchers in other countries;
5. Communication with researchers in other disciplines;
6. Work at home;
7. Work outside of normal office hours;
8. Reading work-related articles and books;
9. Exchanging letters with other researchers;
10. Using the telephone to talk with other researchers;
11. Traveling for discussions with other researchers;
12. Using a method other than letters, telephone, or travel for communicating with other researchers.

ISBN 0-201-03140-X, 0-201-03141-8 (pbk.)

Preliminary results suggest that the following measurable effects on productivity have occured within groups of the energy researchers as a result of using computerized conferencing:

1. "More flexible working hours, allowing better integration of communications activity into the working day." Nearly 40% of the sessions were participated in by the researchers either before or after the normal working hours (judged by their local time of day). About 10% of the sessions took place between 10 and 11 at night.
2. "Increased contact with distant colleagues."
3. Increased efficiency in the use of other communications media, particularly the use of computerized conferencing to plan and follow up on face-to-face meetings.
4. "More precision in communication due to the constant accessibility of a printed transcript."
5. "Opportunity for greater equality of participation within groups."

Will the System be Used?
The I.E.G.'s as a Less Successful Precursor

The closest equivalent of an experiment like the EIES effort described in Chapter 1 appears to be the Information Exchange Groups (IEGs) set up by the National Institutes of Health in the early 1960s. Price and Beaver (1966) have published some data from the first such group (on oxidative phosphorylation and terminal electron transport). This exchange involved the mailed circulation of unrefereed preprints (about 90% of the materials) and of discussions of circulated materials or other general memoranda. A central office photocopied and distributed the contributions, and any scientist who was a bona fide researcher in the field in question could apply for membership. In terms of use, at least, this innovation appeared to be a success for IEG-1. It grew from 32 members in 1961 to 592 in 1965, and the number of memoranda circulated grew exponentially, doubling every seven months, so that at the time of the study there was an average of one a day (Price and Beaver, 1966, pp. 1011–1012).

However, at least one of the groups (IEG-5, on immunopathology) was unhappy with the consequences of this flood of unrefereed manuscripts. At a meeting of the American Association of Immunologists in 1966, nine points of criticism were raised, of which the following seem the most significant (Dray, 1966, pp. 694–695):

(a) No refereeing process is provided for what is, in essence, a form of publication.
(b) The IEG places undue emphasis on priority. It is thus abused by many

ISBN 0-201-03140-X, 0-201-03141-8 (pbk.)

authors who prepublish most or all of their papers in this form, apparently for this purpose . . . .

(c) One of the principal objects of the IEG has not been achieved, insofar as little or no free discussion has taken place in its pages, judging by IEG No. 5.

The group voted (56 to 39) to discontinue this information exchange program.

Though EIES is very different from the IEGs, its evaluation must start from the premise that the communication facility may be used very differently than intended, and will probably be much more successful in some specialty groups than in others. Any evaluation of this project must thus be concerned with two key issues:

1. Will it actually be used? Many communications innovations, such as the Picturephone, have been dropped after the field trial stage. If some groups or individuals use it heavily and others do not, what explains the differences?

2. If it is used, will the effects be desirable or undesirable? What will be the unanticipated or unintended patterns and outcomes of use? Or will there be no noticeable effects at all, with the new communications medium simply substituting for some current forms of communication?

Among the other specific questions that must be addressed are the following.

1.   Can a system like EIES increase the motivation for or probability of a scientist's contributing ideas for a piece of work that another scientist has in the formative stage?

A standard view of motivation in science has the scientists exchanging "gifts" of published results for the reward of recognition.* One of the norms of science is that once something has been published, those who use it are supposed to acknowledge its source. Informally "helpful" information, typically exchanged at conferences or in conversation, however, frequently is not acknowledged; possibly because the scientist who received the insight or advice forgot its source. EIES will provide the date and time and written record of all suggestions or advice; thus it might become much easier and much more expected that the recipient of such material will acknowledge its influence when the results finally are published. The greater probability of this formal recognition for such contributions to the research of others would, in turn, increase the motivation to engage in this activity.

---

*See, for instance, Hagstrom (1965) and Storer (1966). Nicholas Mullins has pointed out that perhaps if the metaphor is to be applied at all, it is more like a potlatch or a frenzied feeding of sharks than a polite exchange.

ISBN 0-201-03140-X, 0-201-03141-8 (pbk.)

On the other hand, scientists may be very reluctant to make detailed suggestions about the research projects discussed by others, because of lack of apparent reward for doing so; or to enter their own research plans and problems, for fear that these may be "stolen" and published by someone else.

2. Is an EIES-like system conductive to productive consensus-building, or does it merely accentuate fads?

In order for the research of separate scientists to have some sort of cumulative impact on the store of knowledge, there must be some minimal agreement on the most important issues in the field and the most promising techniques. As many studies of science have pointed out, "scientists select research problems not only to satisfy their intellectual curiosity, but also because [they perceive that] other persons in the field consider the problems to be important" (Beres et al., 1975). The increased communication flow made possible by computerized conferencing might facilitate this process of "paradigm building," to use Kuhn's term. However, it might also create a "collective folly" or fad in research efforts for the whole specialty field, diverting the best researchers en masse to an approach or set of problems that might later be seen as a dead end or a trivial issue.

3. What will be the effect of computerized conferencing on the sociometric structure of disciplines?

The hypothesis implicitly stated here is that increased communication will strengthen ties and cooperation among the members of a research specialty. However, the lack of nonverbal cues and the tendency toward cryptic messages in computerized conferencing means that there is a good chance for occasional misunderstandings and resultant hard feelings. Sometimes the participants send a message describing the cause and nature of their hurt feelings; sometimes they stop communicating. The long-term effects of this new form of communication on the quality of social relationships among scientists is another problematical issue to be studied.

Besides these specific kinds of possible effects, however, a heavily utilized CC system should have some much more fundamental second-order impacts on scientific specialties, to which we will now turn.

Fundamental Impacts on Scientific Specialties

Field experiments on the use of computerized conferencing potentially offer a "strategic research site" for directly observing the nature of informal communication within scientific communities, and for exploring its relationship to such factors as the social structure, norms, methods and theories, and productivity patterns that characterize a scientific research specialty.

ISBN 0-201-03140-X, 0-201-03141-8 (pbk.)

Impact on Development and Resolution of Scientific Controversies[*]

We are especially interested in seeing computer conferencing available within research specialties in which there are some basic theoretical conflicts or controversies, with the competing theories each having their adherents. Often this will occur when large amounts of new data or new types of data are becoming available.

Studies by distinguished analysts of science such as Kuhn, Merton, and Feyerabend have established that controversies are a perpetually recurring, if not permanent, feature of science. Such studies also establish that controversies are a vital feature of science in the sense that science is fundamentally dependent on them for the interjection of fresh points of view and the challenging of old established beliefs. In other words, it is expected, in the natural course of the development of science, that scientists of different schools of thought, theoretical persuasions, points of view, and disciplines will develop different hypotheses with regard to the same phenomena. It is also to be expected that some of these hypotheses will clash sharply, since they are frequently based on different ideologies. For this reason, scientific groups are especially likely to be affected by the use of computerized conferencing if they are to experience sharp clashes of opinion within the particular group or the discipline as a whole with regard to an important problem area.

An example should help to clarify what is meant by theoretical, methodological, and ethical controversies within a field, and the way in which these controversies may be further developed and/or resolved through the intensified and more complete kinds of discussions of the issues made possible by a CC system. This example will be taken from descriptions of controversial issues in the area of the modeling of large-scale economic and social systems, obtained through questionnaires and depth interviews with members of a pilot group of scientists on EIES[†], representing the econometric and the serious dynamics approaches to modeling.

Systems dynamics represents a challenge to the "establishment" approach of econometrics. As one participant explains, the econometric approach is

> very highly established—econometric models are used by, for instance, the President's Council of Economic Advisors, and are part of the standard education of every economist. There are differences about appropriate data (or

---

[*]We are particularly indebted to the following members of EIES Group 89 for the interview materials on which this section is largely based: Jay Forrester, Dale Runge, Nathaniel Mass, and James Burns.

[†]It should be noted that this pilot group never did activate more than five members and did not achieve the ambitious objectives described here. The quotations included here are meant as an illustration of the possibilities recognized by scientists who have been exposed to the medium, rather than as a description of something that has already been accomplished.

ISBN 0-201-03140-X, 0-201-03141-8 (pbk.)

ISBN 0-201-03140-X, 0-201-03141-8 (pbk.)

inputs); about appropriate ways of specifying a model (equations); about valid ways for testing a model.

The participants in this particular computerized conference explicitly recognized it as a medium perhaps uniquely suited to the discussion of such controversies. For example, one scientist explained:

> These are almost never touched on in the literature. The literature tends to be a shouting match, if you're doing it at all. The turnaround time in the literature is long, and nobody keeps the responses and counter-responses going. On the other hand, in a face-to-face discussion over one or two days, nobody takes the time to think carefully, and the discussion never really gets to the deeper layer of underlying things. I am hoping, therefore, that what we will see emerge is a clarification of the issues that have to do with where the structure and parameters (of models) come from; what is the nature and meaning of validity. What should models be for anyway? That would be much more thoughtful, much more substantial, than anything I am aware of.

A second participant was asked "What are the things, in rank order, that you hope the conference will achieve?"

> Probably the most desirable outcome would be that a number of the methodological issues between the two camps would be resolved. If that weren't possible, the next most important thing would be to have at least an agreement as to what the most important issues and methodological disagreements were . . . .
>
> I guess the third might be simply the added understanding of what the other person's point of view might be—Like, my getting more of an insight into what the basis might be for a certain kind of criticism.

In regard to validation of models, the econometric approach tends to rely on retrospective, data-based validation:

> A great deal of emphasis has been placed in econometrics on how well a given set of equations is able to replicate real world data over a given period of time. One of the many ways an econometric model is used is to run it for a year and then correct it to what it should be to match the data. "The bottom line" of validation is, given inputs over a period of time, that the model should be able to track very closely what the real world actually did.
>
> Much of that is different from what we would even try to do. In the kind of models we try to build, the problem systems are all generated endogenously. We do not have a lot of exogenous data streams fed into the model and have that drive what the model is going to do. It is supposed to be self contained. It generates the problems that we are trying to deal with itself, just like the real world. Consequently, the underlying purpose of this model that we are trying to generate is so different that the kind of test that we would use to justify it would be very different.

Q. What is the approach to validating your model?

There are the two extremes of, one, have the model's validity rest very heavily on the plausibility of its structure. We devote a great deal of time and effort, in contrast to the econometricians, focusing on the equations themselves. So that's face validity, and that's one criterion. The other is that the model, over a period of time, should display the same behavior that the real world does. Say, if we have a model that is purported to deal with the business cycle, it should display, just on its own, fluctuations that look like the business cycle. You should be able to see phase relationships, like leading and lagging indicators; correlations; you should be able to see those things, in a qualitative similarity—but not detailed.

Another participant expands on the differences in approaches to validating models as follows:

We would say that you should start off looking at each piece of the model. Say you have a model of the way in which the corporation makes capital investment. Does each stage of the process conform to the way in which a manager would go about making an investment decision? Does it assume knowledge of future prices, which he cannot possibly have? Perfect knowledge, which he cannot possibly have in making the present decision? What are assumed to be his incentives? What are the types of information he has available?—We would say that each parameter in the model should conform to something for which you could ask a manager, "Does this sound like it's in the correct ball park?"

So you have a whole range of tests along that spectrum. And then you get into tests of behavior. Of which we think some very discriminating tests involve subjecting a model to very extreme conditions or inputs which may never have been encountered in real life, but for which one could assess the results as to whether or not they would be plausible, given the unlikely input . . . .

It can be a very discriminating test, whether or not a model can really hold up outside of the range of observed data. Because any time you implement a policy, what you're trying to do is move outside the range of past data. A model that simply tracks past data may be a very poor guide to the future, or to what happens when you move outside the range of historical data.

There are some theories or views on the scientific process that hold scientists to be advocates for their theories to the extent of rejecting all contrary evidence until it is overwhelming (see Mitroff, 1974a). This view is very contrary to the commonly held view of the logical and rational scientist. If this "advocate" view is correct for some areas of scientific endeavor, then the use of computerized conferencing as a primary mechanism to exchange scientific information is likely to make the evolution of correct results a more efficient process than it is now. Both the fact that it is a "cooler" medium than face-to-face exchanges and the opportunity it provides for fairly neutral observers or moderators for the advocacy of conflicting views may act to bring about quicker and more rational resolution of differences.

ISBN 0-201-03140-X, 0-201-03141-8 (pbk.)

In examining such controversies, some of the questions that need to be treated are

1. How can a CC system help to bring to the surface initially, illuminate, clarify, and eventually resolve important controversies or key problems within the field?
2. What are the most important theoretical or methodological controversies in the specialty, and how do individual users stand on these issues at the outset and after various periods of communicating with one another via a CC system? How much convergence/divergence is there on basic theories and on ranking of the most important areas for breakthroughs?
3. Can a CC system help deviant viewpoints gain a proper forum? Can it help more viewpoints to be considered? To what extent do participants begin raising new ideas or applying findings discussed in the conference to their own work?
4. Can a CC system help to track the development and course of a scientific controversy? Can it make possible the isolation and measurement of critical events (e.g., observations) that are decisive in tipping the scales in favor of one hypothesis or the other?

Impact on the Overall Stage of Development of a Scientific Specialty

With regard to the development of scientific controversies, it is hypothesized that there will be a differential impact of the intensified communications made possible by computerized conferencing upon scientific research communities, depending on the stage of development of the specialty when computerized conferencing is introduced.

Thomas Kuhn has formulated widely used ideas about the nature of the differences among the sciences, which begin with the premise that a fully developed specialty area has a fully developed and fully shared "paradigm." By this he means an "entire constellation of beliefs, values, techniques, and so on shared by the members of a given [scientific] community" (Kuhn, 1970, p.175). Such a fully developed paradigm includes consensus about accepted theories, preferred methodologies, and what questions are important. Kuhn states that the social sciences are at the beginning of developing a paradigm, whereas the physical sciences have well-developed paradigms. In addition, many studies have shown that the formal communication channels differ between the natural and social sciences; the former utilize journals primarily; for the latter group, books are considered the "important" contributions.

Kuhn also finds several social "stages of development" through which a science will pass. During the preparadigmatic stage there is no well-developed social organization (no "invisible colleges"). During the second stage, when a full-fledged paradigm develops, the specialty grows rapidly, almost exponentially. This is called "normal science" by Kuhn, and is characterized sociologically by groups of collaborators linked together into an invisible college. The question that emerges is, which comes first—the paradigm or the invisible college? More specifically, if a computerized conference system can be used to create an invisible college, will this speed the development of a paradigm in the specialty, with a resulting flood of research breakthroughs?

Finally, in a specialty that is already in the "normal science" stage, with a fully developed paradigm and communication links, will the CC system result in a more efficient coordination of communication and research efforts, or will it instead prematurely speed the transition to the Kuhnian "third stage," greater differentiation, the emergence of increasing controversies and anomalies? (This ends with the Kuhnian "fourth stage," exhaustion of the specialty and decline in membership, not exactly a desirable development from the user's point of view.)

Kuhn's conceptual scheme could be used for tracing the developmental stage of user groups, without sacrificing an openness to alternative forms and sequences of development. Indeed, as Whitley (1976, p. 472) points out, "differences between modes of organization in the sciences could be usefully discussed as part of an attempt to understand how scientific change occurs in a variety of situations." A CC system represents a new tool for social organization. It may result in totally new processes and forms of relationships among scientists and specialty areas.

New Interdisciplinary Specialties

In order for a new specialty to emerge, flourish, and become legitimized in the world of both scientific opinion and grant funding, it has been necessary to go through the organizational pains of establishing a new society and a new journal. For the individual scientist who is likely to be engaged in several scientific specialties both simultaneously and sequentially, this can get to be a time-consuming and expensive proposition. In addition, there are considerable investments of time and expense on the part of those who set up and run the professional societies and the journals.

An example of such an emerging specialty area is the study of social networks. Social scientists sharing a common interest in social networks represent the disciplines of anthropology, communications, human geography, information science, applied mathematics, political science, social psychology, and sociology. They are concerned with the forms or structures of the social relations that link persons together into networks and with the processes through which such networks emerge, evolve, and exhibit consequences for human behavior. Freeman (1977 pp. 2, 3, 12–15) describes the emerging specialty as follows:

> The social networks concept was developed independently in several of the
> social sciences. For early users in each of the several fields it was a loose

ISBN 0-201-03140-X, 0-201-03141-8 (pbk.)

intuitive concept designed primarily to sensitize observers to the interpersonal dimensions of behavior. Increasingly, however, as the networks concept has become explicitly grounded in the theory of graphs and in probability theory, there has been a recognition of the fact that the several earlier intuitive traditions all share a common focus. [This development is reviewed in Barnes, 1972; Mitchell, 1969: and Mullins, 1973.] This recognition has provided a foundation for the emergence of a single interdisciplinary social networks perspective . . . .

Recently, there have been calls for both a journal and a professional society in the area of social network studies. The prospective journal editor made the following argument:*

The reason for considering a new journal stems primarily from the nature of the social networks specialty. It is interdisciplinary; it includes work produced by anthropologists, human geographers, political scientists, social psychologists, sociologists, as well as mathematicians and statisticians. It is international, it is increasingly found in research produced in all areas of the world. And, most important, it is just now in the process of emerging as a recognized field of study that ties together a number of traditional concerns of these several disciplines and geographic areas. Students of social networks do not yet have an organization nor even an established identity as a special field of study. The literature is spread out in too many journals representing too many disciplines and published in too many countries to be accessible. A social networks journal could provide the focal point necessary to the rapid and systematic exchange of ideas; it could provide a common forum for a heterogeneous collection of scholars in order to avoid duplication and to maximize the opportunity for building a cumulative body of knowledge. It could contribute in an important way to the development of the coherence of social networks as a single field of study.

In an almost simultaneous attempt to set up a professional organization, one network specialist wrote the following invitation to colleagues:

Many of us have felt that it would be useful to have a coordinating body to serve as a clearinghouse; linking together network analysts and disseminating current information. The Network's first task will be to publish a directory this summer, listing all members, their addresses, and current research interests . . . . We will also publish two issues of a newsletter this year. We think that the best way to keep members informed of each other's work is to invite members to submit extended abstracts or brief working papers (1–5 pages). Such abstracts will facilitate quick communication among interested colleagues without precluding later publication of more formal papers . . . .†

---

*Letter from L. Bergman, Managing Director of Elsevier, May 1977.

†Letter from Barry Wellman, Principal Coordinator of the International Network for Social Network Analysis, May 1977.

ISBN 0-201-03140-X, 0-201-03141-8 (pbk.)

What is interesting about this call for organization is that not only could an EIES-type facility perform the directory and short-paper function, in addition to substituting for professional meetings, but that the fledgling specialty also recognized this. A formal proposal to use the EIES system was submitted at the same time as the calls for the journal and the organization were emerging. The hoped-for impact of EIES was described as follows (Freeman, 1977, pp. 12–15):

> Social networks represents a genuine departure from any traditional mainstream of social scientific thought. Social scientists have always talked vaguely and intuitively about concepts that are probabiy intended to relate to relations among persons—concepts like social structure and social role. But in conducting actual research, they have always concentrated almost exclusively on the traits and characteristics of individual persons. The social networks outlook on the other hand explicitly focuses attention on the attributes of the relationships among people rather than the properties of individual persons; we are finally systematically exploring the *social* part of behavior . . . .
>
> The potential impact of the system of electronic information exchange on the social networks community is very great. Students of networks are currently at a critical point in their development as a community. Their members are geographically dispersed and are grounded in a large number of traditional disciplines. Moreover, they are only recently aware of the identity of the tools and perspectives that have emerged in the context of the several traditions they represent. Social networks scholars are, at this moment, in the process of developing an identity as a single research community . . . .
>
> So far the networks community has developed no formal organization nor standard publication outlets, although both are being discussed. Communication among its members is slow, expensive, difficult and frequently restricted to a small subset of colleagues who happen to be geographically proximate or previously established as collaborators. Individual productivity, therefore, may be expected to increase markedly in a milieu in which communication is immediate, regular and fairly effortless. The stimulation provided by easy communication with one another should result in better morale and more and more successful efforts on the part of most participants.
>
> In addition, the facilitation of regular information exchange can be expected to yield a noticeable improvement in the coordination and cumulation of the efforts of the community as a whole. Conflicts can be aired and independent ideas related in a setting that encourages collective effort and the development of a sense of community. If the proposed project is successful, therefore, it seems reasonable to expect that a major new organized intellectual force will emerge in social science.

A fundamental difference between traditional forms of scientific organization and communication and a conference-based system of exchange is that the latter makes it much easier for scientists to move in and out of a specialty area. We will now turn to some of the ways in which a more fluid social

ISBN 0-201-03140-X, 0-201-03141-8 (pbk.)

structure within specialties might emerge. What is most significant is that a specialty group within a computerized conferencing environment has with its creation all the benefits any other established group has. There is no pressure to have to "progress" or evolve to become a formal society or produce a hard-copy journal. Therefore, groups can be much more fluid in terms of forming and dissolving.

Social Structure of the Scientific Specialty

Another type of general issue that might be explored is the impact of EIES on the size of the scientific group and internal prestige ranking within the specialty. In regard to the size of the group of actively communicating and working scientists within a specialty, for example, will EIES condense the research specialty, so to speak, into a smaller core group, with those not in the system completely cut off? Or will the increased ease of communication within this core facilitate expansion through: circulation of some of the printed output; invitations to "observers" or "visitors" to take part occasionally; the freeing of time to do more letter writing and manuscript circulation to more people; and/or the facilitating and inspiring of specialized face-to-face conferences to which a general invitation is extended?

Scientific research communities are not only networks of communication, they are also stratified social systems that allocate prestige and opportunities. For example, as Price and Beaver (1960, pp. 101–117) described their concept of invisible colleges:

> The basic phenomenon seems to be that in each of the more actively pursued and highly competitive specialties in the sciences there is an "in-group." The people in such a group claim to be reasonably in touch with everyone else who is contributing materially to research in this subject not merely on a national scale, but usually including all other countries in which that specialty is strong. The body of people meet in select conferences (usually held in rather pleasant places), they commute between one center and another, and they circulate preprints and reprints to each other and they collaborate in research. Since they constitute a power group of everyone who is really somebody in a field, they might, at the local and national level, actually control the administration of research funds and laboratory spaces. They may also control personal prestige and the fate of new scientific ideas, and intentionally or unintentionally they may decide the general strategy of attack in an area.

Two interesting inadequacies of the invisible college structure are immediately obvious. First, for those who are "in," the existing communications network is so time consuming, sporadic, and slow that only a few of the many questions, answers, and comments that might fruitfully be exchanged actually are. Second, what about those potentially productive scientists who are "out"? An analysis of productivity patterns of chemists (Reskin, 1977, p. 491)

ISBN 0-201-03140-X, 0-201-03141-8 (pbk.)

suggests that "collegial recognition is particularly important for chemists in contexts that do not stress scholarly publication."

The system as proposed could well make current in-groups more effective. It also would allow the rapid formulation of communities that do not now exist. A group of younger unknown researchers could form their own peer group independent of the "established" in-group. Moreover, the system could allow new modes of interaction between "elites" and "newcomers" (see Mulkay, 1976b, for one view of current relationships).

Thus an issue of interest is the question, which type of scientist can be most aided by such a system, those who are already part of a highly productive elite within a specialty or those who are currently cut off from opportunities for extensive communication and cooperation with others in the field? At present, the academic community is very much a stratified one, with those scientists who are located at the top universities having a much greater opportunity to be productive and gain recognition, because more time, money, and equipment is available for research, and because of the greater likelihood that their academic affiliation will automatically get them included in an existing communication network (see Cole and Cole, 1973; Zuckerman, 1977). This is an example of what Merton (1968) calls the "Matthew effect" in science, quoting from the Book of Matthew:

> For unto everyone that hath shall be given, and he shall have abundance: but from him that hath not shall be taken away even that which he hath.

Crane (1965) provided empirical documentation for the unequal distribution of opportunities for productive research. She concluded that

> scientists at major schools are more likely to be productive and to win recognition than scientists at minor universities, which suggests that universities provide different environments for scientific research . . . [in part] probably because the major university provides better opportunities for contacts with eminent scientists in the same discipline [p. 699].

Blau (1976) suggests that the Matthew effect is particularly strong when scientists relate to one another across international or disciplinary boundaries.

Allison and Stewart (1974) have used cross-sectional survey data to provide further evidence that, at least for chemists, physicists, and mathematicians, getting off on the wrong foot can severely lessen the opportunity to ever have the contacts and resources necessary to be "productive" in terms of research. They summarize their findings as follows.

> The highly skewed distributions of productivity among scientists can be partly explained by a process of accumulative advantage. Because of feedback through recognition and resources, highly productive scientists maintain or increase their

ISBN 0-201-03140-X, 0-201-03141-8 (pbk.)

productivity, while scientists who produce very little produce even less later on. A major implication of accumulative advantage is that the distribution of productivity becomes increasingly unequal as a cohort of scientists ages [p. 596].

It is possible that a CC system can provide equality of opportunity among "unknown" researchers affiliated with small colleges and those at major institutions. It is also quite possible that the researchers at small institutions would benefit more in terms of productivity from the increased stimulation due to improved peer-group communications.

Thus far, we have been discussing impacts on the internal structure of scientific research communities. In studying these, one opportunity that is opened up for sociologists of science who play such an assessment role within a specialty group that uses a CC system is to get "inside" the informal scientific communication process. As an observing member of a group and its conferences, a person does not need to rely on answers to questions about communication behavior, but can actually capture it, with the aid of the computer. In essence, the participant-observer is afforded the opportunity to do a kind of ethnography of a scientific user group on such a system.

This idea of an ethnography of a scientific specialty may seem a bit odd, but it simply entails a long-term exploration of the ways in which members of a developing research specialty think and work, conducted by participant-observation and unstructured interviewing of the members. Arguing for the fruitfulness of this new way of "doing" the study of science, Clifford Geertz has said (1977, p. 27):

> If there is to be any genuine ethnographic knowledge of the disciplines as forms of life or whatever, it is only going to come about by a lot of patient, flexible, responsive, unpressing, and delicate interviewing of those who, though they may not be aware of it, know what it is we want to know.

Geertz has further employed an anthropological metaphor by referring to a community of scholars engaged in studying the same phenomena and using the same approaches as much like a "village," in terms of consisting of a fairly small number of persons, connected by a complex network of relationships, and sharing a whole "way of life." However, this scientific village differs from the village of, say, Moroccans or Balinese in that it is not physically located in one place. Whereas the ethnographer usually studies a group by "moving into the village,"—listening to and watching interactions, and interviewing informants face to face over a long period of time, these techniques will have to be altered when studying a scientific community or "village." He or she who would hope to gain understanding must somehow "move into" the flow of communication, at strategic sites or times when norms, values, and exchange patterns within the community are likely to be made manifest.

ISBN 0-201-03140-X, 0-201-03141-8 (pbk.)

It is in "boundary interchanges" with "outsiders" that the core norms and values of a scientific discipline or a community of any type are likely to be made manifest. Observation of such communications and exchanges can be made through "participant observation" in a computerized conference in which the members of a discipline argue the fundamentals of their theoretical and methodological beliefs with proponents of competing specialties or approaches.

Methodological Weaknesses of the EIES Field Experiments*

As far as the authors are aware, the EIES project is a unique approach to studying factors relating to the organization and productivity of scientific specialties: actually changing the communication modes of several specialties, and then figuratively sitting inside the communications network to observe what happens. It is recognized, however, that this field experiment will distort and fail to measure what might actually occur should computerized conferencing become a "normal" widespread, nonexperimental mode of communication.

A NEW TECHNOLOGY IS LIMITED TO A FEW GROUPS

One analogy that might be made is to the situation when telephones were new and owned by only a few persons. Just as people used to have to shout to be heard over long distance and much static was commonplace, a few technological kinks in the system, which may discourage and frustrate users, can be expected in the beginning.

Second, the scientist-users will have to resort to other communication modes for other roles they play and their associated communications. Eventually, terminals in the home and the use of computerized conferencing might become as cheap and widespread as TV ownership is at present. At that point, a person could belong to many "conferences," corresponding to all her or his roles—a family news conference, for example, and a chess conference. For the duration of this field experiment, however, only the approximately 300 scientists on the system will be able to be reached by computerized conferencing. As a result, use of the system will have to be added on to use of other communication modes rather than replacing much of their use. A related factor is that for system planning purposes, the specialty group's ability to expand to include new members on the system has been arbitrarily limited during the course of the experiment. If computerized conferencing

---

*Several of the ideas in this section benefited from a discussion with Joseph Martino of the University of Dayton Research Institute.

ISBN 0-201-03140-X, 0-201-03141-8 (pbk.)

were a generally available service like the telephone, any number of additional persons might join the network. Still another factor related to the newness and scarcity of the technology is that many of the scientists might never before have used a computer terminal and might not have any other use for it; thus, the learning might be somewhat annoying. Furthermore, since the user will not generally have a terminal both at home and in the office, he/she must take the trouble to carry it around if it is to be available at all times. If the day ever comes when terminals are as omnipresent as TV's, they will always be conveniently at hand without foreplanning, and used with as much frequency and ease as more familiar household appliances are now.

THE HAWTHORNE EFFECT

The scientists in this study will know that they are being observed. They will also know, from questionnaires they receive and from announcements of the project, what variables are being watched. This awareness cannot help but affect the behavior of the persons involved. They may tend to be self-conscious about what is entered into the system, knowing that "big brother" evaluator may be out there somewhere reading the transcript. They may deliberately distort their questionnaire responses to tell the evaluator what they think she/he wants to hear, or conversely to perversely "mess up their results."

LONG-TERM EFFECTS

In current experimental situations, scientific groups are only given access to EIES or other CC systems for a year or two. However, the development of a new scientific concept or the transition from hypothesis to proven "fact" may stretch over time frames of a decade or more. In addition, short-term recognition of the value of a contribution tends to be conferred by peers within an invisible college, but long-term recognition is more likely to be determined by the users from outside the invisible college as well. Thus, it is difficult to determine the impact of the use of a CC system on the quality of work by a group of scientists within a short period of time.

Impacts on Other Forms of Scientific Communication

It might be presumed that teleconferencing, as a substitute for travel, would cut down on the number of face-to-face visits among scientists. However, quite the contrary seems to be the case, based on observations of the initial groups using EIES. Geographically separated scientists who "meet" on EIES and discover common interests tend to plan their subsequent trips so as to provide an extra half-day or day-long extension during which to meet

ISBN 0-201-03140-X, 0-201-03141-8 (pbk.)

with their new colleagues. Likewise, other existing forms of scientific communication are more likely to be supplemented or modified, rather than replaced, by computerized conferencing.

Scientific Journal Production

We have described the possibility of "on-line" journals published on CC systems. In addition, however, these systems might be used to make the publication of traditional journals more efficient, by serving as the communications network for editorial processing centers (EPCs).

Preparing for innovations predicted in papers such as "A Concept for Applying Computer Technology to the Publication of Scientific Journals" (Bamford, 1972), Aspen Systems Corporation and Westat Research Incorporated have conducted a study to assess the feasibility of editorial processing centers (see Berul and Krevitt, 1974; Aspen Systems Corp., 1974). The EPC concept involves a center that computerizes six fundamental areas of the journal production process: authorship, editing, reviewing, redaction, proofreading, typesetting, and business management. Such an EPC would be shared by many journal publishers. The feasibility report recommended an incremental plan whereby the six functions would be brought into full automation over four phases of increased technical sophistication.

The computerization of two of these areas—authorship and reviewing—was conspicuously postponed until the last phase of the feasibility plan. Presumably, this delay is due to an assumption of reliance on optical character-recognition technology for transferring original manuscripts into machine-readable form, and reliance on hard-copy delivery systems to and from the article referees.

A computerized conferencing facility obviates those assumptions. Many of the subfunctions that the Aspen study delineated under the authorship function can be easily implemented on a conferencing system. These are transcription of the manuscript; peer editing and review prior to submission for publication; manuscript proofreading and changing; making copies of the manuscript; and dispatching the manuscript to the required entity (the EPC). Similarly, the distribution of manuscripts to the reviewers, their review and annotation, the return of the manuscript to the editors, and the synthesizing of comments are equally performable via the conferencing system. Save for the special problems of graphics and format standards, the manuscript is then ready for automated redaction, typesetting, printing, and labeling.

Potential Effects on the Grant Process

Informal peer review and formal refereeing of journal and conference articles are not the only systems of scientific review. There are, additionally,

ISBN 0-201-03140-X, 0-201-03141-8 (pbk.)

procedures used to review proposals for scientific funding and advisory boards to government agencies, legislative bodies, and corporate research and development programs. Computerized conferencing could be used to facilitate these processes as well. However, whether this is politically feasible and rational is an interesting question.

As Derek Price pointed out in *Little Science, Big Science* (1962), scientific research is no longer an activity carried out by lone scholars with a few test tubes or rats at their disposal, but rather has become a large-team, expensive process. To make progress in most fields requires research assistants, computer or laboratory facilities, and extensive expenditures for clerical support, travel and communication costs, and supplies. In other words, it requires a grant or contract. The process by which the National Science Foundation, National Institutes of Health, and other sources of support for scientific research select projects and scientists for funding has increasingly determined the areas in which it will be possible for scientific progress to occur.

AN ON-LINE GRANT REVIEW PANEL

The basic argument for peer review processes, whether of grant applications, in the form of refereeing prospective journal articles or books, or in determining scholarly merit for promotion or tenure, is that scientific research has become so specialized and rapidly changing that only other scholars in a specialty are capable of judging its quality. For instance, Ravetz (1971, p. 274) argues that the skills and knowledge vary so much from field to field and the details of the techniques by which good research is produced

> are so subtle, the appropriate criteria of adequacy and values so specialized, and the materials so rapidly changing, that any fixed and formalized categories (for rating quality by outsiders) would be a blunt and obsolete instrument as soon as it were brought into use.

Peer review is seen as the only possible mechanism for maintaining "quality control" in science. The very persons who are the most eminent researchers are also called upon to be the judges of the quality of research carried out by their colleagues. Mulkay (1977, p. 107) describes the social control process by peer review as follows:

> All participants, as they use, modify or disregard the results communicated to them, are continually engaged in judging the adequacy and value of their colleagues' work. As a result of these judgments, recognition is allocated and reputations are created; not only for individual researchers, but also for research groups, university departments and research journals.

In their study of the functions of the referee system for journal articles, Zuckerman and Merton (1971) point out peer review processes also increase "the motivation to maintain or to raise standards of performance" in a

ISBN 0-201-03140-X, 0-201-03141-8 (pbk.)

scientific field, since any weaknesses in an article (or proposal) may result in not only a rejection of that piece of work, but also a decrease in the author's prestige in the eyes of his or her peers in the specialty area. Peer reviewers who conscientiously do their work also serve an important function for authors or principal investigators.

> They can, and often do . . . suggest basic revisions for improving [work]. As for the reviewers themselves, especially in fields without efficient networks of informal communication or in rapidly developing fields, referees occasionally get a head start in learning about significant new work [Zuckerman and Merton, 1971, pp. 96–97].

Thus, there are several functions that the peer review process performs, and that must be examined as criteria for supposed "improvements" in the process:

1. The knowledge that there will be a peer review serves to motivate scientists to pay close attention to developing theoretically and methodologically valid proposals, and to discourage casual submission of proposals that are not well thought out.
2. The peer review should help to maintain the quality of funded scientific research by advising program managers to reject proposals that are inadequate according to the current standards of the field, and by assessing the relative merit of research projects that are competing for the same funds. It should also prevent the funding of studies that have essentially already been done.
3. The review process should result in constructive suggestions to the principal investigators, so that the quality of research that is funded will be higher.
4. Taking part in the process should serve as a channel of communication that helps to keep active researchers aware of the current developments in an area of research.

The peer review process has come under criticism from both within the scientific community and without for its slow and seemingly arbitrary nature (see, for instance, Walsh, 1975; Gustafson, 1975). Putting the reviewers for a grant proposal "on line" for a review discussion (using pen names) that also involves the principal investigator at one stage might resolve some of these difficulties. Our model of the process is that each panel member could first enter her/his individual reviews and ratings in full into a conference; then all would discuss and comment on apparent inconsistencies and oversights among their individual responses and be given a chance to revise their comments or ratings. Then the principal investigator could be allowed into the conference to respond to the criticisms and ratings. Lastly, the reviewers

ISBN 0-201-03140-X, 0-201-03141-8 (pbk.)

would make a final set of ratings and comments, based on the revised and refined proposals made by the principal investigator.

The usual process at this stage is for a number of reviewers to be sent a copy of a research proposal, and to respond individually in writing. Some of the problems that might be solved are the following:

- Each reviewer frequently has expertise in an area that only partially overlaps with the full scope of the research proposal. Reviewers feel qualified to comment only on the sections of the proposal that most match their own areas of expertise. However, they are asked to rate the proposal as a whole. If they could have some communication with other reviewers who claim expertise in complementary areas relating to the proposal, they would feel more competent in making such an overall rating.
- Reviewers tend to be among the best-known and thus busiest persons in their specialty areas. Given their overload, they have a tendency to skim verbose proposals somewhat, and may miss or forget certain points. Sometimes their reviews criticize a proposal for omitting some crucial consideration that is actually included. Their ratings of the overall proposal may be predicated on this oversight; yet, the proposer has no chance to respond and have the rating adjusted accordingly.
- Reviewers frequently make mutually exclusive or contradictory criticisms or demand for revision in the proposal. For example, reviewer A might state that a questionnaire is too long and must be cut. Reviewer B might state that a questionnaire (or other measurement device) is good as far as it goes, and should keep all that is proposed, plus adding another long set of items. The scientist is supposed to come up with a "compromise" that will satisfy both, without the reviewers' being fully aware of the contradictory demands, and without anyone really having a chance to discuss the underlying issues and disagreements.
- For the reviewer, the process is somewhat of a lonely duty. There is no pay for this role, and frequently there is little intrinsic enjoyment in the process, either. If a reviewer could engage in a lively debate with her/his colleagues (fellow reviewers), the process would be much more appealing, and enjoyable, and more likely to be a form of "continuing professional education" for the scientists who take part. The fact that on-line discussion was taking place ought also to increase the motivation to meet deadlines for a timely and speedy review of the proposal itself.
- The total task of a funding agency is not only to rate and respond to each individual proposal to a program, but also to rank-order them and select the best set in terms of overall level of funding and total program objectives. In the past, the only way in which it has been possible to have a review panel engage in this overall ranking of a number of proposals has

ISBN 0-201-03140-X, 0-201-03141-8 (pbk.)

been to call them all to foundation headquarters. It would be relatively easy in the computerized conferencing form of grant review simply to add another conference (another step) to the process. After each individual proposal had been rated and responded to, the panel could then recommend the total set of proposals to be funded by discussing and rank-ordering those that had been found to be meritorious at the individual-proposal-review level.

FUNDING FOR "STATE OF THE ART" REVIEWS

It seems easier to get a grant to do a narrow empirical study within a small specialty area than to try to tie together all of the recent empirical and theoretical advances within a specialty and related areas. One reason that this is so is that no small group at a single location is likely to include the multispecialty expertise needed for a broad review. With a computerized conference that included some strong structuring of the division of labor and interaction process, it would become much more feasible for a fairly large team to produce collective reviews of a field of inquiry, complete with recommendations for the most promising areas for research in the near future.

Professional Meetings

People who have experienced teleconferencing frequently report that it substantially aids them in organizing face-to-face meetings when they eventually occur. Whether the meeting is concerned with administrative or scientific substance, a CC system assists in ironing out the logistics and setting the agenda.

For the professional meeting, a telecommunication system can be used for early and efficient submission, refereeing, and publishing of the proceedings. This also affords the attendees the opportunity to read and reflect on the papers in advance of the actual meeting. Discussions of technical issues can then increase in substantive complexity and allow for wider evaluation and participation.

The prospect of computerized conferencing being used as an adjunct to, or in some cases a replacement for, the scientific conference has wide implications as to how such meetings are structured or who might attend. The very role of the professional conference may be altered within scientific societies and other sponsoring agencies. One particularly promising possibility is that scientific meetings might be opened up to remote participation by scientists from less developed nations, who would then feel less in need of migrating.

ISBN 0-201-03140-X, 0-201-03141-8 (pbk.)

International Conferences for Scientific Exchange and Technology Transfer

One of the most exciting possibilities of teleconferencing in general is the international conference to which participants need not actually travel. Until now, scientific communities have been almost entirely intranational, because of the cost and difficulty of communicating across continental lines. Not only are telephone calls expensive, but time zone differences make it difficult to select times for conversation that are convenient to participants whose days are out of phase by 8 to 12 hours. As for "international" conferences, language differences and the expense of traveling mean that they are in fact dominated by the nation of the host country and surrounding nations.

Use of a CC system would mean that scientists from around the world could be kept up to date on the proceedings of a conference in their field, and could feed in questions and comments to on-site participants. One example of this kind of use was a recent United Nations conference on water, in which a nonprofit organization called Hot Line played such an intermediary role. The potentials in this area of technology transfer were also recognized recently by the U.S. State Department, which has called for proposals to study the feasibility of using teleconferencing to facilitate technology transfer from developed to developing nations—though, strangely, it defines teleconferencing as video conferencing only rather than as computer-mediated transfers of data and communications.

Science, Society, and Computerized Conferencing

There have been many developments during the last several years that signal a new set of demands on science and scientists for social accountability and responsibility. Among these are public debates about potentially harmful research; publicized disagreements among scientists about the correctness of opposing theories; the growth of fields that place science itself as the object of investigation, such as the sociology of science and technology assessment; and the call for such institutional reforms as "science for citizens" and a "science court." In this section, we are going to try to trace some of these developments and to suggest how a CC system might be employed as kind of an "outward-oriented" interface between scientists and the public.

The Growth of Social Studies of Science as an Aspect of the Crisis of Confidence in Science

By the end of the Manhattan Project, which might in retrospect represent the end of the Golden Age of faith in science, scientists themselves, led by men like Oppenheimer, began to question the products of their work. For a

ISBN 0-201-03140-X, 0-201-03141-8 (pbk.)

decade or two after World War II, society at large seemed enamored of "Atoms for Peace," "the race to the moon," and the technological offspring of science, such as overpowered automobiles and color TV's.

By the beginning of the 1960s, however, the student–radical–intellectual movement that was eventually to encompass the environmental, anti-war, civil rights, and anti-big-business-dominated government groups began to gain strength and media coverage. Science began to be viewed by many young social scientists influenced by these movements as an institution that created the technology that was polluting the environment and providing the means for napalming children, and that spent billions to plant an American flag on the moon while our cities became dehumanizing wastelands. Incidents such as the thalidomide babies warned the public that apparent advances often had unanticipated consequences.

The burgeoning literature on science as a socially created, and therefore potentially very fallible, institution is perhaps both an outgrowth of this crisis of confidence and a further contributor to it. Kuhn's *The Structure of Scientific Revolutions* (1962, 1970) is one of the earliest and most influential of an explosion of scholarly studies of the history, philosophy, sociology, politics, and psychology of science and scientists. It is our belief that one reason why Kuhn's book was so influential is that the intellectual and social climate had prepared scholars to shed the myths and awe that had made "hard science" almost a new kind of religious object, and to coldly and critically dissect science and hold it under a microscope to see what it was really made of. From this point of view, Kuhn's book presents an alternative to the positivistic, empiricist view of science. Kuhn argues that there can be no sharp separation of "fact" from "interpretation"; no sense in which newer theories are "right" and older theories "wrong"; they are just based on different shared paradigms.

Other studies that might be seen as reflecting this new view of science* by social scientists and humanists include Reif's 1961 *Science* article "The Competitive World of the Pure Scientist," which attacked traditional stereotyped pictures of the scientist "motivated by intellectual curiosity and immersed in his abstract work, . . . oblivious of the more mundane concerns of ordinary men . . . " and held that "the pure scientist, like the businessman or lawyer, works in a social setting, and like them, he is subject to appreciable social and competitive pressures." It included this devasting quote from the editor of *Physical Review Letters*: "We do not take kindly to attempts to pressure us into accepting letters by misrepresentations, gamesmanship, and

---

*The supposed norms of science (universalism; objective, disinterested weighting of evidence, etc.) are questioned by Barnes and Dolby (1971), Brush (1974), and Mitroff (1974a, b), among others.

ISBN 0-201-03140-X, 0-201-03141-8 (pbk.)

jungle tactics, which we have experienced to some (fortunately small) extent" (Reif, 1961, p. 1957).

Price's 1962 *Little Science, Big Science* helped to replace the image of the isolated, poverty-stricken individual scientist with a view of scientific research as a new kind of "big business," with membership in the "power elite" for its most successful practitioners.

Public Debates

Controversies over the ethics of scientific research have begun "escaping" from debates within the scientific community to editorials, front page stories, and other mass media coverage, drawing the public into the debate and increasing their suspicion of scientific research. Some examples are the research on syphilis, for which black men in the South were left untreated over decades; and the ill-fated "Project Camelot" in South America, which involved social scientists in counterinsurgency research.

The most dramatic and publicized of the recent aspects of both internal and external crises in science has been the controversy over gene splicing. It began within the scientific community, which discussed the dangers of a runaway epidemic as a result of "escaped bacteria" from a recombinant DNA experiment. At a June 1973 Gordon Conference, a letter was drafted calling for the National Academy of Sciences to explore the (potentially dangerous) implications of research with recombinant DNA. The National Academy passed the task along to a committee of scientists that convened at MIT in April 1974. As recently reported (Bennett and Gurin, 1977, p. 48),

> An hour after the group of eight scientists convened at MIT I agreed that an international conference should be held as soon as possible. The conference could not, however, be scheduled before the following February, and by that time many questionable experiments might be under way.

Partly as a result of being unable to schedule a conference when it was needed, they published a letter in *Science* and in *Nature*, calling for a moratorium on experiments that might cause risks, such as improving the antibiotic resistance of bacteria. As a result, a public furor began to break out, with headlines like "Bid to Ban Test Tube Super Germ"; and opposition began to grow, reaching a peak at the hearings in Boston at which Harvard University accepted a moratorium on certain kinds of research in order not to endanger the community.

Public hearings and letters in magazines are hardly the way to enable scientists to manage their own affairs in a timely and efficient manner. If there had been a way for knowledgeable and concerned scientists to exchange ideas, data, and opinions on the issue without having to schedule a large face-to-face international conference in the future in order to accommodate

ISBN 0-201-03140-X, 0-201-03141-8 (pbk.)

busy schedules, then much of the debate that frightened the public and raised opposition to any university research on recombinant DNA might have been avoided.

A Computerized Science Court

At present most scientists realize that very often they will find themselves in the difficult position of taking a stand on scientific matters before a conclusive set of scientific evidence has emerged. Many recent examples come to mind: the debates on the workability of an anti-ballistic missile system, the SST, the effect of fluorocarbons on ozone in the atmosphere, and the fast breeder reactor. In all these situations there were able and honest scientists who could and did lend scientific views and evidence to support contrary positions. It is not that science is any less rational; opposing viewpoints have always abounded. It is a new fact of life that scientific views are now sought for rather pressing decisions, and that controversies are now public knowledge.

This has led to the possibility, often discussed in the literature, of a "science court": a vehicle for the orderly comparison and weighing of scientific knowledge that may support more than one position on a societal issue or policy decision, and which would act like a court to resolve technical issues pertaining to policy.

Although the legal status of such a court is the focus of much controversy, it is easy to see that a CC system would allow for a very comprehensive collection of evidence from geographically dispersed experts at a fraction of the cost of their assembling in a courtroom. This possibility removes many of the arguments against the logistics and economic feasibility of such a court.

Short-term use of scientific consultants might become much more widespread once a group of scientists is using computerized conferencing. It would be very easy for one or more of them to respond to a request to serve as an on-line expert, or consultant, to a government agency or other individual or organization in need of advice. Since a consulting role could be taken on without the time and expense of traveling, it is much more likely that the scientist would become involved in such temporary assignments.

Access to Scientists for Citizens

A related extension is the use of on-line scientists to serve as expert consultants to the public in scientific debates, as well as to business or government. It would be a simple matter for scientists to respond to a kind of bid-and-barter intellectual marketplace. Persons in need of an answer to a scientific question could list it in a particular public bulletin; scientists could respond with one- or two-paragraph proposals of the kind of answer they are prepared to give, and a proposed price for their services.

ISBN 0-201-03140-X, 0-201-03141-8 (pbk.)

There could also be a "public service" bulletin in which citizens who could not afford to pay for such advice or information listed their questions, and a scientist's answer could also be displayed publicly. In responding to such requests from lay people, the CAI-style text with embedded questions about whether the reader wants more detail at a certain point would be particularly useful. At the first level, the scientist could offer a simple generalization or explanation; at the detailed level available on request, the proof or argument in terms normally used by members of the specialty in their communications with one another could be utilized.

Currently, it is very hard for scientists to engage in public service consulting, because citizen groups lack the money to pay travel expenses or to offer an honorarium for full days' services. The public-service-minded scientists available in any local community may not have the expertise needed for a particular problem. A system of this sort means that citizen groups can contact a scientist anywhere in the nation; also, they need not know specifically whom to contact, since the system will provide the contact. Second, the scientist can fit in a few hours of voluntary consulting at his or her convenience without the necessity for travel.

Management of Science and Technology

Attempts at developing a management model and management practices to guide the development of science and technology have more often failed than succeeded, and many times the same approach will fail in one situation and succeed in another. All attempts and their performance still seem to fall back on the knowledge, intuition, and subjective judgments of those in policy and decision positions. However, the dramatic change that has occurred over the years is the increasing "distance" and difficulty in maintaining active communications between those who set policy and make decisions and those who carry out the process of scientific development.

The increasing complexity of institutions, the rules and regulations under which they operate, the increasing need for public justification, and many other factors eat away at the time formally available to maintain the ties needed between those in management positions and those carrying out the work. Computerized conferencing can be a mechanism to gain back some of that lost communication ability. However, it can only be a delaying action if these other trends continue at the pace they have been exhibiting during the past decade. We do not have any way of quantifying scientific advance. There is no sure model that will say X dollars will produce Y amount of advance. It still requires the careful examination of exchange of information between those seeking the advance and those who undertake the research as to the options, the uncertainties, and the details. Contrary to this need, today we see a tendency to move the decisions farther and farther up the ladder in

ISBN 0-201-03140-X, 0-201-03141-8 (pbk.)

institutions, to the point where those making the final judgment are often far removed from having the knowledge of the situation to support a wise decision.

Computerized Assistance to the Research Process

In addition to "pure communication" components, a CC system can enable scientists and engineers to develop and/or utilize models, simulations, data bases, analysis routines, abstract services, and any other electronically storable aids to the research process. We stress "develop" as well as "utilize," since there should be feedback between users and designers of the computer-accessible materials. For example, let us look at how computerized conferencing might be associated with models and with abstract services.

Modeling and Simulation

Computerized conferencing can potentially play three key roles in the area of improving the utility of modeling and simulations. The first is in the formative process, where a wider range of expertise can be brought to bear and the opportunity to provide better interfaces between modelers and decision makers presents itself. The requirement here is for data processing tools to make the specification of model assumptions and structures easier and to analyze, for feedback purposes, inconsistencies and differences of judgments among the discussants. Currently at the New Jersey Institute of Technology one such tool being incorporated in EIES through Hal is Warfield's Interpretive Structural Model (ISM) for producing graphic relationships among interacting concepts.

The second area is in the actual execution of a simulation. The objective here would be to emulate the real-world communication and decision processes associated with the system being modeled. The current assumption that such decision processes can be automated or modeled through such techniques as optimization methods has usually proved to be fallacious in simulations for planning and policy formulation applications. No mathematical simulations today really account for all the factors that humans take into account when they actually make a decision or formulate options.

The third area is the evaluation and assessment of what the model produces in the way of results. It represents both the problem of interpretation and that of validation of the simulation. Once again a great number of potential aids incorporated into the communication process could be useful in various situations of this type.

ISBN 0-201-03140-X, 0-201-03141-8 (pbk.)

Accessible and Evaluated Data Bases

More generally, computer-stored data bases and analysis routines are the basic tools for many scientific disciplines. Computerized conferencing systems through such links as the EIES microprocessor Hal, described briefly in Chapter 1 and in more technical detail later in Chapter 10, can make available to a group of scientists any resource anywhere that is stored on a computer. In addition, it would be possible to use the conferencing component of the system to store comments, evaluations, and suggestions by one scientist, attached as footnotes to parts of the data base, in order to share experience in its use by persons other than those who originally put it together.

A View of the Future

With this base of current development and activity it is useful to project a decade (or two) into the future and see what scientific and technical information transfer will look like. We personally feel that in somewhere between 10 and 20 years the following description will be a reality for a third or more of those engaged in scientific and technical endeavors.

The researcher has access to a version of Bush's Memex. Her personal files are on some sort of storage mechanism (e.g., floppy disk). She can plug into the unit as needed. The unit is designed for the direct composition and synthesis by the researcher of papers from files, but it can, of course, be operated by a secretary. A Memex may also serve as terminal for extracting or sending items to computer systems or to other units of the same type. Costs are less than the mail for communication. Researchers can always borrow a portable terminal to take home. They have access to a couple of conference systems, probably one in their organization and one for communication with individuals elsewhere representing their primary professional peer groups. These latter conference systems may be commercially offered, or they might be run by professional societies or publishing operations. If a researcher has sent in a paper, it is very likely that the complete review procedure would be carried out in a computerized conferencing environment such as provided by EIES. A paper also would have had the benefit of earlier commentary by whatever peer group the authors are commonly associated with through a conference system. The centralized conference system would also have Notebooks, in addition to those in the stand-alone Memex, ensuring the date–time records of ideas and concepts, to prevent plagiarism. The authentication provided by a centralized Notebook facility is potentially a significant factor to facilitate free scientific communication in areas where it

ISBN 0-201-03140-X, 0-201-03141-8 (pbk.)

has become a problem. But as in EIES, the Notebooks and Bulletins are largely for short recent findings rather than longer comprehensive articles, reports, or books.

The abstract and retrieval services would now incorporate a form of communication in terms of reviews contributed on articles and available to the user directly. Because of this, journals will have largely disappeared and publishers will send out individual articles on request. They will, however, print a form of a journal when there is a sufficient quantity of articles on a more specific topic than is common today. The recent emergence of special journal issues devoted to a specific topic by various organizations (e.g., TIMS) is indicative of this future development. Published articles will tend to be larger and more complete, since they will be primarily for students and those interested in the results of research and development in an area in which they are not primarily engaged. Those working in a given area will be utilizing EIES-type systems for their primary mode of exchanging research findings and transferring information.

The professional societies and/or publishers will operate a sophisticated text system allowing for graphics and a wide variety of text fonts. There will probably be centers where an author can go to utilize a sophisticated graphics system to dress up a final version of his or her article.

Microprocessor technology will make these different systems largely transparent to the user since a conference-like system will be the mode he or she uses to gain access to this diversity of services. In essence, the user will see this multitude of services as one system that may be somewhat tailored to a particular field but really represents systems cutting across all fields.

It is hoped that an author will receive credits to pay for the use of these systems as a function of his/her articles, reviews, and commentaries that others have decided to retrieve or order. This would be one of the most effective mechanisms for encouraging relevant article production.

In such a communications environment, it will be much easier for individuals to move into new research areas, for new research groups to form, and for old ones to disband. We also would expect a good researcher to be less dependent on his or her home base. A small college, for example, could attract and retain top-notch people without a tremendous capital investment in library holdings and support staff. Conversely, the researcher at a small institution would not be at as much disadvantage as today in terms of local resources and colleagues to whom to relate.

Professional meetings will orient their programs toward a greater degree of discussion and workshops, and very likely emphasize smaller but more numerous gatherings, with discussion and workshop agendas prearranged through conference systems; and just as there are announcements of scientific meetings and abstracts of papers available, there would be abstracts of

ISBN 0-201-03140-X, 0-201-03141-8 (pbk.)

conference groups and a researcher will be able to check to see if there are some he or she wishes to join.

The Appendix to this chapter is a log of "missed opportunities" in the current pattern of scientific and professional communication, kept by one EIES user, indicating the increased richness and flexibility that might occur if computerized conferencing were available to all scientific and technical personnel.

Barriers and Problems

The image of the university and the scientific enterprise as a "community of scholars" is an attractive one, to which the new technology of computerized communication–information systems could give some reality. However, there are currently institutional forces that militate against the actual widespread use of computerized conferencing for the kinds of scientific communication that have been described in this chapter.

The reward system in science is not currently constituted to measure or give credit for nonpublishing activities. In terms of short-term professional advancement, taking several hours a week to read about the latest research and ideas of a person's colleagues takes away from the time available to engage in the only activity that is defined as "counting": writing and publishing papers.

Likewise, though contributing careful comments and constructive suggestions to the research projects or draft papers of colleagues may in fact be just as valuable a contribution to "progress" in science as publishing a paper, the former is not credited when promotion and tenure are considered in academia. For computerized conferencing, this constitutes a negative motivational force, since the sharing of ideas is the purpose of the communication. In the competitive world of the scientist, the kinds of cooperation and sharing with colleagues made possible by computerized conferencing may simply not be wanted by those who are interested in short-term career advancement. The scientist who does devote a considerable amount of time to communicating with an "invisible college" in this fashion may discover that the contributions made to the building of a common paradigm count for nothing with the local campus administration, because the effort has not resulted in the traditional product, a (nonelectronically) published paper.

A related problem is that computerized conferencing is particularly good for interdisciplinary communications and multidisciplinary projects. Once again, the current opportunity and reward structure of science may work against the successful application of computerized conferencing for such a purpose. The power to employ, promote, and grant tenure is largely controlled by academic departments organized around traditional disciplines.

ISBN 0-201-03140-X, 0-201-03141-8 (pbk.)

Interdisciplinary team research is defined by the gatekeepers as "not our discipline" and therefore "not relevant for advancement within our discipline."

A somewhat different problem is that the current channels of communication in scientific and technical fields have become big business with a heavy capital investment in the current delivery and distribution systems that support both paper publishing and associated data and abstract systems. It is not clear that there is any force at work to overcome the inertia of the current ways of doing things.

Factors internal to many scientific fields also may produce impediments. The competition for research dollars fails to provide the sense of common interests necessary for computerized conferences to be successful.

From a psychological standpoint, in disciplines where there are strong and polarized advocacy positions it is not clear that scientists want to consult, or would enjoy confronting, each other directly, as computerized conferencing provides. The cold detachment of a published paper is far easier to deal with and respond to.

A final observation is that such systems offer the opportunity for frequent communication among research sponsor, researcher, and research user. It is not clear that this heightened state of communication among these three parties benefits the carrying out of research in all cases. It may force the researcher to be too responsive to various outside demands and interfere with the ability to conduct research.

Conclusion

Scientific and technical information can no longer be viewed by researchers as a free good. The past decade has led to both direct and indirect cost increases. Besides the direct costs for reports, journals, etc., the amount of material being generated has led to an increased need for informal communications and associated costs of meetings, travel, telephone, etc. The researcher who cannot afford to maintain active contact with his peer group will be left by the wayside. In most fields hardly anyone can claim to have read all the relevant material. One of the prime mechanisms of selecting material today is on the recommendation of peer-group members. The user will continue to become more selective and only the information sources that can gratify this desire for selectivity will succeed. There is little doubt that the next decade will see dramatic changes in the nature of scientific and technical information systems. The technology is now available and sufficiently cost effective to revolutionize scientific peer-group communications; this, in turn, will alter the nature of the process by which scientific knowledge is produced and disseminated, and will affect the cognitive and social organization of the research communities themselves.

ISBN 0-201-03140-X, 0-201-03141-8 (pbk.)

For Discussion

1. What specific kinds of features or capabilities would ideally be present in a CC system in order to support the scientific work and communication of a specialty field with which you are familiar?

2. How can scientists be motivated to play roles, such as that of the editor of an electronic journal, that initially have no prestige or extrinsic rewards?

3. How might CC systems aid in continuing education and cross-specialty intellectual mobility among scientists? Think of some specific examples (scenarios) of how a scientific career path might be changed by this form of communication.

4. How are the "big names" or "stars" in a scientific research community likely to react to the expansion of their "invisible college" to include persons now outside the "inner circle"?

5. How does the copyright law apply to the scientific material a scientist might enter into an EIES-type system? How do you think the copyright should be handled in such a situation? What law would you design?

6. How do you think authors should be financially compensated for their material utilized in such a system? (For example, it would be possible to pay a royalty for each time a paper is accessed.)

7. How would you design the system to "forget" material? How would you determine when some item is no longer "desirable" to maintain? How would you decide when to delete inactive people?

8. In reviewing scientific proposals, it is felt by some that mail reviews are generally more conservative than when the reviewers act as a face-to-face panel. How would you explain this, and how do you think reviews via computerized conferencing would fare on conservatism relative to these other communication processes?

9. What kinds of cases can you imagine being brought to a computerized science court? What procedures would be necessary for a "hearing"? How could the decisions of the scientific jury be enforced? What would be the analog of judge and jury, and what would be the selection process?

10. What types of scientific projects would you feel are ideally suited to computerized conferencing? Use specific examples. What kinds of projects would be ill suited, and why?

ISBN 0-201-03140-X, 0-201-03141-8 (pbk.)

Appendix:
An Odyssey in the Non-cc World
Robert Bezilla ("213")

Thursday night: I chaired a meeting of our regional professional association. The cocktail hour and dinner hour consisted of about 40 people bringing one another up-to-date on what they had been doing in the previous months. The meeting itself began with a five-minute talk from a man who traveled from New York to Princeton just for that purpose. Unfortunately, the majority of the members for whom his remarks were intended were not at the meeting that evening. Our main speaker generated considerable discussion. For most in attendance it was a new area, and it was obvious that they would have more questions and observations, and equally obvious that they would have no ready means through which to pursue the subject. Some of those in attendance suggested the new topic would be most fitting for the national conference agenda, but since it was full, it was hoped that this might be done in 1979.

Friday: I lectured at the university on a topic that was new to most of the students. They asked some intelligent questions, but I had to wonder if they would have further questions as they later had applications for the ideas presented. Being a guest lecturer, I am not easily accessible to them, so those questions will never be asked or answered. A young statistician saw in one of my remarks a potential answer to a problem that had long troubled her. I did not have the background to deal with her interpretation, but did refer her to two people, one in New York and one in Cambridge, who might help her. Maybe after several months' correspondence she will have her answer.

Returning to my office, I found I had to keep several people beyond the 12 noon closing hour (it was Good Friday). We had received a Request for Proposal two days late because of mail delay. Further delays were caused since it was physically impossible to circulate drafts and estimates to key personnel on time because they were out of town. Now we were down to the final typing and revisions were needed and various errors had to be corrected. Two proofreaders stood by as typists labored to revise the document. They snatched the text one page at a time as it came out of the typewriter. Another person stood by to reproduce the pages one at a time. The proposal had to be delivered by Monday, so we sent one copy by first class mail, another by special delivery. If neither reaches our client by Monday morning, someone will have to make a 120-mile round trip to deliver a copy by hand . . . .

ISBN 0-201-03140-X, 0-201-03141-8 (pbk.)

Weekend: I must prepare for a trip early next week. Some text items will have to be ready for quick typing on Monday morning so that they will be available for me to take to Connecticut when I leave Monday Morning. I hate to do it, but I must sequester myself from my family and weekend guests for long periods of time. While in my study I notice a book and three papers that I wish I had the time to read.

Monday: I spend the morning reviewing text items for my trip and briefing people on what to do in my absence. One of my co-workers must leave his work for about an hour to take me to the train station. The afternoon is spent aboard the train, but I have some time to catch up on my reading.

In the evening I have a dinner meeting with my client to go over the next day's agenda. In the course of the conversation it develops that we would both like to obtain certain materials from each other, but this, of course, is not possible until a later date and that topic of conversation ends abruptly.

Tuesday: In a marathon day I made presentations on five surveys to five groups, a summary presentation, and participated in two planning meetings. In total I saw about 40 people. Obviously, there was little time to answer all of their questions or to carry on much discussion. Some points were repeated three to eight times by me during the day. Those attending more than one session must have been as bored by the repetition as I was. While waiting for my taxi to take me to the train station to begin the trek home, one of the attendees came up to me and presented an outstanding idea. I had no time to talk to him about it, and while riding to the station I had to wonder how many other ideas there might be that will never be presented or discussed.

Wednesday: The morning was spent going through the accumulated mail, making telephone calls, and putting out brush fires. One message tells me to call a client who works a mile down the street; we have been trying to get together for about two weeks now; maybe in another week our schedules will coincide . . . .

Thursday: I telephone a client to get copies of some needed documents. With luck I may receive them by Monday. I send off three letters: maybe I will have responses within two weeks. I receive a questionnaire from a faculty member of the University of Texas. Many of the questions are ambiguous, but I answer them as best as I can. Some dialogue with the researcher might have helped him in his inquiry, but there is no ready means through which to do this. There were some matters that I would have liked to have discussed with two of my senior associates, but both are out of town for the day.

A writer for a trade journal called me and explained that he has a tight deadline to meet and has just learned that I am engaged in some research that might have bearing on his article. As he interviewed me, he discovered some entirely new insights. We talk for nearly two hours on the phone as I

ISBN 0-201-03140-X, 0-201-03141-8 (pbk.)

laboriously read passages from a report, and carefully check his understanding of what I am saying so that I will not be misquoted. I make arrangements to send the full text by special delivery to him so that he can use some of the tables that appear in the report. I am very concerned about how well he understood this very complex topic of research.

During most of my week access to computerized conferencing services would have enhanced my performance considerably. Its absence caused delays lasting from hours to years.

Cost–benefit models undoubtedly could be used to measure the superiority of computerized conferencing over conventional communications. But what model could capture the importance of the ideas that were lost, the questions that went unasked and unanswered? As I and others were traveling, did our families need us? How do we measure this social impact? Where is the model that measures the impact of Whittier's reflection that

> For of all sad words of tongue or pen,
> The saddest are these:
> "It might have been!"

THE BOSWASH TIMES

News Service of the Megalopolis
Science News Section

August 15, 1986

All the News
Fit to Display

ISBN 0-201-03140-X, 0-201-03141-8 (pbk.)

Truth in Polling

Washington D.C. Senator Ted Clark introduced today a "Truth in Polling" bill. If passed, it would require all on-line polls to inform respondents ahead of time of the following aspects of a study.

* Whether any names will be used to identify the source of supposedly anonymous replies

* Whether the replies will be used in conjunction with other available information or previous survey information supplied by the respondent

* Who shall have access to the respondents' replies

* The sponsor of the study

* The mean time to complete the survey for all pretests or previous times the questionnaire has been administered

* What analyses will be used to indirectly infer underlying values or attitudes on the part of the respondent

Dr. Jonathan Hiltz, president of the World Association for Public Opinion Research, commented, "While we support the goal of ethical procedures in survey research, we feel that this bill, if passed, will so distort the replies of respondents and adversely affect response rates that it will become impossible to conduct many important studies. We feel that it is sufficient for the respondent to know the name of the organization conducting a study and whether or not any data will be released or recorded with identifying information."

"Passionate Program" Leads to Hot Suit

Chicago, Ill., August 14, 1986. The nation and many organizations, authors, and journalists await with bated breath what may be the final resolution of the issue of ownership of the content in electronic forms of communication. In federal court here this morning the opening guns sounded in the case of Williams versus Scott Publishers, Inc. The issue is an injunction being sought against further sales of the new electronic book bombshell, *Love Electronics,* by Elaine Taylor. The book is a collection of passionate and revealing messages written by numerous male individuals to "Madam Love," who has been exposed as a computer program simulating an amorous female, designed and implemented by Ms. Taylor as a Ph.D. thesis at the University of Southern Illinois. Mr. John

255

Williams is one of the authors of the letters contained in this book and he and his lawyers have taken the position that messaging is a form of publication; that Mr. Williams retains the copyright on the contents of his letters; and that furthermore, no permission has been given to the publishers of the book to utilize the material. The publishers claim that an electronic message is a letter and under common law precedent a letter becomes the property of the recipient. Furthermore, an electronic book is not a book in the legal sense and "publishing" is just resending the letters owned by Ms. Taylor to others.

Dr. Katherine Amanda Hiltz, president of the American Association of Information–Communication Services, which has separately filed a brief with the court, believes this is an unfortunate case upon which to resolve the information ownership issue. At stake for the nation are issues of more widespread impact than implied in this case. The AAI brief requests that the case be viewed as one of fraud, where a computer program was misrepresented to the "victims" as a real person.

Ms. Taylor has been unavailable for comment but stated in a recent press release that Madam Love's psychology was based on a computer analysis of famous published love letters and bears no relation to her own. Pending the outcome of a University inquiry into charges that she misused computer resources by tying her program into a national network, the granting of her Ph.D. has been blocked. Privately, a number of professors have expressed the view that the quality of her thesis is brilliant, and that the inquiry is a result of political pressure from the state legislature.

Inuit Protest Removal of CC System

Ottawa, August 14, 1986. Parliament was disrupted today by a demonstration of Inuits protesting the sudden removal from their community of the computer and video conferencing system that has been in place for three years as a part of a field experiment.

An Inuit spokesman claimed, "It is immoral and unethical to remove our computer and terminals and other communications equipment just because your experiment has come to an end. We have come to depend on this system for education, for medical services, for contact with other oppressed peoples who share our problems, and for entertainment during the long dark winter months."

A government spokesman replied that the experiment was clearly funded for a maximum of only three years, and that if the Inuit wish to retain the new communications system, they must now pay for it themselves. He also declared that although the system was used for some medical and educational purposes, the emergence of a joint French-Canadian–Inuit–Immigrant Pakistani, antigovernment political movement was not considered desirable by the government, and certainly was not something that they wished to support from general tax revenues.

ISBN 0-201-03140-X, 0-201-03141-8 (pbk.)

256

ISBN 0-201-03140-X, 0-201-03141-8 (pbk.)

CHAPTER 8

Research Imperatives and Opportunities

Introduction

In this chapter we present an overview of high-priority research areas related to computer communication systems. As a communication medium the potential for computerized conferencing extends to almost all human endeavors and situations. There are concerns that can be defined to be within very specific disciplines and there are very specific well-defined questions that can be addressed. However, many of the more significant topics and issues lie in gray areas and the very nature of the most important questions is uncertain. In part this is due to the intimate entanglement of the human users, the application, and the technology.

We will first state our views on the main research imperative, which is the study of the impact of the computer as mediator or presence on human and communication systems.

We will then briefly explore several areas in which CCSs facilitate new research methodologies and kinds of studies. The examples selected are experimental social psychology, public opinion research, and selected areas of computer and information science. Finally, we present a "research agenda" of unanswered questions about the design, policy, applications, and impacts issues related to the implementation of computerized conferencing. This last part of the chapter is actually a generous sampling of excerpts from the report produced by workshops on research issues related to the future of computerized conferencing, which were held on EIES during 1976–1977 and sponsored by the Division of Mathematical and Computer Sciences of the National Science Foundation.

Impacts and Their Evaluation

To evaluate the impact of CC systems properly, we must take a more "holistic" or "systems" view than is usually the practice. Communication is

Starr Roxanne Hiltz and Murray Turoff, The Network Nation: Human Communication via Computer

ISBN 0-201-03140-X, 0-201-03141-8 (pbk.)

the fundamental process by which interaction among the elements of a human system takes place; it is the exchange process by which goals are formulated, decisions made, and "work" accomplished. It is inevitable that changing the form of communication used by the members of a group will affect the goals, interaction, cohesion, productivity, etc. of that group, and their relationship with the rest of society.

In the literature as well as in this book there exist numerous hypotheses and conjectures based on the limited observations that have taken place to date. Some of these issues relate to

1. A specific implementation or application area:
    (a) What are the criteria that make an application appropriate to this medium?
    (b) What are the consequences of group size, length of interaction, and degree of competitiveness?
    (c) What is the impact of various cultural traits, such as politeness and introversion?
    (d) What are the best forms of instruction in what situations?
    (e) What are the appropriate possibilities for charging users?
    (f) What type of message forms and symbols develop?
    (g) How do teams function with this communication medium?
2. The impact on group communication processes:
    (a) How do various factors (software design, social, and psychological) relate to enhancement of the cognitive transmission of information?
    (b) What sorts of norms and rituals evolve?
    (c) How do new friendships or working relationships evolve from communicating on this medium?
    (d) What are the supportive human roles, such as facilitation and gatekeeping?
    (e) What are the characteristics that enhance leadership?
    (f) What are the mechanisms of group control?
    (g) What are the appropriate cuing mechanisms?
3. The impacts on the product or final outcome of group communication processes:
    (a) What are the results of the increased "pace" of communications and how do we best counter information overload?
    (b) What is the impact of individual migration among groups?
    (c) What are the impacts of computerized conferencing on the quality of decisions made by groups?
    (d) What is the impact on risk-taking situations?
    (e) How do characteristics of problem types or group communication objectives relate to computerized conferencing and its design?
    (f) What are the appropriate measures of benefit?

ISBN 0-201-03140-X, 0-201-03141-8 (pbk.)

Although we intuitively know that the use of computerized conferencing might have very fundamental impacts on groups that use it, there has been very little research into the nature and processes by which such impacts occur. For instance, unlike the excellent program of experiments with audio and visual teleconferencing carried out by the Communications Studies Group (see Short et al., 1976), there have not been programs of controlled experiments to date that can help us understand the ways in which CC as a medium of communication may change the amount, kinds, and outcomes of communications. Evaluation has frequently been limited to field trials incorporating collection of some rudimentary statistics on amount and type of use of a conferencing system, and short questionnaires administered to participants after the fact. What is needed at this point is a commitment by the research community engaged in the development and implementation of CCSs to undertake all such efforts only in conjunction with a thorough, multimethod program of research and assessment of the effects of various features or characteristics of CCSs and alternative ways of providing training for users. Ideally, such programs would include a loosely integrated set of controlled experiments, field experiments, observation and automatic monitoring of patterns of use, questionnaires, unstructured interviews, and perhaps simulations.*

There are at least five different levels of the impacts of this new communication medium.

1. *Personality.*   For example, does working on these systems make people feel relaxed or nervous; in touch with others or isolated; "in command" of the system as an able servant, or prisoner of a system they feel unable to understand fully and to control? What unique impact does use of the medium have on the quantity, quality, and "style" of individual communication output, and how does this vary with personality and other individual characteristics?

2. *Small Group.*   Impacts on small groups have to do with the outcome of reliance on this medium as a primary means of communication, and with such factors as the productivity, cohesion, and morale of the individual (work or play) group. For instance, does the system tend to support or to undermine the dominance of a single leader? To increase or decrease equality of participation in discussion and decision-making processes?

3. *Organization.*   The focus is on such factors as the interrelations among the various components of a larger organization. For instance, how does such a system affect centralization versus decentralization of decision making? Does it encourage more or less lateral communication among branches or

---

*See, for instance, the multimethod approach used by The Institute for the Future: Vallee et al., 1974, 1975a; Johansen, 1976.

ISBN 0-201-03140-X, 0-201-03141-8 (pbk.)

staff groups on the same level, or more vertical communication up and down the hierarchy?

4. *Society*. The potential impacts include such things as the substitution of telecommunication for transportation, and effects on the relative power and equality of well being among various groups within the society. It also includes political or policy issues that emerge as a reaction to the effects of the new technology.

5. *World*. Issues such as the effects on the balance of power among nations or the flow of scientific and technical information among nations are raised.

The process of going from individual and group impacts to societal and economic impacts calls for forms of technology assessment. The impacts in this area are going to be the most important ones but they remain 10 to 20 years in the future. This time scale is often referred to as the "no man's land of forecasting" because it usually represents an equalized situation between the elimination of current societal and organizational structures and the emergence of new structures. Neither extrapolation techniques for current trends nor normative approaches work well in this time frame. As a result, only large-scale field trials can provide data with which to gain insight into such areas as work at home; family applications; applications to the disadvantaged; democratic processes; social services; transportation–communication tradeoff; mass media impacts; new employment options; social engineering.

The policy and regulation decisions on the societal level will establish the environment or nutrient in which development of the technology and its applications will either flourish or starve. Until now most developmental activity has been of a research nature. The venture capital and the entrepreneurs for new systems and mass use will only emerge with clear-cut decisions on policy and regulations. We are at a point in time where not technology, but uncertainty in regulation could cause an investment to be wiped out by either a court case or a change in interpretation of existing laws. As long as this situation continues we will remain in a largely experimental environment except in the case of organizations and institutions large enough to afford their own systems. If there is a priority area for study, it is the sort of policy analyses that can provide meaningful information for those in legislative and regulatory positions at all governmental levels. Thus the "political" or "policy" issues should not be seen as a "contaminant" to research in this area, but as a crucial area for inclusion in research.

The particular impacts to be found also depend on a complex interaction among at least four sets of factors:

1. What is being looked for, and how, and for how long. That is, choosing a level of impact and factors within it to focus on probably precludes finding

ISBN 0-201-03140-X, 0-201-03141-8 (pbk.)

other types of impacts. What is found in a study depends partly on how long it goes on; certainly, the behavior of users and the impacts of such use will be much different after five years than after a two-hour experiment. (Though this is merely speculation; no one has collected user profiles for longer than a two-year period.) Finally (and most important for this set of factors), findings are going to be partially an artifact of the evaluation methodology chosen (the controlled experiment; the field experiment; the field trial; questionnaires and interviews with users; participant observation in and/or content analysis of the proceedings of conferences; or simulations).

2.   Features and characteristics of the system itself, and its implementation. This includes the complexity, flexibility, and style of user interface of the system, as well as the print speed of the terminal used.

3.   Application areas, that is, the kinds of groups that are using the system; for what purposes or services; and in what type of environment (e.g., work at home, remote meetings, scientific communication, social or educational services).

4.   Characteristics of the user and the immediate environment. Included here are user attitudes and motivation toward using a CCS; user skills—reading and typing speeds, relative skill and preference for spoken rather than written communication; type of role played by conference moderators or other human facilitators on the system; and the total communication and work load of the user.

To treat all of these factors as simultaneously interacting (which they are) generates a matrix that only a computer can handle! It is perhaps too much to ask any study in the near future to treat such a large number of variables; but those who undertake research in this area should at least be aware of what is excluded by the variables and methods they consciously choose in designing a study.

Each application area has potential impacts at each level. Usually, evaluation studies select data-gathering methods and variables focused on the intended consequences at one specific level, whereas the most important impacts may be unintended ones at a different level. The usual controlled laboratory experiment or the rigidly designed field experiment that does not revise hypotheses and instrumentation to capture measures of such unanticipated influences and outcomes is doomed to methodological failure. Another problem with the controlled experiment is that it takes weeks before a user of such systems gains the skill and comfort with this means of communication that he or she feels with face-to-face interaction; thus, many initial reactions observed among new users in a laboratory will not be generalizable to experienced users in the "real world."

In using a multimethod, multilevel approach to research on the impact of CCSs, it is necessary to iterate between the generation of theories and

ISBN 0-201-03140-X, 0-201-03141-8 (pbk.)

hypotheses from observation and relatively qualitative data, and the testing of specific hypotheses or aspects of the theory with controlled experiments or quantitatively oriented research instruments such as interview guides. It is a mistake to think that we can foresee and instrument ahead of time all of the relevant variables that should be studied and taken into account in a field experiment on computer communication, for it is often the unanticipated consequences of the new technology that will prove to be crucial to the outcome of a long-term experiment stretching over months. For instance, Shinn (1977) states that a researcher should know ahead of time exactly what hypotheses are to be tested in a field experiment, and should then periodically administer predesigned instruments to measure these variables. What is likely to happen in such cases is a finding of very little impact or difference made by the new system because the individuals and groups use the system in unforeseen ways and develop their own goals. Typically, the introduction of a CCS will result in the evolution of patterns of use and outcomes that were not explicitly stated at the outset.

The solution to this methodological problem is that participant observation of the actual system use and unstructured interviews with participants can enable the researcher to modify and add to the initial hypotheses and data collection instruments in order to document and measure what may be very crucial and valuable processes and outcomes. In other words, the qualitative data can be used during the course of a field experiment to modify hypotheses and to suggest changes in or additions to structured questionnaires or the monitoring statistics collected on user behavior.

Another piece of conventional wisdom repeated by Shinn is that an investigator should move "from the laboratory to the field," developing the theory from controlled laboratory experiments and then seeing if the theory holds in the real world. With CCSs, it is likely to be more fruitful to take the opposite approach. That is, the monitoring of field trials or field experiments can contribute to a theory of what variables are crucial to determining the impacts of a computer-based communication system. Specific subsets of this overall, multivariable theory can then be tested rigorously in a controlled experiment. Ideally, the controlled experiment would be conducted not with college sophomores who have never before seen a computer terminal, but with experienced users located in actual organizational units and engaged in realistic tasks.

Thus, the laboratory experiment can best be used to test specific two- or three-factor hypotheses about the interrelationship of variables that form subsets of the much more complex set of variables that determines impacts in actual application of human communications via computers. The laboratory experiment also offers a potentially fruitful area for application of the new technology to research issues in other fields.

ISBN 0-201-03140-X, 0-201-03141-8 (pbk.)

Experimental Social Psychology

In Chapter 3 and in the foregoing discussion, we discussed the ways in which the existing findings and methodological approaches of experiments on group communication and problem solving can give us some insight into the processes and impacts of CCSs. However, this is not a one-way street: social science and computerized conferencing can engage in a genuinely symbiotic relationship. Not only can social scientists produce data that can help to guide the design and implementation of computerized conferencing, but computer-based communication systems can facilitate the design and implementation of studies related to human communication and group processes that would be difficult or impossible without this technology.

One problem in experimental studies of problem solving groups is the impossibility of completely standardizing "treatments" of groups and conditions for each of many separate trials when the experiments are conducted and administered face to face. When used to administer all problems, instructions, etc. automatically, computerized conferencing allows experimenters complete control of certain categories of exogenous factors that have always acted as potentially confounding sources of variation in face-to-face group experiments, such as

1. Variability among several experimenters in their behavior, or even for the same experimenter from day to day (e.g., how much experimenters smile when delivering instructions to the group or how loudly they talk);
2. Variability in the appearance and behavior of the subjects as it affects the reactions and motivations of other group members. Examples are facial and other nonverbal expressions (smiling, frowning, fidgeting while instructions are being given), appearance (attractive versus non-attractive), and verbal mannerisms (stuttering, accent, etc.).

Since the subjects do not see each other and instructions can be delivered over the terminals exactly the same way for each trial, more complete control over all factors other than those that are purposefully being manipulated is possible.

Another methodological technique that can be explored is the use of software associated with computerized conferencing in subject selection, data recording, and initial automatic data processing.

In the computerized conferencing conditions, the automatically generated transcript will completely capture the data; every single thing communicated by each group member is recorded, by time of entry; nothing is "lost." This facilitates later coding and analysis by more than one scheme, if this seems desirable, and eliminates sources of error in the recording and transcribing of the contents and sequence of the communications among group members.

ISBN 0-201-03140-X, 0-201-03141-8 (pbk.)

Another well-known limitation of most existing controlled experiments is that most of the subjects are students. Busy executives can hardly be expected to volunteer to troop into a campus laboratory to play the role of subject. With computerized conferencing, however, the lab can be brought to wherever there is a telephone—to an executive's office, for instance, or a slum storefront—thus greatly expanding the socioeconomic diversity of experimental subjects and the consequent generalizability of results. The portability of the medium would obviously facilitate the conducting of long-term field experiments rather than only short-term laboratory experiments on group communication.

More generally, computerized conferencing's potential combination of automatic administration, data collection, and analysis of experimental runs ("programmed in" as software options), plus the portability of terminals, offers some real opportunities for modifying and expanding existing experiments to test some new hypotheses. In other words, the effects of certain dimensions of communications, such as size of groups, can be examined by computerized conferencing and generalized to all communication media. As far as group size goes, most communication and problem solving experiments have involved very small groups, of seven or fewer members. Yet, we all know that much real-life decision making gets done in staff or committee meetings with a dozen or more participants. It has been too difficult to try to observe, measure, and control medium- and large-sized groups in the usual laboratory setup, but there should be little problem with a computerized conferencing group.

Opinion Research*

On any single day, the chances are one in several thousand that you will receive a visit or a telephone call from a special kind of stranger. If you ask him what he wants, he may reply: "I'm collecting standardized information from a sample chosen to represent the component units of a predefined universe." Or he may say: "I'm taking a poll." In either case the content of his message is the same [Davison, 1972, p. 311].

Public opinion is one area of research in which we believe that there are potentially important future impacts and applications for CCSs. We define "opinion research" very broadly, to include any approach that generates descriptions and explanatory information about the attitudes and behavior of any segment of a population, including elites as well as "the public." When

---

*We are indebted to Robert Bezilla for his discussion with us of many of the issues relating to computerized conferencing and public opinion research.

ISBN 0-201-03140-X, 0-201-03141-8 (pbk.)

the reader begins by understanding that opinion research includes much more than predictions of presidential elections and surveys of TV program preferences, then the potential applications become more obvious and more exciting to contemplate.

This analysis of the functions of opinion research rests on a particular view of the stratification system of modern societies. We picture advanced postindustrial societies as roughly diamond-shaped in form. At the top are a number of loosely integrated elites—the decision-making leaders in politics, business, science, the military, and perhaps the arts. [See Suzanne Keller's (1963) explanation of "strategic elites."] At the next level down are what has variously been termed the "upper middle class" or the "subelites." These are the advisors and assistants to the decision makers at the top, who tend to be "experts" in a fairly narrowly defined field. At the middle levels are the masses, who people the offices and factories and participate in voting and in selective consumerism and in keeping informed of the "news" and of political issues. At the bottom are the disadvantaged, who are unemployed or underemployed, generally undereducated and generally nonparticipants in the institutionalized legitimate political and economic processes of the society. This is not to say, however, that they do not participate from time to time by appropriating a share of the economic benefits of the society for their own (crimes against property) or by expressing their anger and frustration (riots, violent crime).

W. Phillips Davison, in his presidential address to the American Association for Public Opinion Research in 1972, described the functions of public opinion research in a broad overview.

As Davison (1972, p. 312) says, "We can think of public opinion research as part of the communication system of our society and of the world community," serving to connect the various segments of the society to one another by making them aware of the similarities and differences in outlook and behavior. From the elite to the mass go a stream of laws, products, information, etc. Traditional "market research" is concerned with finding the best means to deliver these communications in a convincing way. From the mass to the elite go expressions of needs and desires—the traditional public opinion poll. Such opinion research studies have at least three functions (Davison, 1972, pp. 312–314):

- To explain public opinion on any issue, putting it into context and showing how many people believe what and why;
- To provide a feedback mechanism to decision and policy makers, showing how many people are informed correctly about current positions or situations, and their reactions to recent policy decisions;
- To serve as a *substitute* for general public opinion, in that it may identify incipient grievances, inequities, or sources of problems before they become

ISBN 0-201-03140-X, 0-201-03141-8 (pbk.)

large-scale issues on the basis of which people may be willing to express violent reactions.

Davison also identifies two other, relatively neglected "directions" of communications flow in the form of results of opinion research: from the subelite to the elite (studies of experts, such as most Delphi studies, would fit in here) and studies in which one segment of the population is described for the benefit of others (lateral communications—so that the old may understand the young, for instance, or New Yorkers understand Iowans and Californians).

Computerized conferencing has the potential to overcome two of the major problems with most current public opinion research methods: delays and their noninteractive format. In addition, it can substantially reduce the cost of public or expert opinion studies, and thus extend their use to areas in which they are not considered economically justifiable at present. It can potentially do so while increasing the accuracy of the findings by improving response rate and decreasing response bias.

The most usual formats for public opinion polls are (1) the personal interview with a representative sample of the population being studied, and (2) a mail questionnaire or a telephone interview. All of these take considerable time to administer and tabulate. The mail survey is the cheapest but slowest, since several copies or reminders must be sent out in order to prod the initial nonresponders to complete and return their questionnaires. The telephone is fastest, but even it is fairly time consuming, since each interview must be individually administered. The recruitment and training of a large number of telephone interviewers to be "ready" to do a survey within a two-day period is possible, but in itself takes a great deal of time and foreplanning. It is impossible to rapidly field a telephone survey about an unexpected issue or event for which there has not been considerable warning and therefore preparation time.

A poll or survey conducted over computerized conferencing could quite conceivably have the questions sent out, answered, automatically tabulated, and reported back by the computer within a single day. We can even imagine designing a set of questions in the morning, pretesting them in a period of an hour or so, and later conducting the full-scale survey, all in the same day. There is no need, for instance, to wait for typing and photocopying of survey instruments. They can be composed, corrected, revised, and sent out on line in a matter of hours.

Sources of error in survey or public opinion research that should determine the selection of data collection techniques, in conjunction with associated costs, are completion time, response rate, and response bias. Response rate refers to the proportion of the selected sample of the population for which observations can be secured. Response bias refers to the influence of the data

ISBN 0-201-03140-X, 0-201-03141-8 (pbk.)

collection technique itself on the results that are obtained. Though there are response rate and response bias problems associated with computerized conferencing, we believe that its total effect may be to introduce lower error rates than other forms of data collection.

The main problem that stands in the way of such widespread instant polling via computerized conferencing at present is the sampling problem. A population of interest, or at least a representative sample of it, must have access to a terminal and be in the habit of signing on once a day or more.

They must also be motivated to respond, but that should not be any more difficult than motivating response to other forms of interviews, since part or all of the money normally used to pay interviewers can be used to pay or otherwise reward the respondent. Response rate will suffer from the disadvantage of the lack of personal pressure applied by the pleading interviewer on the doorstep. On the other hand, it should be greatly improved by the facts that there are no fears about "opening the door" to let an interviewer in, and probably less suspicion about the legitimacy of the survey. The main factor that should work in favor of higher response rates for this form of survey is that the respondent does not have to be "at home" and free to talk at the moment chosen by the interviewer. The respondents receive and complete the survey instrument at their own convenience, rather than having to interrupt their activities.

The closest analogy to a computerized survey instrument is the mail questionnaire. Current research indicates that the response rates on these surveys, with proper follow-up procedures, can approach that of the more traditional personal interview. For example, Dillman (1972) designed a "package" of procedures that includes, in addition to the traditional postcard reminder, a personalized follow-up and a final follow-up plea with the questionnaire mailed certified mail (for which the respondent has to sign). House et al. (1977) experimented with the certified mail follow-up and found that it indeed did significantly improve responses (from 56% to 71% among the blue collar workers studied). As they point out (p. 95):

> Rising popular levels of education and experience in filling out questionnaires increase the likelihood of obtaining valid and reliable data from mail questionnaires, thus making them an attractive alternative to increasingly costly personal survey interviews. ... Concerted application of a set of procedures can produce mail questionnaire response rates of 70 to 75 percent in large general population samples.

In other words, there is no longer any problem with the ability of the population to complete written questionnaires, and computerized analogs of the kinds of procedures used in mail surveys can be used to increase follow-ups.

ISBN 0-201-03140-X, 0-201-03141-8 (pbk.)

- Automated reminders, issued one day after the initial questionnaire (which can come complete with graphics and beeps from the terminal).
- Personalized follow-ups in which a private message from a person known to the respondent (such as the head of a union being surveyed, the mayor, or a governor) asking them to please take part.
- A computerized analog of the certified letter, in which the respondent must acknowledge receipt and readiness to answer before the contents of the questionnaire are printed out again.

Another technique to increase participation is to use light-sensitive pens and CRT terminals, so that the respondent need only make pokes and lines on a TV-like surface to respond. This device can be made enough like a game for many survey purposes that it will seem like fun to participate.

Turning to response bias, we find indications that, since CC responses would be entered in private, uninfluenced by a desire not to offend an interviewer, and can be made anonymously, for "sensitive" issues at least, more truthful and fuller accounts may be elicited than by face-to-face or telephone interviewing. One study that is relevant to this question was conducted by Wiseman (1972). He administered the same attitude questions to three carefully matched samples by three modes of survey administration: mail, telephone, and personal interview. In two of the nine questions asked ("Should abortion be legalized in Massachusetts?" and "Do you believe that birth control devices should be readily available to unmarried people?"), a statistically significant technique bias was found, related to religious preference. On these issues, the largest proportion of "socially undesirable" or embarrassing responses was obtained in the mail questionnaire, while the smallest was obtained in the telephone interview (Wiseman, 1972, p. 107).

There are also indications that a small but statistically significant difference in response is associated with whether or not respondents on a mail survey are guaranteed anonymity (see, e.g., Fuller, 1974). To the extent that this is the case, a respondent to a CC questionnaire could choose the option of returning responses anonymously. Similarly, there are suggestions that under some conditions, the sequence in which the response categories are presented can produce a small but statistically significant difference in the results. For example, Carp (1974) found that regardless of the question or the ordering of responses, a response category was chosen a statistically significant larger number of times if it appeared first. Though Powers et al. (1977) argue that the differences tend to be substantively unimportant even though statistically significant, and their results did not replicate this finding, there might be circumstances in which a survey researcher feels that it is important to eliminate this source of bias. It is very expensive and difficult to have numerous questionnaires made up with all the various possible orderings of

ISBN 0-201-03140-X, 0-201-03141-8 (pbk.)

responses when using a traditional printed questionnaire. However, it would be a simple matter to have the computer randomly determine which of five or six or even more versions was administered to each respondent via this medium.

A problem in self-administered questionnaires is usually that the respondents cannot receive any additional information if they do not understand the question or how to answer it. The adaptive-text, CAI-type procedures on a conferencing system could be utilized so that any time a respondent felt that he or she did not understand the question and needed more information, a question mark could be entered and a preprogrammed follow-up would explain the terms or the instructions in further detail.

A related consideration is that the computer can adapt the "space" provided to a respondent to be as large as needed for the response to any question. This is always a problem on a printed questionnaire. Researchers do not want to leave large amounts of space after each question, where a respondent might conceivably insert an additional comment or explanation or objection, for fear of making the instrument forbiddingly thick. On the other hand, to leave no space often means that a respondent will not note such potentially useful qualitative information. In a computerized interview, a respondent might be given the privilege of adding a comment after any answer.

The questionnaire with a "branching" structure, in which the sequence of questions asked is dependent on the answers to sets of filter questions, has been very difficult to administer reliably in self-completion questionnaires or in interviews. A complex set of arrows and boxes that asks the respondent to "answer questions 1–3 if you answered 'a', but question 4 if you answered 'b' and skip to 5 if you answered 'c' " often produces confusion and unreliability among not only respondents but also interviewers. However, this kind of branching structure is very easily handled in a computerized interview, and the respondent need not even be aware that the response to one question conditions the nature of the questions that follow.

A CCS is also especially amenable to forms of opinion research that involve a kind of "dialogue" or round structure. The time between rounds could be as short as one day, so that respondents could themselves suggest additional questions or answer categories or other considerations, which could then be incorporated into a subsequent round of questioning sent to all participants.

Another possibility, once a large population of persons are using CC for daily communications, is to set up a conference using pen names for in-depth exploratory discussion of an issue by a self-selected sample of people (Scheele, 1977). Suppose, for instance, that you wanted to find middle-class persons who define themselves as spending $50 a month or more on wine, in

ISBN 0-201-03140-X, 0-201-03141-8 (pbk.)

order to try to design an advertising campaign that would appeal to such heavy consumers; or suppose you wanted to do a study of persons who define themselves as "living together" without being married. These are rare populations, often found now by the "snowball sampling" technique.* Alternatively, all members of a particular system could be sent a message describing the study/discussion and offering some small inducement to take part.

Finally, many critics of current public opinion research have mentioned that often researchers are not so much measuring public opinion as creating it. When asking about things that respondents have not thought a great deal about previously, and on which they had no strong opinions before the interview began, an interviewer may be inadvertently studying what an informed public would think if they learned about the issue and had to reach a decision. Their education and opinion-formation process could be studied more explicitly and purposefully, with more documentation of the recorded discussion of the questions asked by the respondents and their interchange before making a decision on the issue.

Given these considerations, the most promising immediate areas for public opinion research would seem to be the following:

- Studies of elites and of "experts" who may already have terminals, or who could be loaned terminals for the duration of the study, since a large number would not be involved.
- In-depth explorations of the attitudes of a small group of "representative types" of consumers or voters or whatever, to discuss reactions to potential new products, advertising campaigns, campaign themes, or legislative priorities.
- Permanent "panels" of respondents, chosen as a statistically representative sample of a more general population, who agree in exchange for use of the terminal and other forms of compensation to answer about a survey a week (analogous to mail panels maintained by several opinion research organizations, or the monitoring system for determining TV audiences).
- Simulations, games, or other highly structured communications designed to discover the ways in which different subgroups in the population are likely to behave and feel toward one another should they become involved in conflict over a specific public policy issue.

However, the most promising area for research with this new medium during the next few years is not in electronically replicating existing techniques, but in developing new forms of opinion research that take advantage of its interactive and analytic–feedback capabilities.

---

*In a "snowball sample," the researcher locates a few of the specific types of respondents desired. They are then asked to suggest others who also possess the desired characteristic.

ISBN 0-201-03140-X, 0-201-03141-8 (pbk.)

Computer and Information Science Research Areas

One of the major untapped or currently unexplored potentials of computerized conferencing is the computer's ability to process both the English text and the subjective estimations that flow through any discussion. A host of such processing techniques have been developed and tested on an individual or laboratory basis, or incorporated as a part of other remote communication forms, such as polling and Delphi. Work is needed on modifying, integrating, and testing versions of these approaches in conferencing-type environments for such specialty areas as linguistics, decision theory, psychometrics, distributive processing, information systems, and human simulation.

Linguistics and Language

There is a significant body of literature on the analysis of written communications that tries to identify and characterize writing styles in terms of the conveyance of meanings and emotions to the readers. These include various frequency and correlation analyses applied to the use of words or phrases. Computers have been utilized in experiments to try to identify the unique styles of various authors and poets and even, in some cases, to take text and superimpose a particular writing style or convey certain emotional content through modifications to the style. Other efforts have focused on the grammatical modification of sentences or the codification of grammatical rules necessary to do automatic abstraction.

Some of these experimental efforts and attempts at theory development have never left the laboratories or seen widespread use because of their limitations. No system has even been devised, for example, for the automatic abstraction of free-form English, and it is not likely to be done soon. Most current and past efforts have been directed toward the automated analysis of large volumes of text with little or no human intervention. In the CC environment we are concerned with relatively smaller volumes and with the establishment of computational processes as direct aids to individuals and groups that can take a more direct guidance or correction role in what is taking place. In other words, text-processing systems in the context of computerized conferencing do not have to achieve perfection. They need only minimize the drudgery of the task and can rely on humans for feedback and final correction. What are sought are useful but imperfect systems, with the human an integral part of the process. This considerable change in emphasis from "classical" computer science research leads to distinctly different approaches.

In the mid-sixties, a great deal of research money was appropriated for automatic abstraction and machine translation. The funding sources and researchers involved had much higher expectations than were warranted, with

ISBN 0-201-03140-X, 0-201-03141-8 (pbk.)

the result that there was great reluctance to fund further work. Today there appears to be a renewal of efforts, tempered in many cases by the realization that completely eliminating the human is counterproductive. As we see it, this awareness must be broadened to include the integration of group inputs into such systems, rather than limited to concern with the interface between one person and the machine.

Decision Theories, Utility Theories, and Artificial Intelligence

A great deal of effort is being expended in these areas, especially with respect to understanding the bias and other error sources involved in the process whereby individual humans make subjective estimations of such quantities as expected costs, probabilities, effort, etc. There is also considerably more emphasis on the area of collective group estimation processes (Dalkey, 1975, 1977) and the processing of these estimates to examine consistency and agreement among members of a group.

A closely related area, largely in the management sciences and in operations research, is the work in psychology and mathematical sociology on establishing precise measures of preference, significance, desirability, and other "value" measures.

All this adds up to a steadily increasing storehouse of techniques that may be integrated into CC structures for particular applications. The payoff in this endeavor is the ability to balance qualitative discussions of complex problems with quantitative information summarizing individual and group judgments in a manner that can potentially reduce individual effort needed to bring about a group decision, and simultaneously to improve the quality of that decision as well as the group members' commitment to it.

It is the integration of these techniques that may allow the development of "collective intelligence" systems. In such a system a group of humans would have to meet at least two specific criteria to produce a collective intelligence capability.

1. The group must be able to react to the problems confronting them with the same rapidity as a single individual.
2. The decision arrived at by the group must be at least as good as the best solution offered by any member of the group.

Many experiments can be devised to begin testing the feasibility of these objectives within a CC environment. One example would be the logical extension of what now takes place every year in the artificial intelligence community, where a contest is held to see which artificial intelligence system is the best chess player. Chess would be an appropriate means by which to establish, from an academic standpoint, that the foregoing criteria can be satisfied. In fact, any artificial intelligence chess-playing system can be

ISBN 0-201-03140-X, 0-201-03141-8 (pbk.)

redesigned to be a decision aid for any group attempting to exercise a collective wisdom and act as a single entity chess player. This leads to the further hypothesis, open to argument, that the collective intelligence system should be able to do better than any automated (no human intervention) artificial intelligence system, provided the group has the same processing algorithms available to it.

Psychometrics and Sociometrics

This discipline area has achieved a high degree of ability to expose differences in values and perceptions of concepts among individuals. With methods such as multidimensional scaling (Carroll and Wish, 1975) it is possible to use simple similarity ratings by individuals of different concepts (e.g., rating similarities of countries, people, relationships, jobs, objectives, or tasks) to expose the hidden value dimensions that underly such judgments. What is significant is that these value dimensions may not be obvious to the people making the judgments. This lack of awareness of what dimensions underlie the views of individuals and their relative weights on these dimensions is what causes much of the difficulty for a group in reaching a common understanding. Obviously, it is important to expose these value scales in processes such as negotiation and/or conflict resolution, where various subgroups may represent very different interests and have very different underlying views on the concepts, but may use the same words to represent those different concepts and assumptions.

Most of the work to date in this area is from the classical scientific standpoint of utilizing these techniques as a measurement device. From this standpoint a psychometrician measures the values and/or preference scales of a human group. This measurement may be for the academic objective of understanding a certain value area or for the very pragmatic objective of better advertising. Certainly, the area in which these techniques are heavily applied today is the determination of what underlies a consumer's decision to pick a particular brand of cereal, panty hose, or whatever. From the CC standpoint, however, we are less concerned with measurements as a direct objective. We are far more concerned with utilizing the measurements as a mechanism for feeding back information to a group, and attempting to use this information to allow a group to understand why certain differences exist, as well as to allow the group to make conscious changes of values. To date almost no work has taken place on utilizing such techniques as indirect communication devices. It is a very promising but neglected area.

Distribution and Integration

Even a very large computer completely dedicated to computerized conferencing is capable of supporting an active user population of only thou-

ISBN 0-201-03140-X, 0-201-03141-8 (pbk.)

sands. Advances over the next decade are likely to drive the number into tens of thousands. If developments go in the direction of diversity of systems tailored to specific applications, then we must face up to the problem of interconnect or integration of these systems. Obviously there will be a need for individuals to send information from one system to others and to retrieve information from one or several systems for use in another. An additional facet of the problem is that the user may not be sure which system has the needed information or conference. Interconnection into a single Network Nation, is the problem of distributive processing and the extent to which we can make a network of computers appear as one coherent black box to the user. Distributed processing will require digital transmission networks that are far more intelligent and adaptive than those now in existence. This area deserves attention not only for computerized conferencing requirements but for information systems as well.

It is becoming common for digital* information to be converted to video, and as a result of the home computer market, the device for doing this is now very cheap and widely available. Devices to go the other way, digitalizing a TV picture, are still very expensive and limited to specialized applications. In the years to come we expect significant breakthroughs in the ability of organizations and society to handle information through improved and cheaper devices to automatically convert information from one medium (digital, video, voice, paper, films, analog) to another.

Going from digital to the others is the cheapest process today. For example, digitalizing the spoken word is more expensive than a computer generating the spoken word. Printing text from a computer is inexpensive, but optical readers to feed computers are still very expensive. However, there is sufficient need that a great deal of effort is being devoted to improving the devices.

What are not being examined are the implications, from a systems standpoint, of breakthroughs in this area of conversion devices for the information and communication systems of the future.

Cumulative Information Systems

The emphasis to date in this area has been on the collection of huge data bases and on efficient techniques to allow searching of these data bases. Although this area has seen tremendous successes in the sense that what users now have is considerably more useful than heretofore, it will only be after we merge or integrate communications into these systems that major new advances will be possible.

---

*Digitalization is the use of discrete signal elements to represent data; analog is the use of variations in frequency and/or amplitude of a continuous electronic signal.

ISBN 0-201-03140-X, 0-201-03141-8 (pbk.)

The problem with current information systems is that the content of items is fixed. Users who discover that an index is wrong or that material should be changed have no way of modifying the systems they are using. The element of communications would allow minimizing the intermediaries between those who create information and those who utilize it. Abstracts, data elements, commentary, footnotes, statistics, etc., could all be updated by those who represent the source. The distinction between a draft, preprint, publication, or reprint now turns into the same "paper" or set of information, merely modified by the author as he or she builds on the comments from the readership. Abstracts and indexes can be improved based on the experience of those who utilized them. Raw experimental data are gradually turned into analyzed results and accumulated tested hypotheses, within the context of the same data system. This incorporation of group communication structures into highly adaptable data base structures represents an as yet unexplored challenge in the design of information systems. It carries the implication of information systems specifically tailored for such areas as economics, psychology, materials, anthropology, and construction commodities.

The design and implementation of such systems would revolutionize the way information is dealt with and exchanged in the society, as well as shorten significantly the interval between the generation and the utilization of information.

Human Simulation

We are at the stage today where computers can simulate very particular human reactions. The simulation of the dialogue of a psychiatrist or that of a schizophrenic is such that a significant number of computer-naive, but otherwise intelligent, individuals can be fooled into thinking they are observing what a real human is saying. The objective is not really to fool other humans; rather, the computer's ability to simulate convincingly the dialogue of a human being in specific circumstances means that the computer can be used to carry out a number of interesting tasks for other humans, some of which have been suggested elsewhere as well (Fields, 1977):

- Check the consistency of a writer's style and provide suggestions to the author on improvement of style.
- Change incoming text to the style with which an individual reader prefers to be presented.
- Filter out of text the interesting content for a specific reader, which may differ from that which interests other readers.
- Aid the writer in achieving a particular type, level, or degree of emotional content.
- Remove, when appropriate, material that might be out of place in a

ISBN 0-201-03140-X, 0-201-03141-8 (pbk.)

particular communication process, such as critical comments in a brain-storming session.
* Detect possible deception in bargaining and negotiation processes.
* Adjust the pace of communications to suit the ability of a particular individual to react to it.
* Intensify the making of a point or lend neatness or polish to a presentation.
* Adjust the text to the reading comprehension level, intellectual level, or sophistication of the reader.
* Analyze for an author the reactions to his communications.

The ability to simulate implies the ability to modify actual human communications. This "Cyrano de Bergerac" capability has its dangers as well as its benefits. If successfully developed, it will no doubt be considered counter-productive, possibly illegal, or at least against the norm in some situations. We can foresee a group of people "talking" to one another by turning on their automatic personal simulators tailored to their styles and personalities, as well as by drawing on a large file of concept positions for each person. The possibility is not as farfetched as it may sound. Whether this is an evolving symbiotic relationship between man and machine or the replacement of man by machine would be an interesting philosophical debate.

Summary and Conclusions

The true integration of computer power and communications offers some unique possibilities for new applications in many research areas, including social psychology, opinion research, and information science.

Information science to date has been the weak sister of computer science. In computer science the questions of the effectiveness and efficiencies of hardware and software are at least well-structured problems. In information science the associated question, whether computer systems accomplish anything for a human in a cognitive sense, is less well structured. To a large extent the field of information science is not represented by an agreed-upon paradigm by those that practice it. We suspect that the blending of human communications into information systems will provide the basis for the emergence of a paradigm. We are now presented with the capability of "digitalizing the cognitive process" so it can become part of the research process and be dealt with as an integral part of the system. The realization of the dreams of the information scientists now seems possible.

There are many different morphologies that could be used to provide a context for the range of research potentials that computerized conferencing

ISBN 0-201-03140-X, 0-201-03141-8 (pbk.)

offers. The one we have chosen is based on emphasizing the interrelationships of the factors.* The six major components of the structure can be viewed as elements in a counterclockwise circular impact model:

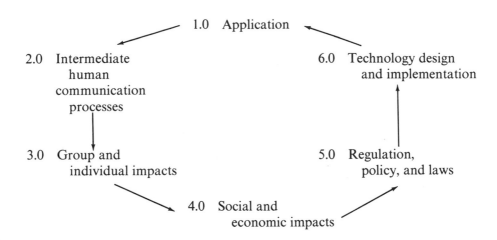

1→2   An *application* (1.0) includes the system, the nature of the tasks, the users, the organization, etc. As a result of the specific application a communication structure is created that can be characterized by various intermediate processes (2.0). We have very little knowlecge about how to optimize the relationships that exist between 1 and 2 and in many cases we do not know the specific impact of a factor in 1 on a factor in 2.

2→3   As a result of these *intermediate processes* we have group and individual impacts (3.0). Some knowledge of the relationships between factors in 2 and their impact on factors in 3 comes to us via studies of other communication media. However, there are sufficient unique aspects of computerized conferencing for there to be sizable gaps here also. This is particularly true where attempts would be made to carry out applications that could only be otherwise accomplished by face-to-face or co-located groups. The psychological process appears to be different enough to prevent extrapolating face-to-face results into the CC environment. Also, a number of potential application areas represent new concepts for the use of communications, so that there is no basis from which to extrapolate.

---

*This structure was utilized as the basis of the 1977 report by Turoff et al., "Human Communication via Computers: Research Options and Imperatives," supported by the Division of Mathematics and Computer Science of the National Science Foundation.

ISBN 0-201-03140-X, 0-201-03141-8 (pbk.)

3→4 The process of going from *group and individual impacts* to multi-group or social and economic impacts is largely one calling for forms of technology assessment. At this point there is general agreement that these impacts in category 4 are going to be large when one looks 10 to 20 years into the future. Beyond itemizing and making gross estimates, the only way to gain a better understanding of category 4.0 is through major field trials.

This seems to be one area of technology where it can be hypothesized that it is meaningful to do technology assessment by experimentation utilizing the technology itself. To date technology assessment has largely been viewed as paper-and-pencil studies, which we do not feel would be as applicable to this area of technology as actual field trials.

4→5 An awareness of *social and economic impacts* or potential ones leads quite naturally to considerations of regulation, laws, and policy (5.0). Here there is opportunity for good analysis since many serious issues appear to be raised by the technology of computerized conferencing that have not yet surfaced in the decision- and policy-making areas. At this stage it may be more of an educational and awareness effort than one of research, except in the area of some new looks at economic impact potentials. Convincing evidence to sway policy makers is going to depend on field trials when it comes to specific application potentials outside of pure business use.

5→6 Over the long term, the regulatory and legal situation may very well determine what is available in terms of the technology. It is the unfortunate circumstance of regulatory uncertainty that is probably preventing sizable investments in commercial research and development into computerized conferencing except for pure business use. This is one of the reasons we must look to government to fund the research and development associated with this area in terms of systems to support various experimental applications. In terms of the variety of potential designs and the flexibility needed for experimental utilization of these systems, there are a significant number of basic computer science problems that arise in this area.

6→1 We now come full circle as the state of technology makes possible new options and the trial of new applications. Without the technology there can be no experimentation to evaluate the effectiveness of the application and determine its consequences. There is a strong feeling by those engaged in this research area that experimentation is the meaningful key to assessment.

Because of the integral role of the human in this type of computer system we need an interdisciplinary approach that integrates the perspectives of both the computer sciences and the social sciences. Imagination is needed in the application of existing disciplinary perspectives and techniques to a new subject area. This imaginative approach is necessary in the design of software to operate CC systems and in the evaluation of the human use of these systems.

ISBN 0-201-03140-X, 0-201-03141-8 (pbk.)

For Discussion

1. What impact would nationwide conferencing and associated polling abilities have on the political process? On the marketing of goods and services?

2. What are the dangers of being able to gather attitudes and data on a mass basis from nonrepresentative samples of subgroups in a society, such as those who bring about change?

3. What is the reciprocal relationship between computerized conferencing design and the formation of user groups for communication?

4. Provide some examples of problem situations for which a true "collective intelligence" capability would appear to be beneficial.

5. How would you judge the attainment of a collective intelligence capability in the situations specified in discussing item 4?

6. In what type of situation would it be wrong for individuals to use simulators to generate their communications? How "wrong"? That is, should it be considered impolite or illegal?

7. What has been your working definition of information science? Has the concept of computerized conferencing caused a modification?

8. Discuss the benefits and dangers of being able to extract underlying value differences among the members of a group from the communications that take place.

9. Describe specific discussion situations where you would like to see value differences exposed to the discussion group.

10. What would be the most startling results of cheap, universally available devices making it possible to speak into a terminal?

11. Can you describe a specific data base system where it would be desirable to integrate communications? What types of communications would you incorporate?

12. What specific new kinds of communication experiments are made possible by computerized conferencing?

13. Will computerized conferencing increase the "pace" of institutional life (i.e., the rate at which decisions are made and actions taken)? What impact will computerized conferencing have on the management of communications?

14. Will computerized conferencing create new human roles in communication processes? What types of data bases would seem appropriate to a general CCS?

15. How could computerized conferencing facilitate the capturing and structuring of "lore"?

ISBN 0-201-03140-X, 0-201-03141-8 (pbk.)

16. What are the potential rules or procedures for CCSs to "forget" things?
17. What potential new forms of etiquette might emerge for such things as leave-taking, getting to know someone, and making introductions? What is the potential for learning via CC discussions?
18. What is likely to be the impact of computerized conferencing on children when they begin to use it?

ISBN 0-201-03140-X, 0-201-03141-8 (pbk.)

PART III

Projecting the Future:
The Technology
and Its Regulation

THE BOSWASH TIMES

The Computerized News Summary Service of the Megalopolis

September 12, 1993

Simulated Psychiatrists Sued

Clifton, New Jersey, Sept. 11, 1993. The Consumer Protection League has filed a formal protest with the Consumer Protection Agency. They charge that a significant number of psychiatrists are using automated computer responses in patient interviews, without informing the patients or families of the patients, and are charging rates for these therapy sessions as if they were personally conducted. The records produced from some conferencing systems show examples of psychiatrists engaging in 20 or more interviews or consultation sessions simultaneously, with such computer assistance.

Sports News

Santa Rosa, California, Sept. 12, 1993. James Harwick today won top place in the Space Wars competition with 169 kills, 33 ties, and only 5 defeats in space battle encounters this year. James, a very typical 17-year-old Space Wars fanatic, explained how he has completely designed his own customized terminal and a more flexible control system than that provided by Space Wars Incorporated. James did not seem too excited by his prize of life-long free access to the Space War Game and is busily dickering with manufacturers for royalties on production of his patented control units.

Legion Protests Un-American Game

Memphis, Tennessee, Sept. 11, 1993. The American Legion and local Chamber of Commerce protested to the School Board today that the "Limits To Growth Game" being used in the public school conferencing system is un-American and should be removed from the computer library. They believe it teaches students to question the values that made this country great.

District to Disband the "Dis-Dis"

Washington, D.C., Sept. 11, 1993. Congress has quietly requested that the new "Dis-Dis" (Disagreement Distiller") analytic package added to the Congressional Computerized Conferencing System be removed or deactivated. This package automatically scanned the text of *Congressional Record* entries, press releases, and published newspaper accounts of speeches of members of Congress, with the purpose of classifying and distilling their positions on all issues. One of a family of new text-oriented analytic programs

ISBN 0-201-03140-X, 0-201-03141-8 (pbk.)

285

meant to stimulate dialogue, the academics who designed the system thought that it would be a great boon to the legislative process. A Congressperson would use the system in either of two ways. An issue might be entered, and then a printout would be given of all other recorded positions on that issue, classified by whether they were in agreement or disagreement. Alternatively, a name could be entered, and the printout would show the extent to which the second Congressperson agreed or disagreed with the first, issue by issue. In either case, the user could request the key sentence or two and date and source, for the recorded position on the issue.

Apparently, reporters have been able to gain access to the system and have used it in an unintended way. Past speeches of individual Congresspersons have been scanned in order to turn up inconsistencies, with the conflicting quotes then presented in critical articles. The view on Capitol Hill is that Dis-Dis has turned into a destructive political tool for "computerized mudslinging" (in the words of one irate victim), and has bred considerable animosity. There is no doubt that it will be removed from service. However, the new headache for politicians may not be cured so easily. It is rumored that a consortium of leading newspapers is now negotiating with the original contractor to acquire their own copy of the system.

Computerized Trials

Spokane, Washington, Sept. 12, 1993. This city has become the first to mandate that trials of all criminal cases be conducted via a computerized conferencing system structured to emulate courtroom proceedings. While many localities have offered this on an optional basis, it is expected that the action by Spokane will be tested in higher courts. The differences of opinion among lawyers and judges are well known. However, many scientific studies now support the position that justice is more equitably administered in a non-face-to-face environment.

Equal Opportunity: The Home Terminal

Shaker Heights, Ohio, Sept. 11, 1993. The City Council was shaken recently by a class action suit filed on behalf of poor segments of the community. It asks that if the City Council continues to hold its monthly public meetings over the local computerized conferencing system, it must pay the expenses of those citizens who wish to observe or participate, including the rental of terminals. The Council argues that the presence of public terminals in the library fulfills its obligation to provide public access. The suit argues that the limited hours at the library and the long waiting times for use of a terminal, in fact, give poor people an unequal chance to participate in comparison with those who own their own home terminals.

ISBN 0-201-03140-X, 0-201-03141-8 (pbk.)

CHAPTER 9

Structured Communications

Introduction

The fragrance of the future of computerized conferencing emanates from its ability to provide structure to enhance the human communication process. Specification of such factors as the number of participants; the roles that they play; who may communicate with whom, how, when, and under what conditions; are aspects of structure. Even where a communication structure is not explicitly designed and imposed on a group, there will be an implicit or emergent structure.

There exists an obvious need for structure as the size of a group increases; hence we have evolved highly structured parliamentary systems for large face-to-face groups. Such large-scale events as a parade could also be thought of as a particular structure for communication and interaction. The "paraders" are to move down the center of the street at a pace determined by the leader and are not to move backward or wander off on side streets. The "watchers" are to stay on the curb or behind ropes or barriers set up for the occasion. Forms of communication are to be limited to certain kinds of nonverbal display by the performers; and the watchers are not to try to interact personally with the performers.

From this point of view, the democratic voting process with modern voting machines can be seen as a way of structuring communications within a very large group (the nation) in order to assure equality of opportunity to participate and freedom to oppose rather than conform to those in power. The written input into a voting machine with closed curtains provides anonymity and the opportunity for each person to carefully consider input and to have an equal chance for input. It can also be seen to be something like the "final round" of a Delphi study, in that primaries and public opinion polls have given the public an idea of the opinions of the group on the candidates and issues before voting. Likewise, a bureaucracy can be viewed as a structure for

Starr Roxanne Hiltz and Murray Turoff, The Network Nation: Human Communication via Computer

ISBN 0-201-03140-X, 0-201-03141-8 (pbk.)

the flow of information and communication for decision making and implementation in large-scale organizations.

There are many small-group interaction processes that have been implicitly recognized as needing structure because they involve communication in a group marked by heterogeneity, conflict, competition, or lack of any shared norms for interaction. An example is courtroom procedure, which is a very complex structuring of who may say what to whom. This includes the possibility of a judge's telling the jury to disregard any communication that falls outside the prescribed structure—that is "out of order."

Our presumption is that for many situations, an explicitly designed communication structure will be more likely to produce a fruitful group process in terms of both the efficiency of the process and the quality of the results. As far as a generality can go, we believe this applies to a majority of group communication situations. It is also our conviction that the prime research objective associated with the development of CC systems should be increased understanding of what communication structures and protocols are appropriate to various human group communication situations.

There are an infinite number of different human communication structures that can be mediated and facilitated by computer. In this chapter, we will review some of the most successful kinds of attempts at structuring the human group problem solving process in other communication modes, and suggest how these might be implemented in computerized form. It is hypothesized that the computerized mode would present certain advantages over the face-to-face mode or written mode in which such processes as Delphi, "nominal group technique," or brainstorming are now conducted. However, none of these designs have yet been implemented or tested in the environments to which they apply. So in a significant sense this chapter represents a set of opportunities, ideas, or concepts for designers and experimenters to consider and improve on.

For the first two areas covered, Delphi and nominal group processes, significant work on structured communications has taken place, but not in a computerized conferencing environment. Both have in common the incorporation of the following kinds of "structuring."

1. Anonymity;
2. Independent generation of ideas or judgments, by assuring that all participants have an opportunity to think and record their ideas or judgments before receiving the ideas of others;
3. Specification of modes of communication for some or all the communication, i.e., the use of written communications;
4. Mechanisms for assuring equality of opportunity to participate;
5. Appointed facilitator(s) to assure the flow of communications in the prestructured manner (rather than reliance on informal leadership from within the group itself);

ISBN 0-201-03140-X, 0-201-03141-8 (pbk.)

6. Specification of allowable subjects of and forms of communications (example: voting or discussion segregated by time period);
7. Some sort of organized feedback to the group of the "input" of each member and the aggregate "group decision" that is emerging;
8. Specification of allowable "who-to-whom" patterns of communication (i.e., no private communications).

These are the key techniques that constitute "structuring" of group decision-making processes rather than leaving these processes to "occur naturally."

The main purpose of such structuring is to try to elicit all the ideas and expertise available in the group, by creating conditions that prevent conformity and elicit input in a systematic manner, rather than simply according to who speaks first or most. Since Asch's classic (1951) studies on group conformity pressures, it has been acknowledged that people tend to "go along with the group" rather than state their own independent judgments. The conditions that seem to produce this inhibition among group members are "surveillance" (others knowing the identity of the person stating the opinion or making the judgment, rather than anonymity); ambiguous situations for which the participants have little confidence that they are making a correct response; and being faced by a unanimous majority. (See, for instance, Deutsch and Gerard, 1955; Hardy, 1957; Allen, 1965.) On the other hand, it has been experimentally demonstrated that when these conditions are removed, there is no tendency toward conformity (Tyson and Kaplowitz, 1977).

The introduction of anonymity is thus one of the strongest techniques to prevent conformity to group pressures. Another is to lessen group pressure by changing the mode of communication from face-to-face discussion to a medium that creates some "distance" among the members. Audio-only communication and written communication have this effect. For instance, Stanley Milgram's famous experiment on obedience to authority studied the conditions under which a person would conform to an order to give painful and potentially fatal shocks to a second person. Among his findings were that when the "authority" gave the order over the telephone rather than in person, only one third as many subjects complied with it (see Milgram, 1965).

Another problem addressed by structuring may be to make explicit differences in underlying assumptions and epistemologies among participants in the process. People employ different ways of assigning validity or of measuring the degree of truth in some conclusion, observation, or "fact." Some of these have been characterized by Churchman (1971) as the Leibnitzian, Lockean, Kantian, Hegelian, and Singerian inquiry processes. The implications of these and others for structuring communication processes have been discussed by Mitroff and Turoff (1975) and by Scheele (1975). What it amounts to is the belief by many that a great deal of the difficulty in getting a group to come to grips with a complex problem in a collective manner is that

ISBN 0-201-03140-X, 0-201-03141-8 (pbk.)

members of the group will often take very different views on the nature of "truth" or validity for a given "fact." This is further complicated when these fundamental differences in perspective are not recognized at the conscious level.

It is therefore necessary to evolve communication structures that will promote the exposure of differences that exist at fundamental levels. These communication structures may even incorporate analysis tools to aid the process. How often do two people disagree only to later (if at all) discover that the disagreement existed because the same phrase or concept was being defined differently by both of them? With the variability possible in language, free-form discussion by any group that does not have a history of communication or share the same disciplinary background can almost be predicted to exhibit such problems. If we can accept as a valid objective of computerized conferencing the goal of creating collective intelligence capabilities, then these can only emerge via structures within which a group can effectively demonstrate an ability to produce results and to make better decisions than any member of the group acting as an individual.

The Delphi and nominal group processes represent hundreds of communication designs yet to be translated in some appropriate form to a CC environment. Because of the potential here we shall treat these in some detail. Following that we move into a series of speculative situations where no real experimental knowledge exists on the benefit of alternative structures. However, the processes discussed are very real in terms of typical group objectives. We hope the reader will come to understand the rich potential that exists for computerized conferencing and the host of open-ended researchable issues in human cognitive communication and information processes that are introduced by the incorporation of the computer as an integral part of the communication process.

Delphi

Represented in the literature on Delphi (Linstone and Turoff, 1975) are many designs for structured communication processes that have proven successful in addressing a particular type of problem. For illustrative purposes we will discuss only two examples from this literature.

Forecasting Delphis

One of the more common applications of Delphi is to forecast problems. How do you, for example, allow a group of planners in the steel industry to come up with a joint forecast (Goldstein, 1975)? The basic structure, subject to some variation in different applications, is the following.

ISBN 0-201-03140-X, 0-201-03141-8 (pbk.)

ROUND OR PHASE ONE

The members of the group are first asked to make estimates of the future values of certain key variables usually used to measure the situation. For each forecast they are asked to supply qualitative statements of two types:

1. Assumptions they have made in order to arrive at an estimate.
2. Uncertainties—things they are not assuming but that would cause them to change their estimation if they did occur.

The monitor team collects this information and will usually discover that some participants' uncertainties are others' assumptions, and vice versa. Duplications and excessive language are eliminated, and both assumptions and uncertainties are provided back to the group as possible assumptions as well as a distribution of the forecasts.

ROUND OR PHASE TWO

The group is asked to vote on the confidence or validity of each assumption or assertion, utilizing an appropriate scale that ranges from high confidence through "not determinable" to complete disagreement.

These are collected and for each variable the potential assumptions in ranked order from highest confidence on down are listed. The summary of the original group estimate is also provided for the respondents.

ROUND OR PHASE THREE

The group is asked to reestimate the variables now that they have seen the exposed underlying assumptions for the group as a whole. They may also be asked to revote those assumptions exhibiting prolonged or highly spread distributions.

The result of this process is a set of estimates that are supported by a collection of explicit accepted and rejected assumptions, representing the collective wisdom of the group. Those assumptions that fall in the nondeterminable category provide an explicit qualitative expression of the uncertainty in the estimate.

An exercise of this sort may involve 20 to 50 people considering 10 to 100 variables, and generate 20 to 50 assumptions per variable. It represents an extensive pooling of knowledge in a highly efficient manner in terms of the time of the experts involved. The end user is provided a much clearer perspective on the foundation for the forecast than is the case for most other forecasting techniques.

Policy Delphis

Another type of Delphi structure is utilized to look at policy issues. The objective here is not obtaining consensus in the form of a single pooled

ISBN 0-201-03140-X, 0-201-03141-8 (pbk.)

judgment, but the development of the strongest arguments for or against particular resolutions of a policy issue. The purpose is to expose the underlying factors bringing about the different resolutions of the policy issue (Turoff, 1975c).

ROUND OR PHASE ONE

The group is given the more obvious resolutions of a policy issue and asked to suggest others. Members are then asked to designate how desirable and how feasible each is. They are also asked to specify the arguments in support of their position on each possible resolution.

The monitor group will organize this so that pro and con arguments are matched as much as possible, and this becomes the input for round two.

ROUND OR PHASE TWO

The group is asked to vote on the importance and the validity of each argument. In addition, they are requested to offer any stronger arguments and to explicitly fill in the counterpoints for arguments that are not matched to others.

ROUND OR PHASE THREE

The group is now provided with the ratings on all items and given the opportunity to revote on any as well as to fill in any missing votes on new arguments.

Usually items fed back are reordered by the voting process into categories of desirability, feasibility, importance, and validity. A typical example occurred in a Delphi done in 1974 for the Canadian government. The respondents were given the detailed internal bureaucratic procedure for determining how much money is to be spent on a new office building. They were asked to itemize the strong and weak points of the process and to suggest changes. All three item categories were summarized and processed through the foregoing round structure. The result was a shopping list of potential changes to both policy and procedures, itemized and rated by desirability and feasibility and cross referenced to strong and weak points of the current processes.

Computerizing Delphi

Because individuals often believe that no intelligent people would make a forecast or resolve a policy issue differently than they themselves would, strong differences are often not perceived until the second round of a Delphi. This means that some people do not realize the need to express well-thought-out arguments until it is obvious there really are other people who disagree with them. As a result, much good material does not enter the process until the second round, and four or five rounds could be necessary for an issue to

ISBN 0-201-03140-X, 0-201-03141-8 (pbk.)

be treated properly. This is one reason why the introduction of computerized conferencing into this structure would probably create a more favorable policy analysis process than does the paper-and-pencil process utilized in the standard Delphi exercise.

Besides the reduction of elapsed time to carry out a Delphi via the computer, the other significant impact is the ability for the process to flow steadily and incrementally. In other words, forecasting of one variable or one policy resolution can be examined first in computerized conferencing and carried through the whole process. Alternatively, different items could be in different phases of the process according to the wishes of the group. This provides a greater ability for the group to focus its effort and should result in raising the quality of the result.

Delphi is a process that has received very little attention in terms of formal evaluation of the process other than the careful work of Dalkey, Martino, and a very few others. For short-term Delphi forecasts, accuracy can be checked against subsequent reality. For example, incorporating Delphi into sales forecasts has been reported to reduce error to less than 1% (Basu and Schroeder, 1977). Today most Delphi work is applied or result oriented, since Delphis are largely done and paid for by organizations with a need to study some specific topic. We really have very little understanding of its internal dynamics as a human communication process. As a result, Delphi design is largely an art form and there are far more people who have tried to do one Delphi than people who regularly design successful efforts. There is a strong need for careful experimentation in this area, which computerized conferencing may make a more feasible process. It is hoped that there will emerge guidelines that will better enable future designers of Delphi processes to avoid errors rather than have to learn from a trial-and-error process. With Delphi studies normally costing from $20,000 to $300,000, some mistakes can be very costly.

Nominal Group Technique

Direct comparisons of the usual face-to-face interaction mode for group decision making with more structured and less "intimate" modes are very important and suggestive of a promising area for research with computer conferencing. Van de Ven and Delbecq (1974, p. 606) have developed and utilized what they call the nominal group technique for group problem solving:

> The nominal group technique (hereafter NGT) is a group meeting in which a structured format is utilized for decision making among individuals seated around a table. This structured format proceeds as follows: (a) Individual

ISBN 0-201-03140-X, 0-201-03141-8 (pbk.)

members first silently and independently generate their ideas on a problem or task in writing. (b) This period of silent writing is followed by a recorded round-robin procedure in which each group member (one at a time, in turn, around the table) presents one of his ideas to the group without discussion. The ideas are summarized in a terse phrase and written on a blackboard or sheet of paper on the wall. (c) After all individuals have presented their ideas, there is a discussion of the recorded ideas for the purposes of clarification and evaluation. (d) The meeting concludes with a silent independent voting on priorities by individuals through a rank ordering or rating procedure, depending upon the group's decision rule. The "group decision" is the pooled outcome of individual votes.

Note that the kinds of operations performed by the participants could be done by computer conferencing, without the uneasiness that sometimes accompanies sitting around a table and looking at one another without talking.

The effectiveness of the NGT mode of decision making was compared with a "normal interacting" group communication process and a Delphi process, conducted as described next (Van de Ven and Delbecq, 1974, pp. 605–607).

> The format followed in interacting group meetings generally begins with the statement of a problem by the group leader. This is followed by an unstructured group discussion for generating information and pooling judgments among participants. The meeting concludes with a majority voting procedure on priorities, or a consensus decision . . . . Unlike the interacting or NGT processes where close physical proximity of group members is required for decision making, participants in the Delphi Technique are physically dispersed and do not meet face-to-face for group decision making . . . . While considerable variance exists in administering the Delphi process, the basic approach, and the one used in this research, is as follows: Only two iterations of questionnaires and feedback reports are used. First, a questionnaire designed to obtain information on a topic or problem is distributed by mail to a group of respondents who are anonymous to one another. The respondents independently generate their ideas in answering the questionnaire, which is then returned. The responses are then summarized into a feedback report and sent back to the respondent group along with a second questionnaire that is designed to probe more deeply into the ideas generated by respondents in the first questionnaire. On receiving the feedback report, respondents independently evaluate it and respond to the second set of questions. Typically, respondents are requested to vote independently on priority ideas included in the feedback report and to return their second responses, again by mail. Generally, a final summary and feedback report is then developed and mailed to the respondent group.

The task chosen for systematic comparison of communication structures was one that was meant to represent a subjective "real-life" human relations type problem. There was no clearly "correct" solution and there was emotional involvement and different vested interests represented among the

ISBN 0-201-03140-X, 0-201-03141-8 (pbk.)

participants. Specifically, the problem was to define the job description of part-time student dormitory counselors who reside in and supervise student housing.

Sixty group sessions of seven members each were conducted, with heterogeneous members (student residents, student housing administrators, faculty, academic administrators) representing different points of view. Dependent variables were

1. Mean number of unique ideas generated;
2. Mean total number of ideas;
3. Quality of ideas (in terms of originality, practicality, and creativity, as rated by judges);
4. Satisfaction of the group members with the process and outcome of the decision-making process, as measured by the following five questions.
    (a) To what extent did you feel free to participate and contribute your ideas?
    (b) To what extent did you feel your time was well spent in this meeting/completing these questionnaires?
    (c) How satisfied were you with the quantity (number) of ideas generated by your group?
    (d) How satisfied were you with the quality of ideas generated by your group?
    (e) To what extent do you feel the group meeting/series of questionnaires you just participated in is an effective way to deal with the problem?

(This summary of dependent variables is from Van de Ven, 1974, pp. 14–27.)

In terms of quantity of ideas, NGT groups generated 12% more than the Delphi groups (difference not statistically significant). Delphi generated 60% more ideas than the interacting group process (significant at $p < .01$). In terms of satisfaction, the NGT groups were significantly higher than Delphi and interacting groups, whose scores were practically identical.

A content analysis of feedback generated by open-ended questions on what was liked most and least about the meeting or Delphi resulted in a summary of the qualitative differences among the three processes as conducted in this experiment (shown in Table 9-1). The authors conclude that:

> This research suggests that when confronted with a fact finding problem that requires the pooled judgment of a group of people, the practitioner can utilize two alternative prodedures: (a) the Delbecq–Van de Ven nominal group technique for situations where people are easily brought together physically, and for problems requiring immediate data, and (b) the Dalkey Delphi technique for situations where the cost and inconvenience of bringing people together face to face is very high, and for problems that do not require immediate solution. Both

ISBN 0-201-03140-X, 0-201-03141-8 (pbk.)

Table 9-1

Comparison of Qualitative Differences between Three Decision Processes Based on Evaluations of Leaders and Group Participants[a]

Dimension	Interacting groups	Nominal groups	Delphi technique
Overall method	Unstructured face-to-face group meeting; high flexibility; high variability in behavior of groups	Structured face-to-face group meeting; low flexibility; low variability in behavior of groups	Structured series of questionnaires and feedback reports; low-variability respondent behavior
Role orientation of groups	Socioemotional group maintenance focus	Balanced focus on social maintenance and task role	Task-instrumental focus
Relative quantity of ideas	Low; focused "rut" effect	Higher; independent writing and hitch-hiking round-robin	High; isolated writing of ideas
Search behavior	Reactive search; short problem focus; task-avoidance tendency; new social knowledge	Proactive search; extended problem focus; high task centeredness; New social and task knowledge	Proactive search; controlled problem focus; high task centeredness; new task knowledge
Normative behavior	Conformity pressures inherent in face-to-face discussions	Tolerance for nonconformity through independent search and choice activity	Freedom not to conform through isolated anonymity
Equality of participation	Member dominance in search, evaluation, and choice phases	Member equality in search and choice phases	Respondent equality in pooling of independent judgments
Method of problem solving	Person-centered; smoothing over and withdrawal	Problem-centered; confrontation and problem solving	Problem-centered; majority rule of pooled independent judgments
Closure decision process	High lack of closure; low felt accomplishment	Lower lack of closure; high felt accomplishment	Low lack of closure; medium felt accomplishment
Resources utilized	Low administrative time, and cost; high participants' time and cost	Medium administrative time, cost, and preparation; high participants' time and cost	High administrative cost
Time to obtain group ideas	1 1/2 hours	1 1/2 hours	5 calendar months

[a]From Van de Ven and Delbecq, 1974, p. 618.

the nominal group technique and the Delphi method are more effective than the conventional discussion group process. [Van de Ven and Delbecq, 1974, p. 620].

The nominal group technique in the face-to-face condition has also been observed to foster participation on an equal and nonconflict basis among members of a very heterogeneous group. For example, a health care administrator gives the following account.

> In my experience, this technique seems to work in groups made up of members from any socioeconomic level or culture. We have used it successfully with members of poverty communities and also with some ... English speaking participants and those needing translators. It has also proven to be effective and acceptable in groups where backgrounds are mixed and ... it allows what in other settings might be conflictual or potentially offending statements to be brought out, clarified, and ranked without strife or discord. It creates situations where those who might not participate for reasons inherent in ideology or in the social situation become full partners. In Gura, for example, the NGT groups contained Chamorran women from the villages who spoke little English along-side women from the capital city who were active in civic affairs. It became quite clear after the first few hours of the day that both groups took deep satisfaction in the work they accomplished [Parker, in Delbecq et al., 1975, p. xi].

One variation of the nominal group technique that makes it closer to the Delphi structure and provides a process for enabling the group to be "self-corrective" is to introduce a second iteration in the voting. In other words, in first-round voting, each person might privately rank-order the options generated on a scale of one to five. These could then be posted on a flip chart or other display mode for each item, and a "resulting mean ranking" could be obtained. In the iteration mode, there is then a brief discussion of the distribution of votes for each item, with persons who gave exceptionally high or low rankings able to explain their reasoning. Often, it will become obvious at this point that the extreme votes are attributable to a different interpretation or kind of evaluation of the item by some members—based on misunderstandings, misinformation, or unequal distribution of information among the group members. The last step then becomes a second round of voting, which usually reduces the dispersion in voting and tends to "improve the accuracy of the group output and create a greater sense of completion among group members" (Voelker, 1977, pp. 2-3, 2-4).

It is important to note that neither unstructured computerized conferencing nor Delphi conferencing need have the disadvantages attributed to the Delphi process as conducted by Delbecq and Van de Ven, and may have all or most of the advantages attributed to their NGT process. For example, there is no need for such a time lag. The conferencing may be synchronous, or in the case of Delphi conferencing, all rounds may be completed within a few weeks (see Turoff, 1971).

Another major inhibitive characteristic found in the two round mailed Delphi was that "there is no opportunity for social–emotional rewards in problem solving. Respondents focus all efforts on task-instrumental role activity, derive little social reinforcement from others, and express a feeling of detachment from the problem solving effort" (Van de Ven and Delbecq, 1974, p. 619). This is not characteristic of the computer conferencing mode, as anyone who has examined a conference record can attest. The second major criticism found by the authors for their Delphi was that "the absence of verbal clarification or comment on the feedback report generated by anonymous group members creates communication and interpretation difficulties among respondents" (Van de Ven and Delbecq, 1974, p. 619). Likewise, a computer conferencing mode can provide ample opportunity for such clarifications or comments.

To summarize, nominal group techniques have been found to produce better results than unstructured face-to-face interaction. Modification of NGT by conducting it on a CC system and by introducing a Delphi-like second round of voting appears to be very promising.

Structured Group Experiences

More generally, any group goal can be reached more efficiently and with a better outcome if thought is given ahead of time to the goals of the group and to how communication can be structured so as best to reach these goals. This forethought can vary from a structured replacement for the usual haphazard and dissatisfying process for introducing group members to one another (see Figure 9-1) to workshops organized to rethink organizational goals and strategies.

The structured "getting to know each other" process described in Figure 9-1 is a very important kind of information and social–emotional exchange to use for a CC session in which many or most of the participants have never met one another. It is important that group members get to feel comfortable and knowledgeable about each other before trying to attack a specific problem solving task. There are other ways to do this. For example, a conference moderator can ask each person to introduce herself or himself. An "on-line party" can be held, as described in Chapter 3.

We agree with Parsons and Bales that problem solving groups must satisfy both their social–emotional needs and their goal attainment needs. Within the goal attainment process, we see four phases. In each of these phases, specific structures or aids might be provided by the computer to enable the group to move toward attainment of their goal.

ISBN 0-201-03140-X, 0-201-03141-8 (pbk.)

Figure 9-1 A structured group experience: Getting acquainted (from Pfeiffer and Jones, 1973, pp. 7–8). (The Pfeiffer and Jones version is for face-to-face meetings. Computer conferencing adaptations are in square brackets.)

Goals
I. To help group members get acquainted quickly in a relatively non-threatening manner . . . .

Group Size
Even-numbered groups of 8 to 20 persons. [The group moderator may participate or merely observe, in order that there be an even number of participants.]

Time Required
Minimum of ten minutes plus two minutes per group member. [Probably 15 plus 4 for new computer conferencing users, or conduct asynchronously over a two-day period.]

Physical Setting
Room large enough for participants to converse freely in dyads, with a circle of chairs for each group, one chair per member. [Each participant at his/her own terminal.]

Process
I. Group members are paired in dyads, and the facilitator instructs participants to "get to know your partner" for the next few minutes (5–10). Participants take turns interviewing each other (without note-taking). They are instructed to avoid demographic data (where you are from, what is your job, etc.) and to try to find out what kinds of characteristics the interviewee has. The individual who is being introduced should "hold his comments" for the discussion period. This is an exercise in active listening, so participants should paraphrase often to make sure that they are hearing what is intended (e.g., "What I hear you saying is . . . "). The person being interviewed should not volunteer too much; the interviewer should have to work at finding out who he is.

II. After the interviewing phase, group members reassemble, seated in a circle. The facilitator indicates that they now have the responsibility of introducing their partner to their group. Each group member, in turn, is to stand behind his or her partner, with his hands on the partner's shoulders, and introduce his partner by speaking in the first person, as if he were that partner. There should be no rechecking between partners during this phase. Both the interviewer and the inverviewee should have eye contact with the "audience." Optional: The person being introduced may be asked immediately afterward to share his reaction to the experience.

Problem Solving Phases

The following represents a structure of phases for a generalized problem situation, in which a group utilizing computerized conferencing might spend days, weeks, or months going through the four-phase process. The particular situation is one in which a group has a specific problem to resolve and the members are assumed to be cooperative. A CC system might be so designed

ISBN 0-201-03140-X, 0-201-03141-8 (pbk.)

as to facilitate each of the following phases by providing appropriate capabilities for each phase in the process.

These phases are quite similar to the processes groups carry out in Delphis or nominal group processes, but are expanded here to incorporate various forms of computer augmentation in terms of aids to the pooling and analysis of the wisdom provided by the group.

1. Phase One: Creativity for Factor Enumeration and Exploration of the Topic

This is the phase in which the group provides a pooling of knowledge as to the factors underlying the issue or the problem or the possible options for problem solution.

A number of specific approaches have evolved for structuring face-to-face groups to enhance creativity. All these can be implemented in a CC environment. Without digressing into opinions about how effective these techniques are, we present a summary of four of them in order to point out the feasibility of translating these methods to a CC environment. Although the structures in these face-to-face processes are often implicit and guided in a subtle manner by the facilitator, they are well enough defined so that some aspects of the structure or process can be automated. It is our feeling that there are many people who cannot work creatively in face-to-face groups, and that computerized conferencing would offer a better atmosphere for enhancing creativity.

BRAINSTORMING

Perhaps the most widely known of face-to-face structures (see Stein, 1975; Osborn, 1963; Parnes, 1967), brainstorming is based on the principles of deferring judgment of an idea and that quantity breeds quality. The rules under which such a session is run are (Linstone et al., 1978):

1. Criticism is ruled out;
2. Freewheeling association is welcomed;
3. Quantity is wanted;
4. Combination and improvement are sought.

A checklist of ideas is gathered for later evaluation, as an input or a source for creative solutions. The makeup of the group is considered quite crucial to the exercise.

A CC system designed to optimize brainstorming would probably limit text items to a small size, might censor items containing negative words and phrases, and utilize stored profiles on individuals to suggest group members. It might also use automated indexing techniques to group and organize items.

ISBN 0-201-03140-X, 0-201-03141-8 (pbk.)

OTHER TECHNIQUES

Among the other techniques that have been employed for creativity enhancement are synectics, brainwriting, and generative graphics. Synectics means "the joining together of different and apparently irrelevant elements" (see Gordon, 1961). The method is based on the use of metaphors and analogies within a systematic framework. Group participants are encouraged to take an "excursion" along a different topic and then try to apply the excursion to the problem at hand.

Since synectics requires training and facilitation, it is likely that the introduction of computerized conferencing would require automating a significant number of the aids and procedures used to stimulate the process. Certain simple techniques like the random choice of words to say something about are quite easy to have incorporated. Others would approach the use of artificial intelligence techniques to achieve the goals of making "the familiar strange and the strange familiar."

Brainwriting (see Warfield et al., 1975) is essentially written brainstorming: each person writes an idea down and passes it to a neighbor, who must add to it. These pieces of paper are passed around until everyone has commented on every piece of paper. With slight changes, in most CC systems this would mean passing a comment to each participant in turn, to make a required addition before incorporating it into the conference.

Generative graphics, developed by Joseph Brunon at the Neuropsychiatric Institute of UCLA, relies upon the graphical organization of text, words, and phrases (see Linstone et al., 1978). Clearly this augmentation of computerized conferencing would require graphical capabilities and the facilities inherent in Nelson's hypertext (see Chapter 10).

We believe that an important factor in stimulating creativity is the breaking down of human and group conditioning and habits that prevent reexamination of operational implicit assumptions. Omar Khayyam Moore in a classic article (1957) pointed out that the Naskapi Indians of the Labradorian Peninsula would utilize the shoulder blade bone of a caribou to guide the choice of the locale of the next hunt. The bone was placed in the fire, and the resulting cracks and burns were interpreted to provide the direction of travel for the hunting group. As Moore pointed out, the use of experience to guide this decision, returning to where previous successes had been made, would have resulted in depletion of food supplies in this area and a probable worse result than the random pattern of hunting that resulted from "reading" the bone. The computer is an ideal mechanism for playing the same role as the burnt bone in many human group processes. This potential to introduce randomness and to shake up established beliefs or foster their reexamination is still largely unexplored.

The other key factor that plagues current creativity processes is the capture and structuring of the results. The key output of the initial phase of a problem

ISBN 0-201-03140-X, 0-201-03141-8 (pbk.)

solving group is the itemization of the causal factors and components underlying the problem. In the next phase, something must be done to order and
integrate all of these items.

This may mean that a number of separate lists are enumerated by the
members of the group, and the computer can collect and track the items
enumerated by providing some sort of auxiliary data base capability for the
easy review of the material in the lists. There may also be some sort of
approval or voting process for the incorporation of an item into the final list.
The subject of each list would depend on the particular problem and might
represent areas such as goals and objectives; decision options; environmental
or uncontrollable factors; consequences of actions; influencing roles (e.g.,
stakeholders), etc.

The degree of complexity of the problem determines the scope and number
of these factor lists that the group may wish to develop. The design of the
computer aids here must be such as to make it easy for a group to develop
and organize such lists, as well as to reach agreement on the items and their
meanings.

As a result of the computer presence, the initial creativity and exploration
phase can become an integral part of all the phases of the problem solving
process.

Phase Two: Evaluation and Consensus Exploration

This is the phase concerned with weightings, ratings, and rankings of items
within a given category. Its purpose is to determine to what extent there is
agreement and/or similarity among the ratings and to provide sufficient
insight for the group so that disagreement can be identified and discussed.

The design problem is to ensure that the actual reason for disagreements is
brought out. For example, the question "Is it a real disagreement or is it a
different interpretation of the meaning of the terms?" represents one of the
typical sources of confusion that arise in any communication medium.

It may be desirable that the analysis of individual ratings be utilized to
stimulate, on an automatic basis, requests for certain individuals in the group
to further define certain items or support their views prior to the analysis's
being made available to the group. It is very difficult in, for example, a
Delphi process conducted by paper and pencil to have the flexibility to utilize
the analysis of what has transpired to selectively solicit additional information on an individual basis. This constraint can sometimes lead to situations
where premature exposure of the group view can foster polarizations in the
group.

Also in this area is the problem of clustering a long list of items into a
shorter list of major items representing groupings of the original items. In

ISBN 0-201-03140-X, 0-201-03141-8 (pbk.)

some cases these clusters may exhibit a structure according to levels of preference or importance ratings. Interpretive structural modeling (ISM) and cross-impact analysis, for example, represent two of a host of morphological, classificatory, or clustering tools that may be useful for this process. The goal would be to develop a library of such tools so they can be applied to the particular situations for which they are best suited.

One can easily hypothesize structures for the explicit purpose of investigating the underlying nature of disagreements and providing a mechanism for resolving them. An example of such a structure follows to illustrate this point.

CONFLICT RESOLUTION

Practically every professional field involved directly or indirectly in some form of human communication or problem solving has its own ideas about this area and how to approach it. Unfortunately, most individuals and groups having real conflicts to resolve do not often lend themselves as subjects for experimental study. The best we can do is to try to provide the tools to aid this situation and encourage groups to utilize them in field trial type atmospheres. Here is one hypothesized CC structure that might be characterized as such a tool. There is some issue that divides a group into three subgroups: those considered to be pro, con, or neutral about a particular view on the issue. Each individual joins a private conference whose membership consists of each of these three subgroups. The conference subgroups attempt to formulate the justification for their view if pro or con, while the neutral group tries to formulate the questions it would like to ask of the other groups in order to decide which is the best position on the issue. A comment generated in one of the conferences may only be passed to another conference (group) if a sufficient majority (perhaps 80%) of the members of the group agree to passing the particular comment.

In addition, any member of a group may decide that he or she wishes to change positions and join another group. To do so, the member can message the members of the group he or she wishes to join and must convince that group to vote in a new member. This procedure has the interesting feature of eliminating group pressure to stay, since there is no knowledge that the "turncoat" negotiations are taking place, and the group will be notified of the change in membership only after it is an accomplished fact. It also means that the person changing sides must have good enough reasons to convince the group that he or she desires to join.

Obviously this process will be dependent on the facilitation and negotiation skill of those in the neutral group if it is to bring about meaningful conflict resolution in a polarized situation. However, the medium and a structure of this type could allow a greater chance for bringing this about than does a face-to-face situation.

ISBN 0-201-03140-X, 0-201-03141-3 (pbk.)

Phase Three: Relationship Judgments and Model Formulation

This phase is concerned with the identification of the primary or key relationships among items in the various lists developed in the prior phases of the communication structure:

What actions will be most likely to produce which consequences?
Who has what power to influence what action?
What objective is fostered by which consequence? etc.

This area represents the challenge of getting a group to focus on only the most significant relationships. For this task the information has to be gathered in terms of some sort of generalized relationship model that can provide the group a structure and an analysis of their judgments.

The degree of sophistication in the model is likely to be a function of the complexity of the task. Some examples of typical modeling capabilities that can be provided are simple binary causal models; Bayesian probability models; decision tree models; Markov chain models; cross-impact models; system dynamic models.

What is sought here is to use model formulations that are not problem dependent but represent the problem according to an appropriate choice in parameters made by the participants in the conference. A significant amount of work has been done in this area to allow individuals who are not necessarily quantitatively oriented to be able at least implicitly to define quantitative relationships. The work in Delphi, gaming, operations research, and subjective judgment analysis or decision theory can be drawn upon for the design of this capability.

Perhaps the most important aspect of introducing the computer in this phase is the potential for examining group consistency on relationships proposed. As models get more complex, the need for incorporating the knowledge of a wider group of experts increases, as do the associated problems in consistency of assumed relationships. This raises questions for research in the area of compatibility of cognitive models and computer models.

The classic problem in utilizing computers to aid the cognitive transfer and use of information and data is that there is a fundamental difference in the way computers "prefer" to structure relationships and the way human beings view relationships. Computers are very efficient in dealing with binary (or yes/no) linkages among data and information items. Most designs for data base structures and attempts at generalized routines in this area usually presume that a relationship between any two items exists or does not exist in terms of typical on/off status situations. In other words, event A is the cause of event B or the class of items X is a subcategory of the class of items Y. The inherent pressure on the design of data structures operating on computers is

ISBN 0-201-03140-X, 0-201-03141-8 (pbk.)

toward pure directed graphs with no "fuzzy" relationships among the nodes. However, when humans deal with the relationships among items, there are many more options.

One example of a typology of relationship options is expressed in Table 9-2. Each entry in this 3×3 table represents a different type of relationship between any two or more items. Each of the nine types has its own representations (quantitatively or qualitatively) and methods in human discourse.

Thus, by merely taking the dimensions of sufficiency and necessity, we see that there are a host of alternatives for the way humans may express the relationships among items. There are also particular discipline areas that have had to pay special attention to a particular class of relationships (shown in parentheses in Table 9-2).

Adding to these dimensions is the problem of determining the degree or sensitivity of the relationship within any one of the subcategories in the table. Problems here deal with degree of linearity, range, quantification, and feedback or direction. All these lead to other possible aids and needs for consistency checks.

Human beings dealing with complex problems in a communication process will often express relationships in a qualitative manner, with no explicit quantitative clarification of the specific nature of the relationship along such dimensions. Potential communication problems develop because different

Table 9-2
Possible Relationships among Two or More Items[a]

	Necessary	Undetermined necessity	Not necessary
Sufficient	Ateleological Cause–effect (Physical sciences)	Causality	Genetic (Biological sciences)
Undetermined sufficiency	Dependency	Circumstantial Probabilistic Fuzzy	Circumstantial causality
Not sufficient	Teleological Producer–Product (Social sciences)	Circumstantial dependency	Accidental Coincidental (Law)

[a]To what degree is event or item A necessary and/or sufficient to the occurrence or observation of item or event B? Is, for example, the statement by a leading group of economists at the beginning of a year that this will be a bad year a self-fulfilling prophecy? Is it due to the reluctance of industry to invest, as a result of the statement itself? Is it merely coincidental that the two events both occurred? Or were the economists merely stating the cause–effect result of a scientific truth, without any impact resulting from the statement itself?

ISBN 0-201-03140-X, 0-201-03141-8 (pbk.)

individuals have unconsciously viewed a relationship among two items differently, without properly communicating the nature of that relationship. This discrepancy is further aggravated by computer models and data base structures that confine individuals to expressing judgments within one classification scheme for relationships. Another aspect of the problem is that many individuals do not consciously understand the distinctions inherent in different forms of relationships. In areas such as planning, forecasting, and evaluation, individuals and groups must deal with related hypotheses about various categories that span the relationship matrix in Table 9-2.

We believe that this is one of the key problems in going from a pure CC environment, where free-form text is utilized to express relationships among items concerning a complex problem, to the augmentation of the process by computer aids to provide structures for the information developed in the discussion. We believe this can be accomplished by developing appropriate data structures that will account for varying types of relationships among items. Such a system might utilize a question–answer input system that allows the person to specify a relationship by its properties without prior knowledge of the logic or philosophy of the classification scheme that underlies the data structure. In other words, data structures and modeling aids could be designed that provide the mechanisms for producing explicit quantitative estimates of subjective expressions of relationships.

By fitting these data structures into a CC environment, a group can see explicitly how they arrive at similar or dissimilar views. In itself, the exposure to a group of people of the extent to which disagreements are the result of underlying differences in the nature of assumed relationships would present a major cognitive augmentation capability.

There are a number of computerized inference systems in which an individual using a restricted vocabulary can build an associated data base capable of inferring the answers to certain questions. To the best of our knowledge none of these inference systems have been implemented in an environment where a group of people are contributing information on a single topic to the same associated data base, with routines incorporated to make the group aware of inconsistencies in the information provided by various members. Such an inference system can be viewed as a restricted form of communication or an augmentation tool to a free-form-discussion-based computerized conference. It is also possible to augment these inference systems with probabilistic weights on the definitiveness of the relationships to reflect the span of relationships exhibited in Table 9-2.

The crux of the concept here is that in certain situations it would be useful for people to communicate indirectly at times, by contributing their premises or assumptions to a model and data structure that evaluates the consistency of the accumulated group wisdom. Disagreements may then be resolved in

ISBN 0-201-03140-X, 0-201-03141-8 (pbk.)

free-form discussion. The likely benefit is much quicker detection by a group of underlying differences that usually do not see the light of day.

When a person enters a relationship into a system of this type, such as a statement that the completion of project A will impact on project B, he or she would perhaps be asked questions leading to the establishment of the degree of sufficiency or necessity implied. Then the system would perhaps discover that someone else has established project C rather than project A in a similar relationship to B. The system would then produce feedback to both parties and perhaps others to aid in determining if each party is forgetting something, or if a real difference of views existed. There is actually a large number of types of inconsistencies for which such a computer-aided system could check.

Phase Four: Comprehension and Decision

A process of the sort we have been describing is likely to leave a group in a position, after the first three phases, in which it may feel satisfied about all the pieces it has contributed and agree that those pieces represent the items and relationships that govern or concern the issue at hand. However, it may also leave the individuals or the group members without adequate comprehension of how to integrate their contributions for what may be a fairly complex problem. We cannot state, at this point, that we have more than a belief as to what is a feasible approach to developing a Gestalt for the group as a whole. This is an especially rough problem when the group is not quantitatively oriented. For certain problems and groups it may require the explicit use of graphic aids. In some cases this can be handled by allowing the group to vary initial conditions on the model they have composed and observe alternative consequences. In some contexts this can be provided in such a manner that the original group or later groups can selectively go back to earlier phases to update or improve the model represented.

GAMES AND ROLE PLAYING

The area of games is one of the most promising exploratory subjects for CC investigations and deserves elaboration. At present, we feel that what will be a desirable tool in many situations is the use of a game in which individual members of the group can assume player roles for both supplying decisions and engaging in dialogue with other players. We find that the work in games to date seems to support the view that games can help players comprehend the nature of a complex situation. For that reason, it is quite common to use them for training and education in urban planning and similar areas to provide insight into the multiple factors involved in such environments.

Once the group has compiled item lists (such as roles, action options, consequences, and environmental factors) and the relationships among these, we can incorporate the design of a generalized game controller that will allow

ISBN 0-201-03140-X, 0-201-03141-8 (pbk.)

the generation of an event-sequenced scenario-game in CC form. This means that the group can play out the "world" model or Gestalt that resulted from their contribution of judgments and views. Such a result also becomes a helpful vehicle for conveying to others what the group has arrived at and discovering if others agree or disagree.

We have already mentioned that four people at terminals could engage in a bridge game with the computer dealing the cards, tracking the play, and regulating the communication structure. Obviously the computer can also aid bridge players in finding one another when they feel like playing. A wide range of board games can also be implemented in computerized conferencing, from Monopoly to business games or any of the popular battle games. Instead of the popular Star Trek, which computer-types play against the computer, there can be fleets of spaceships, each under the control of a different human group, with the added dimension of teams.

However, games have objectives other than recreation, and business, urban planning, and other educational or decision-oriented games can be incorporated into computerized conferencing (Scher, 1976). The major defect that most games exhibit, especially educational ones, is that the communications actually used in the face-to-face game environment usually do not reflect the real world. By putting the game into a computerized communications environment, we can program the structures for communications that the game implies. This may include which players in the game can talk to whom and in what circumstances; costs or resources that must be expended for communications; leaks of communication; rumor simulation and unanticipated breakdowns or busy signals. The computer can act as the game controller, scheduling the events to occur and providing the outcomes based on the actions the role players take. One very significant aspect of this flexible degree of control is the ability to control the clock. Because of this, the game can be played in a regulated time manner (such as every week of play representing a year) or in real time. There are many games where playing in real time rather than accelerated time would be beneficial to enhancing the realism, including some of the disaster type games designed to educate people on how to deal with crisis situations. Since people can interact at a time of their own choosing, a computer-based game can go on over days, weeks, or months, just as for a computerized conference.

Most computerized business games are very limited and usually deal with running a business over many years. This macro approach provides no insight for students into the multitude of small short-term decisions that will occupy most of their time in the "real world," such as:

Trying to estimate costs for a bid on a major project;
A labor–management contract negotiation;

ISBN 0-201-03140-X, 0-201-03141-8 (pbk.)

Arriving at a decision to take over a company;
Deciding to launch a new project.

require a high degree of communications structuring and role playing. The role playing could probably be done more realistically through the computer than in some of the face-to-face acting games used, especially if the student were not able to tell which of the other players were students, faculty, or real-life jobholders playing at their convenience from their own terminals.

Every single game represents a different and probably unique communications structure for incorporation into a CC environment. Very likely the most popular ones in this environment have not yet been invented. This area holds tremendous opportunity and will probably have a large public market in the late eighties.

COMPUTER-ASSISTED INSTRUCTION

Another method of comprehension more tied to conveying the results of knowledge outside the group that developed it is, of course, the educational process. We believe most work in computer-assisted instruction (CAI) has suffered from a lack of the incorporation of structured communications. Although we utilize the CAI label, the form and function of CAI for comprehension of complex situations implies to us an integration of CC concepts.

The value of structuring communications to facilitate learning has been implicitly recognized, of course, by the evolution of the structure of formal education. The classroom and its norms traditionally define the teacher as the originator of the subject to be discussed and process to be followed by all participants. The movement toward "open classrooms," CAI, self-paced or programmed instruction, and "free universities" (or experientially based learning) can be seen as a recognition that the traditional structure is not always optimal.

For all the work that has gone on to date in CAI, we find that a report published by the Educational Research Group of the Institute for the Comparative Study of History, Philosophy and the Sciences, Ltd. (1967), on the subject of structural communications makes the most sense to us and comes closest to our own views on the limited utility of current generations of CAI systems. This document is a lucid thesis on the need for structured forms of communication as an integral part of any educational process and the inadequacy of programmed learning of any form (audiovisual, manual, or computerized) to satisfy this need. This can be emphasized by the following quotes:

> We are familiar with the benefits of centrally prepared audio-visual aids: but these do not allow for interaction between the creative mind of the author and

ISBN 0-201-03140-X, 0-201-03141-8 (pbk.)

the receptive minds of students. At first sight, it would seem impossible to combine the advantages of central preparation of very high quality teaching material with genuine dialogue between the author and the recipient. The attempt to do so by "programmed texts" has given results very limited in scope . . . [p. 187].

In structural communication the aim is to evoke understanding, not to convey facts, except as a by-product. Facts are always necessary, but they are seldom of much value unless they are placed in a framework of understanding. This is true not only of advanced studies, but even at the elementary level . . . [p. 191].

By inserting structure into the act of communication, the author makes it possible for the recipient to reconstruct the mental process behind the message [p. 195].

The document identifies seven stages of what is believed to be the structure behind the educational communication process and illustrates these for various topics.

1. Acceptance—The recipient decides to accept the communication.
2. Intention—Primary orientation toward the author's purpose.
3. Presentations—Assimilation of content of the message and the associated challenge for the receiver to investigate the content (e.g., problems or questions posed).
4. Response—The indicator of the degree of transition from knowing to understanding.
5. Discussion—Recipient is brought to a state of understanding.
6. Comment—Expansion of the original theme.
7. Viewpoints—Integration and review of the communication.

The current generation of CAI systems deals only with stages 3 and 4. As a result, these systems can substitute only for the development of skills through drill and practice on the assimilation of facts. It will require the integration of CAI systems with CC systems to allow the encompassing of the total educational process. For a teacher to control the dialogue in order to promote understanding, very particular communication structures would be necessary. For example, being able to review and decide ahead of time which student comments on questions the teacher should deal with in what order would provide more flexibility than teachers now have in the classroom environment. By allowing the teacher to review pupils' comments before the class as a whole sees them, the inhibition of students about asking possibly naive questions or making embarrassing comments can be reduced, since the teacher can choose to respond to some questions or comments on a one-to-one basis. Also, what a student may consider naive can often generate a significant insight for other class members.

ISBN 0-201-03140-X, 0-201-03141-8 (pbk.)

As yet there has been no design work on what appropriate educationally oriented communication structures should look like in a CC environment. However, there is both the need for such a capability and the promise of benefits to both the educational process and citizens' comprehension of the complex problems that must be dealt with in the future.

TRANSLATION

The decision and implementation phase will often need to incorporate "translation" processes. A structure must be provided for arriving at comparable terms in the language of various cultures or subcultures. Although the type of system deserving immediate attention is one to deal with the translation processes described in this subsection, the long-term research potential is for systems that will check for violations of cultural norms between members of different societies who are attempting to communicate. For example, this would ease the problems that occur between Japanese and Americans because of the Japanese norm that it is impolite to admit that what has been said has not been understood and the American assumption that silence means agreement. In this sort of situation, an explicit structure could put the computer in the position of compensating for such a problem.

A multilingual CC capability is one of great significance and appears to be rather straightforward in nature. In such a conference a separate copy of the proceedings would be kept in each language. The conference structure involves a group of human translators as well as the actual conferees. When an individual writes something in one language, the following sequence of operations might occur:

1. A slot is reserved for the item in all the transcripts in the other languages.
2. If the individual writing uses a particular translator, then the item is delivered to that translator when the author is ready.
3. If no specific translator is designated, then the computer seeks the next available translator qualified to translate to a particular target language.
4. If the writer has used a key word title, this can be automatically translated and put into the proceedings immediately by the computer, using a stored file of keys and their translations, so that the conferees are aware of what is "on the way."
5. When the translation is completed, the next action depends on which of a number of options the writer chooses. The translation might go directly to the transcript, or if the writer can read in that language, he or someone he designates may review it. A more demanding criterion could also be that the translation only goes to another translator, to be put back into the original language so the writer can compare the original with the retranslation.

ISBN 0-201-03140-X, 0-201-03141-8 (pbk.)

Accompanying translation-oriented computerized conferencing should be a data base where the conference group and the translators can accumulate a file of words, phrases, and concepts for which appropriate translations have been agreed on. While this file can be a service to the translators, the proper design of this data base can allow the group to accumulate a documented set of understandings. This means that the discussion would not be overly sensitive to changes in either individual conferees or translators, as is often the case in face-to-face meetings. In other words, the data base structure that underlies this type of conference is not necessarily so straightforward as the communication structure and there is opportunity for some careful experimentation in this area. Such a CC system could effectively be used for real-time discussion if there were a sufficient number of real-time translators to keep up with the conferees. Because of time zone differences, participants located in different countries would tend to use it in an asynchronous manner.

Example and Summary

The four phases described earlier amount to a rather straightforward process:

1. Gathering of the factors underlying the problem;
2. Assessment of those factors;
3. Formulation of a model to relate those factors;
4. Interpretation of the behavior of the resulting model.

The phases suggest a rich list of computer-based analytic aids that could be brought to bear. The analytic aids would be designed to improve the process of cognitive transmission within the group by enhancing understanding and improving the precision of the information as well as exhibiting degrees of agreement in an explicit manner. One can visualize such a system as largely providing a library of tools and protocols from which the facilitator of the process can pick and choose to fit the group and its concerns.

It is possible, however, to design simpler systems that incorporate all four phases within one communication structure. This is done, of course, by restricting the processes to a high degree through focusing on very specific questions and specific measures of relationships. An example of this, at this point, provides a top-down view of the process we have been discussing throughout the chapter with bottom-up specifics.

Thus, in contrast to the generalized system hypothesized earlier, we present in the Appendix a specific communication structure designed in 1971 by Turoff as a "technology assessment workshop" for use by engineers in

ISBN 0-201-03140-X, 0-201-03141-8 (pbk.)

developing technology assessment considerations about a particular area of technology (contributions to an earlier version were made by Steven Maimon). This workshop has been conducted with a number of face-to-face groups and was developed for use by the Committee on Technological Forecasting and Assessment of the IEEE. It incorporates a mix of nominal group processes with Delphi-like processes to form a communications structure that can be implemented quite easily in a CC environment. The workshop design is meant to be a specific illustration of the concept of structuring communication around a specific objective. The reader is also invited to utilize it as a vehicle to experience a structured group communication process.

In this workshop structure a group is divided into four separate face-to-face working groups of 5 to 15 individuals each. These groups each have a specific objective related to the examination of a well-defined area of technology, such as antennas. The groups and their objectives are

1. Technology Group—To describe specific and significant possible developments.
2. Applications Group—To describe specific new application potentials or major shifts in current application areas.
3. Societal Group—To hypothesize societal changes that may have significant impact on this area of technology and its use.
4. Policy Group—To identify key policy or decision issues whose resolution can affect this area.

When a group has focused on an item they believe to be "significant," they fill out a well-structured form providing certain details on the item (see Appendix) and the form is routed to the other groups. Certain of the other groups must provide specific responses to such a form before they can continue with their own objective. For example, the Application Group must itemize possible applications of a technological development sent to them by the Technology Group. If this stimulus provided by the Technology Group causes them to develop an application they believe to be significant, they may start their own form on that specific application. The Societal Group must respond to an application form with a list of societal consequences of that application.

As can be seen, each face-to-face group acts as a single respondent in a four-respondent Delphi process. Although each group starts independently, the structure is such as to force the focusing of all groups on those items that emerge as significant. The format was originally evolved so that each working group could meet separately and the intergroup communication process could be carried out over a long period of time and by mail when necessary. Even for such a specialized area as antennas, the considerations evolved are usually

ISBN 0-201-03140-X, 0-201-03141-8 (pbk.)

quite interesting to the participants; for example:

1. Low-cost home satellite receiver antennas;
2. International CB equivalents;
3. Reduced work week;
4. Regulatory decisions on frequency management.

In general this workshop has been used to sensitize engineers to the considerations that must occur in a technology assessment process. Social scientists and others with the appropriate background were usually included in some of the working groups.

The point of this illustration is that the whole structure translates easily to a situation in which four separate computerized conferences are set up to reflect each of the groups and the appropriate communication protocols are introduced to provide for:

1. Voting on significance within the groups, with the outcome of those votes governing what may be communicated between groups.
2. The introduction of special forms generated by the computer to be filled out as the mechanism for standardizing communication between the groups.
3. Introduction of a data structure to track and compile all of the relevant information on a particular item.

The current version of the workshop itself incorporates no specialized analysis tools and merely provides an organization and structuring for the gathering and communication of information. Each significant item accumulates its own file of associated considerations and the result is a well-structured compilation of the knowledge of the group. However, it is easily seen that many of the analysis tools we have suggested as aids to problem exploration are applicable to this task. In principle, the translation of such structures to the CC environment would make it far easier to conduct this type of exercise in terms of the possible convenience for the participants, and would be likely to produce a higher-quality product because of the basic reflective nature of the undertaking.

The examples of structured communication systems in this chapter are meant to illustrate the rich variety of options available for highly structured computerized conferencing. No doubt the concepts explored in this chapter will seem rather elementary, naive, and nearsighted when compared with what is actually available 10 or 20 years hence. Even in today's simpler systems, experience seems to indicate that designs, once implemented, generate among their users a host of ideas for improvement that somehow escaped the original conceptual process and insight of the designer. This is not to disparage the designer: rather, it is part of the phenomenon that a human

ISBN 0-201-03140-X, 0-201-03141-8 (pbk.)

communication system inherently is, or should be, self-modifying. To survive and serve adequately it must allow the communicators to take over the design role. Traditionally in computer systems this is a role the original designer is reluctant to delegate, which accounts for the demise of many information systems. This adaptive nature or requirement for CC systems also indicates that we must view this area as an evolutionary one with many divergent and useful paths to explore. As you have gathered, we take strong issue with those who seem to feel that there is ultimately one pure communication system best for all to use, analogous to, say, the single universal telephone system. We see and hope for a future that presents a wide variety of systems from which people and organizations can pick and choose in order to fill a variety of group communication and problem solving needs.

For Discussion

1. Consider some games with which you are familiar. How would they change if incorporated into a computerized conferencing environment?
2. One of the authors has often been accused of viewing the whole field of computer and information science as a subfield of computerized conferencing. Do you agree or disagree, and why?
3. By adding robotic systems, we can introduce the ability of members and groups to incorporate remote physical activities. What CC systems would this make possible?
4. Beyond translation, what features or designs might be utilized to compensate for cross-cultural differences?
5. Pick a problem area of concern to you and design the tools you would like see incorporated into a conference system to aid a group in examining that topic.
6. With a world rich in alternative communication structures and overflowing with communications, what sort of "future shock" phenomena do you foresee, and what groups would be most subject to it?
7. What type of system might you design to carry out a religious ceremony? Which religions might be adaptors of this technology for holding services?
8. Would you consider an on-line play as a feasible system? Would people pay to listen in on group discussions of a dramatic nature? What about the audience influencing the dialogue?
9. Would there be a job market for new forms of on-line employment for facilitators, judges in conflict situations, leaders, jokesters, poets, dramatic writers, game controllers, etc?

ISBN 0-201-03140-X, 0-201-03141-8 (pbk.)

10. Given a choice of a course in some subject at your local community college or from a renowned leader in the field over a CC system, what cost differences would sway you one way or the other?

11. For what human communication endeavors would you not be able to use computerized conferencing? Are you sure?

12. How would the nature of a legal proceeding change if done through a CC system?

13. What are some examples of cross-cultural differences that can lead to communication bottlenecks? Do you foresee communication structures or protocols that can be used to overcome these?

ISBN 0-201-03140-X, 0-201-03141-8 (pbk.)

APPENDIX

Instructions for Conducting a Technology Forecasting and Assessment Workshop[a]

Murray Turoff

Introduction

This workshop may be conducted with two to four groups of from 5 to 15 individuals each, allowing anywhere from 10 to 60 individuals to participate. Ideally each group should consist of around seven people plus a group leader. It may be conducted at one three- to- eight-hour sitting or be phased over a number of three-hour sittings occurring at different times.

A given workshop should focus on some reasonably specific areas of technology (e.g., computer networks, sensor systems, mass transit, process control, or computer terminals). It may also be desirable for certain applications of the workshop to focus on societal consequences and policy issues to a narrower range than society as a whole. For example, if the workshop is for a group of antenna engineers, the focus might be on the consequences and policy issues related to frequency management and regulations. In addition, the workshop should focus on a time scale that is somewhere between 5 and 20 years in the future. The exact range should be a compromise between the particular topic and the background of the group. The farther one projects into the future, the more difficult it will be for most individuals to make reasonable estimates or to conceptualize, unless they have had some practice at this.

Each of the possible four groups making up the workshop start off on their own tasks. What they do is communicated to the other groups (via written forms), which require particular responses from particular groups. As the exercise progresses, each group will find less time to originate new considerations and will find it must spend more time to reflect on what the other groups have done and meet the requirements or particular responses to particular items. In essence, the structure of the communication process imposed for intergroup communication is intended to force a consistency in

[a]These guidelines were developed in 1971 for use by the IEEE committee on Technological Forecasting and Social Change; they were modified on August 1, 1977.

317

and focusing of the scope of considerations coming out of the workshop as a whole.

In addition to the primary goal of assessing a particular technological area, there are some significant secondary features of the workshop. One of these is education; that is, if the workshop is successful, the individuals in any one group will come to realize that to a certain extent, in explicitly doing their specific tasks, they were implicitly doing some of the work the other groups were responsible for. A second feature is the realization that the forecasting and assessment function is not as easy as it may sound. Even the rather straightforward set of procedures laid out for this workshop should prove a very demanding mental framework for the individuals involved. Finally, the workshop is so structured as to reflect the breakup of various functions (groups) as they usually occur in society or organizations. However, in the real world the communication among the groups reflecting research and development, marketing, budgeting, and social impacts is not usually as precisely structured as this exercise. It is therefore hoped that through the combination of these factors the individuals participating in the workshop will gain an understanding of the technological forecasting and assessment process that will benefit them in their own endeavors.

Because the written forms preserve the information generated by a single group, it is possible to have each group meet separately at regular intervals and exchange forms and briefings on a regular basis. While this will slow the process down somewhat, it may be the more practical alternative to a larger workshop gathering in some instances. The forms proposed here are not sacred and provide for only the minimum of information to make the result useful. Users should feel free to augment these as may be dictated by their particular area of concern.

Group Responsibilities

Each group is responsible for originating a certain type of item which will be augmented by information generated by one or two of the other groups. The organization of the responsibilities for this process is presented in Table 9A-1 and clarified in the following outline.

Technology Group

1. Develop specific *major present* and *potential* technological developments.
2. Assess, for the Application Group, the technological developments required to support specific major applications they have generated.

ISBN 0-201-03140-X, 0-201-03141-8 (pbk.)

ISBN 0-201-03140-X, 0-201-03141-8 (pbk.)

Table 9A-1

Technology Forecasting and Assessment Workshop Structure: Responsibilities of Groups[a]

Group	Items originated by group	Responses of other groups to originated items[b]			
		Technology	Applications	Societal	Policy
Technology Group	Major potential technological developments	X	Possible applications of technological developments X	X	X
Applications Group	Major potential applications of technology	Required technological developments for application X	X	Societal consequences of application X	X
Societal Group	Significant and possible societal changes resulting from or affecting technology and its use	X	Applications affected by societal changes X	X	Policy issues raised by societal change X
Policy Group	Significant and possible policy or resource allocation decisions affecting technology and its application	X	X	Consequences of policy or resource allocation decisions or actions X	X

[a] Actual workshop may use all four groups or any neighboring (on chart) two or three-group subset. Interaction of resulting subsets may be carried out at different meetings of workshop. Single group size: 5 to 15 individuals. All communications between groups via written forms for the information indicated in the chart.
[b] "X" means "no action required."

319

Application Group

1. Develop specific *major present* and *potential* applications of technology.
2. Assess, for the Technology Group, possible applications of major technological developments they have generated.
3. Assess, for the Societal Group, what applications are affected by potential significant societal changes.

Societal Group

1. Develop specific *major present* and *possible* societal changes affecting technology and its use. These may be value or goal changes as well as behavioral changes in such things as work, education, or recreation (i.e., sociological, demographic, and psychological changes).
2. Assess, for the Application Group, the social consequences of the major applications they have generated.
3. Assess, for the Policy Group, the possible consequences of a significant potential policy or resource allocation decision or action.

Policy Group

1. Develop specific *major present* and *possible* policy and/or resource allocation decisions or actions affecting technology and its use. These may be such items as major research and development investments that *change* current trends, the introduction of new training or educational programs, new regulatory policies, or changes in laws. These items should reflect a change in the current set of related decisions or resource allocations.
2. Assess, for the Consequence Group, the policy or resource allocation issues raised by the projected significant societal changes.

The foregoing group responsibilities are largely self-evident from the communication forms provided each group at the start of the exercise. In addition, there is a special form for Questions/Comments/Suggestions that any group may utilize to request information or clarification not provided by the other forms.

It has been our experience that when the participants represent a very homogeneous group in a single specialty area, it is advisable to invite some additional people to seed some of the groups with expertise not represented in the original group.

Table 9A-1 summarizes the responsibilities of each group.

ISBN 0-201-03140-X, 0-201-03141-8 (pbk.)

Specific Procedures for the Groups

1.   Each group should have a table it can sit around, a blackboard, an "in box," stapler or paper clips, and scratch paper. Distance between the groups should be such that the conversations in one group do not disturb another group's deliberations.

2.   Each group should begin by having all the individuals introduce themselves and their specific interest and areas of expertise.

3.   The group should pick a chairperson. This and other jobs may be rotated by the chairperson's passing his responsibility and/or reassigning tasks.

4.   The chairperson's responsibilities are to

   (a)   Direct discussion on a topic so as to foster a relaxed and informal atmosphere. However, should disagreements or prolonged delays occur, the chairperson may halt discussion, restate the issue, and call for a vote at any time.

   (b)   Assign, as needed, the preparation of various items (such as question responses) to particular individuals. The resulting items will be presented to the group for an approval or revision.

   (c)   Select an assistant who will receive and keep track of items coming from other groups and summarize them verbally for the group. This person will also remind the chairperson, as needed, of the priority of consideration of items pending before the group.

5.   Each group, as it generates a major item form, should assign a sequential log number to the form. The forms have a self-routing checkoff list at the bottom that will ultimately route them and their attachments back to the originating group.

6.   It will take a group an hour or more of general discussion before it is probably ready to begin generating items.

7.   A group should not expect to generate more than two major items per hour on the average. However, if it is not called upon to respond to other groups for a time, it is possible to generate four items per hour.

8.   The following forms (samples of which appear at the end of this Appendix) are to be provided to each group in prestapled packages. Approximately four to five packages of forms will be needed by each group for each hour of workshop time planned.

Technology Forecasting and Assessment Workshop Participant Profile
   Technology Group: Major Technological Development Form; Applications List Sheet; Questions/Comments/Suggestions Form.
   Application Group: Major Application Form; Technology List Sheet; Societal Consequences List Sheet; Questions/Comments/Suggestions Form.

ISBN 0-201-03140-X, 0-201-03141-8 (pbk.)

Societal Group: Major Societal Change Form; Policy Issues List Sheet; Applications Affected List Sheet; Questions/Comments/Suggestions Form.

Policy Group: Major Policy Decision Form; Consequence List Sheet; Questions/Comments/Suggestions Form.

9.   At the start of the workshop, each group is to determine the meaning of the descriptors in Tables 9A-2 to 9A-4 in terms of its area of responsibility. These definitions should be provided to the workshop monitor at the end of

Table 9A-2
Significance Scale for Application or Policy Sheet

Descriptor	Abbreviation		
Very significant/strong effect[a]	VS	or	1
Significant/significant effect	S	or	2
Slightly significant/slight effect	SS	or	3
Not significant/no effect	NS	or	4
Not determinable at this time	ND	or	5

[a]A very significant application or policy item developed as a response item deserves treatment in a separate form as an item to be originated by this application or policy group for a full analysis.

Table 9A-3
Desirability Scale for Consequence Form

Descriptor	Abbreviation		
Very desirable[a]	VD	or	1
Desirable	D	or	2
Not determinable at this time	ND	or	3
Undesirable	UD	or	4
Very undesirable	VUD	or	5

[a]A very desirable or very undesirable item deserves treatment as an item to be originated in a separate form by the consequence group for a full analysis.

Table 9A-4
Current Development Status for Technology Sheet

Descriptor	Abbreviation		
No significant effort	NE	or	1
Too little effort	LE	or	2
Effort about right	RE	or	3
Accelerated effort	AE	or	4
Too much effort	TME	or	5

ISBN 0-201-03140-X, 0-201-03141-8 (pbk.)

the exercise. No more than 30 minutes should be spent at this task. The group is free to devise its own scales if need be but should notify the other groups of its change via the Question/Comment/Suggestion Sheet.

The object of rating the items you generate in response to major items of the other groups is to determine if they are significant enough for your group to consider originating as a major item in itself in order to receive more in-depth examination.

10.   The priority of consideration for items before a group is, *first*, Questions/Comments/Suggestions; *second*, attachments for which the group is responsible, and responding to items generated by other groups; *third*, generating the new items the group is responsible for.

11.   Questions on procedures should be brought to the exercise monitor. The exercise monitor may also be asked to offer advice on any problems that may arise within a group or between groups. The exercise monitor is responsible for seeing that the required responses are handled in a timely manner and has the power to take over chairmanship of a group for a period sufficient to clean up a backlog.

12.   When voting is carried out by a show of hands, there is a tendency for members of the group to "follow the leader." This biases the results and is to be avoided. When voting on numerical quantities, therefore, it is strongly urged that people write estimates on slips of paper and that these be collected and tabulated. In particular, for the date of occurrence the chairperson should ask everyone to write down *the year by which they expect the event to have occurred*. Then these should be sorted from earliest to latest and the earliest year that accounts for 50% or more of the votes occurring in that year or earlier is "Earliest Year, probability > 0.5" and then the year that accounts for 80% or more of the votes is the "Latest Year probability > 0.8."

13.   If the exercise is completed at one sitting, approximately a half hour per group should be allowed at the end of the workshop for presentation of each group's results to the workshop as a whole. A separate meeting of the workshop allowing time for discussion and an assessment of the workshop itself is desirable; this also affords the presenter of the groups' results some time for reviewing and organizing the material.

14.   If the exercise represents a first trial of the workshop by a group, it is a good idea to prohibit the start of any group to group communication for the first hour of the exercise.

15.   Members of each group should fill out a Participant's Profile sheet, which should be used to generate a composite profile of the group that produced the information. The individual profiles can be filled out anonymously without affecting the overall profile.

16.   In some situations it may be desirable to allow individuals to change group memberships after a period of time. How this is to be handled is left to the exercise monitor.

ISBN 0-201-03140-X, 0-201-03141-8 (pbk.)

TF & A WORKSHOP
PARTICIPANT PROFILE

Primary Area or Product of Employment:

      Academic
      Manufacturing
      Services (Nongovernment)
      Government
      Consulting/Research (Corporation)
      Consulting (Private/Self-employed)

Area of Personal/Professional Expertise: _____

_____

_____

Job Title: _____

Job Function (if different): _____

Education:
      Degree _____

      Year _____

      Field (Specified) _____

Age: _____
Sex: Mr.   Ms.
Workshop Group: TG   AG   SC   PG

ISBN 0-201-03140-X, 0-201-03141-8 (pbk.)

MAJOR TECHNOLOGICAL DEVELOPMENT FORM

Title:

Description:

Current Development Status	Characterize Current Effort (Manpower. $. organizations. etc.)
No significant effort ————	
Too little effort ————	
Effort about right ————	
Accelerated effort ————	
Too much effort ————	

Certain to occur with sufficient effort/resources, OR
Dependent on fundamental breakthroughs in R&D, such as:

Year Expected	Based on Current Effort	Based on About Right Effort (If different from current)
Earliest Date (Year in which $p > 0.5$)		
Latest Date (Year in which $p > 0.8$)		

ROUTING CHECK LIST: Date/Time
SENT TO APPLICATION GROUP ————
APPLICATION SHEET FILLED OUT ————
SENT TO CONSEQUENCE GROUP (For Information Only) ————
SENT TO POLICY GROUP (For Information Only) ————
SENT TO TECHNOLOGY GROUP ————
LOG NUMBER   T-————

ISBN 0-201-03140-X, 0-201-03141-8 (pbk.)

APPLICATIONS OF MAJOR TECHNOLOGY DEVELOPMENT
(to be attached to Technology Form)

Log number on Technology Form T-_____

Date/Time Started_____ Date/Time Completed _____

APPLICATION TITLE	EFFECT OR SIGNIFICANCE

ISBN 0-201-03140-X, 0-201-03141-8 (pbk.)

MAJOR APPLICATION OF TECHNOLOGY FORM

Title: _____

Description:

Market Evaluation:

Type of Uses At Specified Cost Levels (From first user/ highest cost to last user, lowest cost)	Relative Market Cost on Today's Basis (e.g., TV, car, etc.)	Available to User	
		Earliest Year $p > 0.5$	Latest Year $p > 0.8$

ROUTING CHECK LIST: Date/Time

    SENT TO TECHNOLOGY GROUP   _____

    TECHNOLOGY SHEET FILLED OUT   _____

    SENT TO SOCIETAL GROUP   _____

    CONSEQUENCE SHEET FILLED OUT   _____

    SENT TO POLICY GROUP (For Information Only)   _____

    SENT TO APPLICATION GROUP   _____

    LOG NUMBER   A-_____

ISBN 0-201-03140-X, 0-201-03141-8 (pbk.)

TECHNOLOGY SUPPORTING A MAJOR APPLICATION
(to be attached to Application Form)

Application Form Log Number A-_____

Date/Time Started _____

Date/Time Completed _____

TECHNOLOGY DEVELOPMENT TITLE	EARLIEST* YEAR EXPECTED $p > .5$	LATEST* YEAR EXPECTED $p > .8$

*Based on Current Effort.

ISBN 0-201-03140-X, 0-201-03141-8 (pbk.)

SOCIETAL CONSEQUENCES OF AN APPLICATION
(to be attached to Application Form)

Application Form Log Number A-_____

Date/Time Started _____

Date/Time Completed _____

CONSEQUENCE TITLE	DESIRABILITY

ISBN 0-201-03140-X, 0-201-03141-8 (pbk.)

MAJOR SOCIETAL CHANGE

Title: _____

Description:

Affects Evaluation:

Affected Societal Groups	Nature of Effect	Desirability

ROUTING CHECK LIST: Date/Time
   SENT TO POLICY GROUP   _____
   POLICY SHEET FILLED OUT   _____
   SENT TO APPLICATION GROUP   _____
   APPLICATIONS SHEET FILLED OUT   _____
   SENT TO TECHNOLOGY GROUP (For Information Only) _____
   RETURNED TO SOCIETAL GROUP   _____
   LOG NUMBER   S-_____

ISBN 0-201-03140-X, 0-201-03141-8 (pbk.)

POLICY ISSUE ARISING FROM A MAJOR SOCIETAL
CONSEQUENCE
(to be attached to Societal Change Form)

Consequence Form Log Number S-_____

Date/Time Started _____

Date/Time Completed _____

POLICY ISSUE TITLE	SIGNIFICANCE

ISBN 0-201-03140-X, 0-201-03141-8 (pbk.)

APPLICATIONS AFFECTED BY A MAJOR SOCIETAL
CHANGE
(to be attached to Societal Change Form)

Consequence Form Log Number S-_____

Date/Time Started _____

Date/Time Completed _____

APPLICATION TITLE	EFFECT OR SIGNIFICANCE

ISBN 0-201-03140-X, 0-201-03141-8 (pbk.)

MAJOR POLICY OR RESOURCE ALLOCATION DECISION

Title: _____

Description:

CURRENT TREND	TREND INDICATORS
In this direction _____ Opposing this decision _____ None or not determinable _____	

Is there now a policy or decision body to implement the policy or decision?　　　　Yes　　　　　　No

If yes, name: _____
For the decision or action to occur, is a major change in the structure of that body necessary?　　　　Yes　　　　　　No

If yes, describe: _____

_____

_____

_____

YEAR POSSIBLE	BASED ON CURRENT TREND	WITH NECESSARY SOCIETAL OR POLICY BODY CHANGES
Earliest Date (Year in which $p > 0.5$)		
Latest Date (Year in which $p > 0.8$)		

ROUTING CHECK LIST:　　　　　　　　　　　　Date/Time

   SENT TO SOCIETAL GROUP　　　　　　　　_____

   CONSEQUENCE SHEET FILLED OUT　　　　_____

   SENT TO APPLICATIONS GROUP (For Information Only)　_____

   SENT TO TECHNOLOGY GROUP (For Information Only)　_____

   SENT TO POLICY GROUP　　　　　　　　　_____

   LOG NUMBER　　　　　　　　　　　　P-_____

ISBN 0-201-03140-X, 0-201-03141-8 (pbk.)

CONSEQUENCES OF A MAJOR POLICY DECISION
(to be attached to Policy Sheet)

Policy Form Log Number P-_____   Date/Time Started _____

Date/Time Completed _____

CONSEQUENCE TITLE	DESIRABILITY

ISBN 0-201-03140-X, 0-201-03141-8 (pbk.)

QUESTION/COMMENT/SUGGESTION SHEET

NOTE: This sheet may be routed to more than one group but must ultimately be returned to initial sender. Responses should be timely. Use reverse side of sheet as necessary.

COMPLETED BY INITIAL SENDER		COMPLETED BY RECIPIENTS	
FROM	TO*	DATE/TIME RECEIVED	DATE/TIME SENT
TG	TG		
AG	AG		
SG	SG		
PG	PG		

Question / Comment / Suggestion:

Replies (Denote Responding Group(s),

ISBN 0-201-03140-X, 0-201-03141-8 (pbk.)

---

*Use numbers to indicate routing sequence. Return to sender at end of routing. Staple additional written pages if necessary.

THE BOSWASH TIMES

The Computerized News Summary Service of the Megalopolis

August 12, 1986

State Capital a Ghost Town?

Boston, Mass. August 12, 1986. The Massachusetts State Legislature voted today to cancel all face-to-face meetings of the legislature for the remainder of this years' session. Senator Williams said that the legislature's conferencing system now provides all the voting and legislative tracking of activities needed to carry out the business of the legislature. Plans are also being made to augment electronic subcommittee meetings with roving subcommittees that would travel around the state for public meetings when needed.

One Representative who was contacted, McKendree of Cape Cod, says that next year he is no longer going to maintain an office at the state capital, but will conduct his legislative responsibilities from his residence. He feels that the legislature buildings in Boston should be turned into a museum.

Privately, a number of professional lobbyists expressed concerns about their ability to function if the members of the legislature no longer meet physically in Boston.

New Orleans in National Playoffs

New Orleans, August 11, 1986. The city's Star Trek team won a place in the nationwide star battle today. The team of 400 crew members drawn from all walks of life has been together for only two years.

Star Trek Enterprises is classed as the most elaborate of all communication game simulations in existence. With over 100 teams and team members or their sponsors paying an average of $3000 a year (per player) in rental fees for the Crew terminal, this is one of the more profitable of game companies. While many people have objected to the commercialization of computer games, Jay Scher, the president of Star Trek Enterprises, said that "Our objective is to put communication games in the same league as baseball. This is a progressive form of participatory recreation which does not require the wasting of energy resources to travel to the site of the game."

Controversy Plagues New Art Form

Omaha, Nebraska, August 12, 1986. Mr. Jerome Hendricks of Omaha has been awarded the Variations Prize this year for his adaptive book version of the classic movie Star Wars. Mr. Hendricks's on-line book has available 575 distinctive plot vari•tions from which readers can choose dynamically as they

ISBN 0-201-03140-X, 0-201-03141-8 (pbk.)

read the book. For those with graphic terminals, Mr. Hendricks has included automated space battles.

Some literary critics have claimed that because this book treats the reader as a character of his or her choice, there is some difficulty in distinguishing the book from a computer game, and it therefore should not have been eligible for the prize. The award committee, however, has decided that books with game-oriented characteristics and automatic graphics are still books.

The National Gaming Association has charged the Adaptive Book Writers Association with trying to muscle in on its territory. Mr. Hendricks commented that his readership did not care much about the debate over the distinction, and that he was quite willing to accept prizes from either association.

ISBN 0-201-03140-X, 0-201-03141-8 (pbk.)

CHAPTER 10

The Human–Machine Interface: Design, Dilemmas, and Opportunities

Introduction

The design of human–machine interfaces for the use of computer and information systems is still more of an art form than a science. Because humans are so adaptable that they can cope with a wide range of situations, there has not been as much experimentation on and evaluation of interfaces as might be expected. Most of this generation's designers of terminals and software interfaces were trained when computer hardware was expensive relative to people costs. Many user communities represented captive audiences, such as employees who had to use whatever system was provided. As a result, human factors and optimization of the human interface were often sacrificed for machine efficiency.

Today we have reached a point where the tradeoff of human optimization in favor of machine efficiency is no longer valid. In fact, it hardly ever has been, except that most cost–benefit models applied to computer and information systems do not measure user costs. Furthermore, now that we are dealing with human communications, it must also be realized that there are alternatives to the use of a CC system. We are not in a situation in which users must use this form of communication. Another dimension of this problem is that a goal of CC systems is augmentation of communication processes by the presence of the computer. Many of the potential benefits of computerized conferencing are contained in these added capabilities. We strive to bring to the user capabilities far more flexible and richer than a telephone, but which require an investment of several hours of time to learn how to use. In these circumstances the design of the human–machine interface becomes crucial to the success of the system and deserves very careful attention.

Starr Roxanne Hiltz and Murray Turoff, The Network Nation: Human Communication via Computer

ISBN 0-201-03140-X, 0-201-03141-8 (pbk.)

Dimensions of the Problem

There is no single target group for which a CC system interface can be designed. Very often, for other computer systems, a designer can count on only a specific kind of user, such as secretaries, nurses, managers, students, or chemists. Such homogeneity allows the designer to pinpoint minimum skill levels and assume certain standard motivations of the users. Furthermore, the designer in these situations can often count on required training programs for prospective users. Even when a CC system is introduced to a specific organization with the intention of utilization by a single specific group, it often happens that its use as a communication system spreads. With a system designed to be used by professionals, a significant number will be found to delegate the mechanics of interacting some or all of the time to secretaries, assistants, spouses, or children. As a result, the degree of intelligence or general education of the actual terminal users may be quite varied.

The degree of computer sophistication has become another significant factor. A person experienced with computers comes to expect certain capabilities and ways of doing things that are alien to a person with no computer experience. In practically all situations today the designer has to satisfy both the experienced computer-type and the neophyte.

Many designers make the mistake of assuming that all users want to learn all about their beautiful system. However, in the real world there are a significant number of users who want only to turn the system on, do the minimum necessary to accomplish their specific task, and turn it off. They do not wish or do not have the time to learn more than is absolutely necessary and really do not care that the system has capabilities they do not need at the moment.

Some designers also feel compelled to have the user understand why the system works a certain way, based on its internal design. This is like expecting an executive to understand the mechanical or electronic makeup of her/his dictating machine. Yet there are users who do want to learn a great deal, and some who will feel comfortable only with some understanding of the internals. Hence, the designer is faced with a wide spectrum of attitudes, from "I couldn't care less" to "I want to know everything." The latter can often be the most difficult when voiced by a person lacking the background to understand "everything."

The final problem is of a more subtle nature and therefore more difficult to deal with. Through the ages we have associated human identities and characteristics with our domestic animals, our tools, and our machines. Perhaps no machine personifies human characteristics or roles more than a computer. It is the first machine that has really taken over certain functions that used to require a human mind. We even attempt to make it appear "intelligent," and

ISBN 0-201-03140-X, 0-201-03141-8 (pbk.)

at least some schools of thought hold that to be an ultimately attainable goal for computers. Social scientists (Goldstein, 1975; Sterling, 1975) who have looked at the roles computers play in the minds of humans have found the computer to be viewed as an evaluator, magician, helper, entertainer, companion, challenger, mentor, foe, criminal accomplice, producer, overseer, pressurer, priest, dictator, servant.

Computers can give the impression that they are testing the knowledge of the user and often lead to a feeling of success or failure on the part of the user. The machine can also be nonthreatening in that it does not ridicule or has infinite patience and the ability to respond as told. These are traits that other humans do not possess. The computer can present a challenge and a feeling of competition or of being a foe. The computer can give the impression of surveillance or of being a competitor at doing a job. It can pressure its user and present many inconveniences with which the user must deal. Worst of all, it can leave the impression that there is only a certain logical way to think about things—the way the computer does it. All of these characteristics encourage the average person to view the computer as having human traits.

Computer designers sometimes unconsciously aid this process by programming in such "human" responses as "Hello, John" or "Thank you, John." We wonder about the ethical issue here and some of the implications of this approach for systems designed for youngsters or other new computer users. As for CC systems, we believe they should be as neutral as a telephone, with the human element left to the humans involved in the communication process. However, because many people approach computer systems with preset inclinations of fear, boredom, confusion, awe, timidness, frustration, embarrassment, impatience, tension, pressure, disinterest, distrust, etc., the friendly, "human" touch is tempting to employ. This problem is somewhat easier to deal with for CC systems than for other computer systems. The sooner a person is actually communicating with others and enjoying that communication, the quicker his or her negative preconceptions will evaporate. For the designer, this means that a minimum of learning should be necessary for the user to start a straightforward communication process.

Other aspects of the design problem that should be seriously considered are the atmosphere and institutional policies surrounding the introduction of a system. How private are the communications? Are your errors at the terminal being tracked and reported? What statistics on your behavior are being collected? The sooner that users can forget the computer and adopt the attitude of communicating with other human beings, the better able the designer will be to introduce users gradually to the many other conveniences that the computer can contribute to the communication process.

In summary, the designer of the human–machine interface must satisfy, or

ISBN 0-201-03140-X, 0-201-03141-8 (pbk.)

compensate for differences in,

1. degree of intelligence or education;
2. sophistication with respect to computer systems;
3. amount of time available or inherent interest in the process;
4. emotional attitudes toward computers.

This means in effect that there is no unique interface to satisfy all users, and somehow the interface will have to provide different faces to different users of the system.

Folklore

Although interface design is still largely an art form, there is a considerable body of conventional wisdom or folklore that has evolved from observation of users on interactive computer systems. Much of this wisdom can be traced back to the earliest efforts in this area by people like Baker, Shaw, Neisner, Nelson, Orr, and Culler and Fried, and initial systems like JOSS and MAC.* However, not all designers draw on this resource, and some seem to prefer to rediscover this knowledge by making the same mistakes again.

Perhaps the best approach is to start with an itemization of the problem manifestations that typically occur in interactive computer systems. We summarize these in the following subsections.

Fishbowl Effect

The impression a user can have of being watched or monitored and that "stupid" errors are obvious to all is described as the fishbowl effect. It becomes increasingly significant at high levels of social status or rank of the individual user. To counter this effect, the user must be provided with a clear statement of the personal data the computer keeps and with an understanding that even experienced computer people make mistakes at terminals. Most important is the provision for the use of terminals in privacy (put them on wheels to roll into offices) so that no one can look over a user's shoulder unless invited.

ISBN 0-201-03140-X, 0-201-03141-8 (pbk.)

---

*References from which this folklore has been gathered and expanded upon include Bennett, 1972; Cherry, 1966; DeSio, 1966; Martin, 1973; Neisner, 1964; Orr, 1968; Shaw, 1964; Sheridan, 1974; Turoff, 1966, 1969; Walker. 1971.

Peephole Effect

The terminal is analogous to a peephole into a rather large ballroom. In regard to content, a single comment in a conference may be one of a thousand making up the whole discussion. For this reason the peephole must be fitted with a zoom lens with which the user can view summaries or organizational outlines or focus down to specific discussion threads. In other words, even for a communication system the retrieval tools a computer can make available become extremely important. In regard to discussion partners, in computerized conferencing the user has potentially available a tool to provide a tremendous increase in the number of people with whom he or she can actively communicate. However, this is only possible if the appropriate mechanisms are provided for finding others with similar or complementary interests and for dealing with the volume of material generated by multiple discussions.

Bully Effect

The rapid response of a terminal and sometimes the wording of questions will make the user feel that just as fast a response is required. A primary advantage of computerized conferencing is to be able to work at one's own pace. Thus interaction with the computer should not create the impression that a user must type something instantly whenever the terminal requests an input. Tied into this situation is the practice of charging a user according to the time the terminal is turned on, since this method can give the impression that a person's "think" time is costing money. For this reason, charging by quantity of material generated or retrieved is preferable to charging by time on line.

Concrete Effect

Very often users come to feel that they must adapt to the way the computer does something rather than to adhere to an individually preferred way. A lack of flexibility or variety in the interface design is one primary cause of this problem. The common impression that it is very expensive to change the system adds to this feeling. The user should not be given the impression that what the computer does and how a person goes about using its capabilities is cast in concrete. Rather, the designer should convey the impression that the system can evolve to fit needs and preferences as users gain experience.

Clutter Effect

Providing the user with more information than is actually needed is known as the clutter effect. Too many instructions, options, etc. will cause the user to

ISBN 0-201-03140-X, 0-201-03141-8 (pbk.)

ignore many of them with the result that the really important bit of information is not grasped. Instructions should be available on a selective basis, fitted into some generalized structure providing an overview of the system.

Computer Angst

Fear of breaking the machine or the system, often reinforced by phrases such as "In no circumstance are you to use the following keys," is known as computer angst. To dispel it the designer should strive to encourage a trial-and-error approach to learning the system. Therefore the user should not be inhibited from trying anything by intent or by accident. Only a naive designer assumes that users will never make a mistake or not do something they were instructed not to do. There are always a few who will want to see what happens when you break the rules. Another aspect of this problem is the attitude that the user cannot comprehend the system because computers are terribly mysterious and complex. This syndrome has been furthered by many salesmen and computer-types who like to be thought of as possessing esoteric knowledge, but it should be considered basically unethical. Enough skepticism has been generated about computers for this point of view to be more likely to diminish the utility of the system than to keep users "in their place."

People Angst

Fear of going to other users for help in using the system is people angst. The most favorable situation a designer can foster is that in which the user community becomes a cooperative group that helps its members to improve their knowledge and command of the system. It is usually desirable to designate a member of the user community as an agent for aiding new members and informing them whom they can talk to about doing special things on the system. Important for encouraging this approach is having advanced features in the design that are convenient for certain tasks. Some users enjoy the role of helping others and will be very quick to learn all there is to know about the system. Others who need an advanced feature will often learn it more easily with the help of another person than from a reference manual.

Rorschach Blot

Users unfamiliar with computers will assume that the system is very much like something else they know—in this case, letters and/or telephones. They may also unconsciously infer that what is easy for them is easy for the system and what is hard for them is hard for the system. The best way to overcome this outlook is by the very careful choice of examples of what the system is capable of doing. Optimal use of the system will not be immediately evident

ISBN 0-201-03140-X, 0-201-03141-8 (pbk.)

to such people, who may, for example, assume it is used only in the same format and for the same functions as internal memos. In fact, some message systems have been purposely designed to replicate the exact format of an internal memo, creating an atmosphere that discourages innovative ways to do things. This problem is also related to the syndrome of viewing the use of computers as just the automation of what is done manually.

Dictating to the User

Some computer designers mistakenly feel they should dictate to users exactly how to use these systems and, based on their study of user needs, the actual user applications. The appendix to one user manual for a commercially available message system contains the following checklist for users of the system, with instructions to consider each question carefully when sending a message:

Is this message necessary?
Is this system the best way to communicate this information?
Is the distribution appropriate?
Is the message too long?
What impression does this message give the recipient?
Is the message pertinent?
Have you properly disposed of all messages?
Who else should see this message?

None of the foregoing belongs in a user manual! Such questions presuppose a very formal atmosphere for the message system that could well inhibit certain applications even in an informal organization. The prerogative of determining the matters queried in this list should rest entirely with the users and their own peer or management structure. In the history of the development of management information systems it has been common for computer people to preempt user prerogatives through actions such as determining what information is collected and how it is to be displayed and organized. In CC systems, users should be free to establish their own norms, such as the degree of formality in discussions.

Commitment

Designers tend to become so committed to their designs that they do not recognize a need for change. Users are frequently unable to recognize if a difficulty they are having can be overcome by an improvement in the design. Usually this problem implies the need for someone to act as an ombudsperson between designers and users. A fortunate situation in a CC system is that it can be used to involve the users in a process of self-design of the system to

ISBN 0-201-03140-X, 0-201-03141-8 (pbk.)

their own specifications, with the official designer stepping in only when an exception occurs—"design by exception."

The Three Witches of Interactive Systems

The three most common fallacies that computer people fall into when designing interactive systems for noncomputer types are the following:

1. That users wish to interact with the system utilizing natural languages (such as English-like sentences), so a user should be able to type: Give me all new messages after February 2, 1977.
2. Users should learn to use a programming-like language because this provides the greatest degree of power (ultimately) to the user.
3. The design should be built into hardware.

Most user communities develop their own abbreviations and other techniques to minimize effort in communication. Consider pilots, taxi drivers, police, ham and CB radio operators as examples. The user will need the capability of entering on the keyboard the minimum typing strokes to accomplish a job. For control of the interaction, natural language would become too lengthy for users. Using a programming-like language requires training, practice, commitment, intelligence, and a certain way of thinking about problems that may be quite foreign to many users. Although such a language may be made available for users who have the aptitude and time to learn it, it should not be required of all users.

There is no better way to create irremediable problems in the interface design than to incorporate it into hardware. This approach has created a built-in obsolescence phenomenon in many earlier information systems. Fortunately, new technology in the form of programmable or intelligent terminals has eliminated the need to do this for the sake of economy.

The Four Phases of User Development

The first phase a user goes through in learning a new system is the *uncertainty* phase, during which he or she has to overcome hesitancy and anxiety. The designer is, in essence, given a line of credit by the user. The user will invest only so much time getting through this phase, after which he or she will reject the system without having any real idea of what is being rejected. The designer's goal is to minimize this time period. The sooner the user can do something personally useful, the better off the designer is. A half-hour to an hour is the time the designer should shoot for with respect to CC systems.

The second phase is the *insight* phase, in which the user understands the general concept of the system and can make at least limited use of it for his or her own purposes. The user should also be able to gain new knowledge of the

ISBN 0-201-03140-X, 0-201-03141-8 (pbk.)

system during this phase. In this phase the user can see shortcomings of the system. If the user is willing to complain during this phase, it means there is a genuine interest and some recognition of potential utility. The worst situation is one in which users do not have complaints or suggestions and merely say it is "all right." This shows a lack of interest and ability to relate the system to their needs and motivations.

The third phase is *incorporation*. In this phase the mechanics of the interaction have become second nature and the system is utilized as a part of the user's normal environment and for his or her purposes or applications. The user is hooked and comes to depend on the system.

The fourth phase is the *saturation* phase, in which the user begins to develop needs the system cannot satisfy. Not all users will reach this phase. For those who do, if the system cannot evolve to satisfy the new needs, then it begins to die or become outmoded from that point on.

Each of these phases requires that different capabilities be inherent both in the internal system design and in the interface design.

Communication and Information Domains of Users

There is a general set of increasingly complex communication and information requirements users may have that lead to specific demands related to the degree of computer power a CC system should offer users. Being able to determine which domain applies to which users or user community provides the designer with the ability to segment the design and the resulting learning requirements. In essence, these domains are the expectations users have for their potential use of the system.

The first domain is *simple inquiry or message exchange* and represents the desire of the user to keep in active contact with a number of individuals. This requirement is usually satisfied initially by straightforward message-type systems, although the computer may offer such embellishments as delayed messages, routing for approval, and group messages.

The second domain is *consensus formulation and report generation* and represents the ability of a group of individuals to discover their degree of consensus on a topic and to formulate options and group reports. This process of pooling knowledge requires the sort of joint or group writing space typified by a conference or joint notebook. It also demands a human moderator and the design features to support that person's role. Obviously various computer capabilities such as indexing, retrieval, reorganization, and voting can be useful here.

The third domain is *discovery and analysis*, or the objective of a group attempting to develop new knowledge by joint investigation and exploration. This goal can impose a diversity of requirements that may be a function of the particular topic for which the system is used. It may very well imply

ISBN 0-201-03140-X, 0-201-03141-8 (pbk.)

tailored communication structures, especially if the group is heterogeneous. Other computer tools such as data bases, models, simulations, and analysis routines might also have to be made available to provide information for the discussions.

The fourth domain, *planning and decision analysis*, involves the process of examining and evaluating alternatives. Because computerized conferencing is a reflective, or "cool," communication medium, the process may have to be highly structured in order to control the behavior of the participants, as discussed in Chapter 9. A group might very well operate under a version of a Robert's rules of order imposed by the computer, but oriented to the delimitation and evaluation of specific options. There are also a host of data processing tools designed to aid the individual formulation and group consistency of subjective judgements.

The fifth domain comprises *management, monitoring operations, and command and control*. This occurs in a day-to-day operational environment or for urgent action and decision-making exchanges, as in a crisis management situation. To support such applications, users must treat the system as part of their everyday communication environment. It also implies an integration of the communication process with the information process in an organization.

Difficulties with Users

The designer is not responsible for all the problems that arise. However, it is true that users will exhibit certain undesirable behavior that the designer has to accept and to allow for in the design of the system.

- Users will fail to notice even the most explicit instructions.

- There is no single way of describing anything that will satisfy every user.

- Users will do the unexpected, the unanticipated, and the forbidden.

- Users will disregard or forget instructions.

- Users will formulate opinions based on inadequate knowledge.

- They will not always ask for help when they need it.

- They will not appreciate the system for its complexity, sophistication, or elegance of the design, but only for the use they can make of it.

Design Principles

Although the difficulties the designer faces seem endless, there is the saving grace that humans are fairly adaptive and if something is useful to them they will tolerate some degree of imperfection. Any single design will, in essence,

ISBN 0-201-03140-X, 0-201-03141-8 (pbk.)

be a compromise, because different users will have completely opposite views of what is right or wrong about the system. The designer must also consider the efficiency of the system, so that users have good response. Sometimes this consideration will conflict with providing some users with features they desire. Users of CC systems will expect the instant response of a telephone and not be very tolerant of delays or willing to trade this feature for others. However, response can be varied according to function, so that a user may be willing to have a delay to do a search but not to have a message delivered. Delays when they do occur should be consistent. The result is that the designer is a compromiser or negotiator rather than an optimizer. In any case, there are certain general principles of design that have evolved to deal with the problems we have been discussing.

Forgiveness

Basically, this is the concept that there should be no error or incorrect input by the user that cannot be recovered from or corrected with a minimum of effort. For example, in some "unforgiving" systems, pushing the wrong key can cause the user to be transferred out of the user program and into the executive software* of the machine. Most users would not know what to do in such circumstances. A more elaborate example is where a user wishes to delete an item and is asked to supply a short identification such as an item number. In this situation more complete identification, such as a full title, should be provided back to the user for verification. If it is the right item, a choice should be offered as to whether to complete the deletion process. This cautious approach can be taken even farther by actually leaving the deleted item in existence for a period of time in case the user has a change of heart or mind about the action. In general, an error on the part of the user should require about the same amount of effort to correct as was involved in making the original error.

One way of attaining this objective for single errors (as opposed to sequences of errors) is by designing a fairly comprehensive interface routine between the application program and the user. A routine of this sort can be told by the application program what the allowed options are for the next user response, and as a result be able to check the input and ask for a new response if the one provided is wrong. Depending on the application, all user inputs can usually be grouped into 10 to 20 general types, including (1) free text of a certain maximum length; (2) a set of numeric choices; (3) a number within a certain range; (4) one of a set of special symbols; (5) a date; (6) a time; etc. Such a routine can be generalized for error correction by keeping a

---

*The "exec" of a computer is the software package that manages the computer's resources and allocates those resources to the different tasks users are demanding of it.

ISBN 0-201-03140-X, 0-201-03141-8 (pbk.)

set of buffers, which represent the last set of separate inputs of a user, so that mechanisms can be provided for users to step back to correct previous input.

Complete forgiveness would require storing a complete audit trail of the user's inputs during a session and in most situations is prohibitively expensive. However, 90% of the problem, which is usually single rather than sequences of errors, can be handled by a good insulation routine between the user's input and the program that actually carries out the operations.

"Forgiveness" was coined for this context by Charles Baker, one of the designers of the JOSS system at RAND (Baker, 1966). Computer people still tend to use the term "bulletproofing," which is more reflective of an attitude of protecting the program from things the user can shoot at it. There is a definite, perhaps unconscious, psychological difference inherent in the choice of which word is used to guide interface design. One shows a concern for the system; the other for the user.

Escape

Users must have the freedom to change their minds. There should be no action they can bring about that they cannot cancel or escape from. For example, if the user asks that a set of items be printed out, and the items turn out to be too long or not relevant, the user should be able to hit a single key to cancel the rest of the printout.

Related to this capability is the incorporation of warnings to the user on what to expect next, like: "you found 30 items comprising 638 lines of text." These warnings provide the user with sufficient information for her or him to make an intelligent choice on what to do next so as to minimize the need to escape from actions. Frequent need to escape discloses that a user has significant problems in using the system as presented. A frequency record of the use of the escape function is a significant indicator of a mismatch between the user objectives and the design of the interface and should be tracked in any on-line system. The use of escape, after all, is an expression by the user that the system is doing something for him (or to him) that he really did not want done and was not able to anticipate.

Generalizability and Segmentation

In terms of human–machine interfaces, this is a specific tradeoff between the holistic and reductionist view of a system. Ideally, the user should grasp the totality of the system or recognize its capabilities from the most general or comprehensive perspective. On the other hand, this comprehension may require far more learning effort on the user's part than he or she has time for, in terms of the specific practical uses to be made of the system by that particular person. We have, as a result, the artistic tradeoff a designer must

ISBN 0-201-03140-X, 0-201-03141-8 (pbk.)

make between the following two objectives:

Segmentation of the capabilities of the system such that a particular user can learn just enough to accomplish his or her specific task.

Making the user aware of the most general capabilities available in the system, so that they can be learned and utilized as the need arises.

There is apparently no single solution to this tradeoff, and it is one of the factors that leads to having a variety of ways in which the system can be used to accomplish the same task, with the result that the system itself can be viewed by users from a number of different perspectives. For example, a CC system may be viewed as one tremendous file of pages of text, certain pages of which might be messages, others conference comments, and others pages not committed to either use. This perspective makes it possible to deal with everything a particular user has written and to treat that collection of text items as an entity.

From another point of view, it is possible to deal only with messages, "in" boxes, "out" boxes, and other analogies familiar to people who communicate by memos. In this view the user is required to learn only a few simple operations in order to deal with message traffic. Usually, this approach of presenting the system in segmented pieces made analogous to nonautomated concepts is followed in educating novices. However, unless care is taken to edge the user gradually toward more general perspectives, the user may never reach the stage of comprehension at which he or she can grasp or develop completely new uses for the system not possible in the nonautomated physical analogies used.

The more limited or segmented the user perspective, the less likely that users can ever aid in evolving the use of the system. Underlying this concept is a fundamental view that with respect to human communication in particular, the standard computer approach of studying user applications and then designing a specific system to fit those applications is fundamentally wrong. The real advantages for these systems lie in new ways of doing things, not in automating the old ways. Therefore, the design goal should be to provide human communication capabilities that are not problem or application specific. This goal is not always attainable, but is one of the objectives for such systems. As long as the communication system is not application specific, there exists the opportunity to incorporate new applications.

This does not exclude designing a system for a problem class such as "negotiation." However, the designer should not go to the level of specificity of, say, "labor–management negotiations." Unfortunately, it is often easier to sell a system to an institution when it is a specific system tailored for the XYZ organization, rather than a system to handle a class of problems for any institution. This has been the standard approach to management information

ISBN 0-201-03140-X, 0-201-03141-8 (pbk.)

systems. The result was very rigid MIS systems that often failed because they could not evolve as the problems and approaches to those problems evolved and were outdated by a year or more between the analysis and actual delivery of an operating system.

Part of the utility of segmentation is to be able to introduce segments of the capability to the user and let user feedback evolve further elements of the design. Systems must be designed so that change and evolution of both the interface design and some of the capabilities constitute the rule rather than the exception. Once again, this flexibility is foreign to most institutional practices in implementing computer systems. Managers think in terms of a completed system being delivered that will require no further effort except for "routine" maintenance. In actual practice, most successful computer and information systems require a level of "maintenance" effort indistinguishable from the original development effort. Such systems are undergoing evolutionary development regardless of what they call it. The inherent difficulty in not calling a spade a spade in this environment is that for evolutionary development you can set up the evaluative and feedback mechanisms between user and designers, whereas for maintenance you have no justification to invest effort in this activity.

One significant concept in segmentation for CC systems is that text processing is a generalizable concept that applies to any structure or form of communication used to characterize the particular conferencing system. As a result, all the operations associated with text processing (copying, merging, editing, filing, searching, composing, relating, etc.) can be handled in the same manner, whether you are dealing with messages, conferences, notebooks, or any other communications structure. Thus, once the user has learned some of these capabilities for messages only, no additional learning should be necessary to apply them to other classes of text items. This generalizability also provides an indirect way for users to begin to realize that there are different perspectives from which to view the system as their knowledge of the components increases.

Variety and Flexibility of Interaction

The interaction must actually exist in a variety of forms to satisfy the variety of talents on the system. The simplest form for beginning users is to present a numbered "menu," or list of choices, and to pick one to determine what happens next. The next option is to provide "commands" as input, such as CNM (compose new message), which allows the user to direct that a certain thing should occur regardless of what was done previously. Users will tend to learn and remember the commands for the things they do most frequently rather than step through a set of menu choices. However, the

ISBN 0-201-03140-X, 0-201-03141-8 (pbk.)

average user will do this only after discovering what represents the frequently done things. New users who do not know computers and intermediaries (such as secretaries acting for their bosses) will often rely on the menus.

A significant enhancement to menus and question–answer response design is to allow the user to supply a set of answers in anticipation of questions. This device enables the user to avoid sitting through a set of questions and challenges the user, in a nonthreatening way, to remember what is asked in order to do something. To take advantage of this device the user must have a capability to permanently store sets of responses and give them short labels. A personal set of commands can then be evolved to make it very easy or quick to do things that the user frequently does. The general principle the designer has to strive for is to minimize the time a user spends carrying out the mechanics of the interaction, so that a maximum of time can be spent actually communicating.

Finally, for advanced users, the ability to store sets of answers can be expanded to include looping and test operations, so that the user has his or her own programming language with which to tailor the interface on an individual basis. As some users reach this level, they will in fact have learned a programming language without realizing it. In terms of flexibility, this ability to place a programming language or procedures between the communication system and the user allows the designer to tailor the system closely to the demands of the user with a minimum of programming effort. The designer can declare those features evolved by individual users that prove to be popular as "global," or available to all users as system commands.

Guidance

It is mandatory in such systems that there be very easy capabilities for users to obtain selective pieces of information or features they wish to learn about directly from the terminal. Furthermore, the material on a specific feature must have a short concise explanation, as well as a more detailed explanation with examples when necessary. There must also be people who are designated as available to provide direct guidance on the use of the system.

After teaching a user to send a message, the next important lesson is how to find out something more about the system when he or she needs it.

Leverage

The greatest challenge for designers of future systems consists in providing users with the fullest possible set of advantages in regard to the computer's potential for human communications, with a minimum of knowledge needed to invoke the capabilities. To what extent can other data processing functions

ISBN 0-201-03140-X, 0-201-03141-8 (pbk.)

be incorporated and provided to the user as a part of the communication process? The goal of the interface design is to provide the longest possible lever. Even in the area of handling text, the computer makes possible things the user cannot initially grasp because of being conditioned with largely sequential text mediums, such as reports and books. The manner in which the user can evolve text handling in computerized conferencing can lead to completely new forms of writing style.

Terminals

Computer people have a tendency to design terminals for other computer people. Computers utilize many special characters internally that the average user will never need to know about; however, designers sometimes label these characters as a third level of shifting on the keyboard, so that some keyboards look quite complicated. The layout of special keys is not often optimized to prevent errors. Somehow engineers do not seem to realize that people make mistakes at keyboards. The feel of console keyboards is usually different from that of typewriters, and some have no sound to indicate that a key has been activated. Very often there is no signal warning the user that the end of a line is approaching.

In terms of handling text, most display units (cathode-ray tubes, or CRTs) are quite limited (around 2000 characters) in the amount of text they can display, and the motion of the text as it is produced precludes rapid reading or skimming. However, now that the price of memory has come down considerably, new terminals are appearing that allow large amounts of material to be fed into the terminal and reviewed or manipulated locally in the terminal.

Hard-copy portable terminals are extremely useful in this environment. However, print mechanisms utilizing special heat-sensitive paper do not produce very satisfactory output. Vibration-free and light, high-quality print mechanisms for portable terminals must still be developed. The technology of the "ink jet" may be the solution at some future time.

Without special expensive modems* or conditioned communications lines, text cannot be printed faster than 30 characters a second on ordinary telephone lines at present. This rate is well below reading (let alone skimming) speed, and slows down a group communication process considerably. It is hoped that a print speed of 300 characters per second will be available with standard equipment soon.

---

*Modem is a word formed from "modulate and demodulate" and represents a device used to interface and convert back and forth between telephone line transmission and digital signals such as those produced by a computer terminal.

ISBN 0-201-03140-X, 0-201-03141-8 (pbk.)

Most visual terminals allow text to be placed on the screen only in the sequential manner of one line at a time. Of more interest is the ability to place information selectively in any order at any position on the screen. For example, the key phrases of a paragraph can be placed at such a location that the reader perceives them before the rest of the text is filled in. It would be interesting to see if this deployment improves comprehension of complicated material. The other current limitation is the lack of variability in type fonts as a mechanism for highlighting information. Although all these limitations will gradually be ovecome in the next few decades, the terminal currently is the prime "peephole" limitation on the use of computers for human communications.

Terminals represent the point of contact between the user and the system. Hence, they must be chosen with special care when they are expected to be used by noncomputer-types. People who work with computers regularly become oblivious to the inconveniences of some of the terminals they use and the difficulties others might have. For example, visual display units will give people with bifocal glasses stiff necks unless they invest in trifocal lenses.

Most of all, users have to have convenient access to terminals and be able to utilize them in private, as they would a telephone.

Specific Short Items of Wisdom

- The most efficient structures for computers are not necessarily efficient for users.

- Effectiveness of an interface design is measured by the extent to which the user can forget that there is an interface.

- Users should not have to cope with details extraneous to their tasks.

- Error messages should be informative.

- Charges should represent resources that the user can understand or that relate directly to the user task.

- Every designer assumes his or her system is easy to use.

- Designers should have reasons, or at least a rationale, for design choices.

- User behavior changes as a result of using the system.

- Designers should not evaluate users; user evaluation is their own management's job.

- The user should be involved in the evolution of the system.

- A system that does not evolve begins to die.

- There is not much new in interface design, but a lot of rediscovery goes on.

ISBN 0-201-03140-X, 0-201-03141-8 (pbk.)

- In science, theory leads experiment; in interface design the opposite is true.
- Users tend to fall into habits or specific things they always do, even if these patterns constitute the "hard way to do it."
- Systems should not trick, deceive, or manipulate users.
- Users learn best by doing—trial and error.
- Users should find it convenient to get to a system when they need to.
- If instructors are used, they should not be knowledgeable about computers.
- Users need to know they cannot "break" the system.
- Designers should provide a variety of forms of instructional material.
- On-line instruction is better than off-line instruction.
- Users will assert minimal effort to obtain benefits.
- Users will evaluate systems in their terms, not in computer terms.
- Users who have spent years learning their jobs should not be expected to spend years learning about computers.
- What you can count on for at least some users is that they are highly intelligent, too busy for any training course, very impatient, and intolerant of inconveniences; they also require useful results from the system.

Design Principles Pertinent to Computerized Conferencing

The previous discussion applied for the most part to any terminal-oriented computer system. However, there are a number of considerations that are very specific to computerized conferencing. One fundamental characteristic of CC systems is that they handle large amounts of free-form text. The first level of concern with this feature is the ability of users to edit and correct the fragment of text that they are currently composing. This immediate text-editing capability must be such that:

- Learning only a very few commands requiring a minimum of typing will allow correction of most typographical errors.
- The user can make corrections to all parts of the text in one operation, and can transpose, insert, and delete various portions at will.
- Correction of text that is being composed is clearly distinguishable from other, more advanced text-processing capabilities.

ISBN 0-201-03140-X, 0-201-03141-8 (pbk.)

Competing with nonautomated methods for handling written forms of communication is not as easy for the computer as might be expected. A human being handling written materials can utilize notebooks, card files, file cabinets, index tabs, file folders, scissors, paste, Scotch tape, typewriters, correction fluid, Xerox machines, desks, and cubbyholes or piles of material. When one considers how these materials enable a human being to manipulate, organize, and synthesize text, it becomes evident that to compete electronically with paper is not an easy task. Also, we must remember that users have had many years in which to learn how to use the tools for text processing they are now accustomed to and are unlikely to invest the same effort in relearning a new electronic set of tools. The user will only be willing to spend a limited amount of time learning what appears to offer an advantage over what can now be done.

At least since Vannevar Bush's 1945 article on Memex, it has been recognized that it would be possible to augment the composition and communication process with a machine to do all these things and more.

Computerized processing, to truly be beneficial, must allow users to handle a large body of text representing the proceedings of conferences, group and individual reports, and papers. The art of composition is one of rearrangement and reprocessing of fragments of text in many versions, from crude drafts to polished finished product. The possible complexity of the process is reflected in the components that a group might use in putting together a report: manuscript fragments; chapters; pages; footnotes; outlines; summaries; tables; figures; sources; points to develop; quotations; notes; correspondence; points to check; glossaries; indexes.

A group or an individual can well break up a composition effort into an unpredictable set of groupings. The problem arises when a fragment of text in one grouping of text relates to other fragments in other groupings. The design challenge is the problem how to provide users with a smooth and flexible manner of recording, reviewing, and reorganizing these links between text fragments. This concept of associated text fragments has sometimes been referred to as *hypertext* and is due to Theodore Nelson. (See, for instance, Nelson, 1965, 1970, 1972, 1973(a), 1976.) Various attempts to design and implement hypertext have been made on a number of text-processing systems at such places as Northwestern University and Stanford Research Institute (Englebart et al., 1976). Hypertext represents the tantalizing but ellusive concept that a computer can allow people to develop text in a more parallel and associative framework, which would seem more compatible with basic cognitive processes than normal paper composition modes. Providing these capabilities within a system that is easy to learn and master remains a challenge to designers of CC systems.

Many of the ideas that go along with hypertext still make extraordinary demands on the available technology but will not do so in the future. They

ISBN 0-201-03140-X, 0-201-03141-8 (pbk.)

include

Complete audit trails on all drafts of a piece of text.

Joystick control of text: up is forward in the text; down is backward; left is summarization to outline; right is footnotes and more detail.

Viewing multiple streams of text, like two or more separate scrolls, where the relative speed of each is determined by the linkages of text fragments between two streams.

Inference propagation of changes in links where appropriate.

Graphical manipulation of text fragments where text condenses before one's eyes to key fragments, or expands to fill in details.

A concept related to hypertext is that of *adaptive text*, which allows the writer of text to ask questions of the reader and has a number of different purposes:

To ask the reader such questions as, "Do you want more or less detail on this topic, or wish to move on to another subject?" The answer would then be used to determine the next item of text the user receives. An "adaptive story," where the user chooses the outcome of a critical event in the plot, can be imagined.

To incorporate the answer or answers in the body of the text for the benefit of subsequent readers of the item.

To compile the responses of the different readers into a form of return communication to the author, as in a poll or Delphi exercise, or for the collection of information in an organized manner.

For providing question–answer-type lessons in an educational environment.

Inherently computer-assisted instructional (CAI) systems have limited forms of adaptive text in the programming language used to create the instructional lessons. However, the goal is to check the reader's response against the correct answers and merely record whether an answer is right or wrong. In a CC environment we are more concerned with recording the answer (which might be a text item itself) and utilizing it in subsequent communications. The challenge here is to provide these capabilities in a manner in which a user does not have to learn a programming language in order to take advantage of these capabilities.

Associated with the collection of information in an organized manner from users, including filling out of forms, is the desire to process the information. The simplest form of processing is voting and estimation of variables, such as cost, by a group of people. However, there is another level of processing—the provision of analysis aids to check consistency and to structure complex related sets of subjective judgments, on both an individual and a collective basis. We know from Delphi work and other efforts with complex problems

ISBN 0-201-03140-X, 0-201-03141-8 (pbk.)

that individuals employ certain common methods or steps in trying to deal with complex problems. For example, it is quite common first to itemize a list of options, such as goals and objectives; possible actions; events that might occur; important factors; influencing factors; requirements; resources; tasks.

Once such a list is generated, individuals and/or groups can make judgments about how any two or more items are related according to any one of a number of measures: preference, priority, similarity, causality, influence, etc. Once this is done, there are a host of analytical methods by which to summarize the resulting structure, consequences, or consistencies of the judgments, so the level of complexity of the results can be reduced to more tractable subsets. Techniques such as interpretive structural modeling, cross-impact, or multidimensional scaling are representative of tools currently available. To enhance the use of computerized conferencing for dealing with these and other forms of complex problem approaches, another goal of design should be to provide mechanisms for communicating the judgmental information to computer programs designed to provide these types of analyses and to pass the results back in the form of communications. About all the designer should require of the user is to fill out some standard form to collect the information; the user should not be expected to learn to operate these tools directly. Included in these tools would be the models and existing data bases used by any organization or group, so that they could be utilized to produce results for review in conferences on a selective basis.

Another wide open design area is the use of the computer to adaptively facilitate the communication process. One example in the original EMISARI was the collecting of those keywords readers were searching for and not finding, and returning them to the individuals responsible for authoring material in the particular file. Once upon a time, before libraries made themselves internally more efficient, you could ascertain from the card in the back of the book who else in your organization was interested in the same topic. We can visualize procedures and analysis routines that would allow CC systems to accumulate profile information on users and automatically generate messages like; "so-and-so has been looking for material on topic X and that appears to be an area of concern to you," or "Are you sure you want to use that phrase because this other group seems to infer different relationships for it?" The dividing line here between having the computer do facilitation and having it do surveillance is very fine. This area will require much experimentation to carry beyond fairly simple processes in an acceptable manner. Some of the analysis routines available today can infer basic values of individuals without any awareness on their part that they are providing information of that informative a nature.

Another example in EMISARI of using the computer to allow addressing a message by content was the ability of the user to send a message to a piece of

ISBN 0-201-03140-X, 0-201-03141-8 (pbk.)

data, such as a table supplied by one of the conferees. Therefore, the requesters of the data, the suppliers, the analysts, and the users could all build up a commentary associated with a single item of data. When one went to look at the data, the new communications associated with it would then be made available.

The final area of concern, as the costs of graphic terminals come down, is the development of graphically oriented conferencing systems. There is a large class of applications in which a group of people must be able to work on a common diagram, just as they would in front of a blackboard if they were in the same room. In many fields a diagram conveys information far more concisely than does text. Typical are design, construction, engineering, and planning activities in many areas. In addition, many successful Delphi exercises have solicited the collection of information in a graphical representation or form. Since, in principal, it is possible to have a group of people making simultaneous erasures and additions to the same diagram, a complete audit trail must be maintained until the group reaches agreement on the final form of a particular diagram. The mechanics of such a conferencing capability are still an open question, since no such system has yet been implemented.

Once design aspirations go beyond the straightforward passage of free text between communicating individuals, a host of issues are still open to development and experimentation.

The Psychology of the Design

Perhaps the most subtle design issue for computerized conferencing concerns features that foster the atmosphere or psychology of a communication process among the system's users. Among these are letting a person know (1) that someone has received his communications, either by confirming a delivery of a message or by indicating who has read what in a conference; and (2) who else is on the system or is participating in the same conference at the same time, or when others enter or leave the conference.

Another aspect of this issue is the meaning of a confirmation. On most message systems a user may review a title and reject looking at the text of a message. In the EIES computerized conference system the choice was made that a user must accept his or her message text before going on to the next new message. The reason for this feature is that the system is designed to facilitate a group communication process. Therefore, allowing people to reject messages means that those who write verbose or irrelevant messages may never realize that their messages are not being received. Hence, the group need not deal with such problems as a group. In this particular case, giving individuals too much of the computer power available can adversely affect the communication process.

ISBN 0-201-03140-X, 0-201-03141-8 (pbk.)

Pen names, anonymity, and voting represent extensions of the communication process that have definite psychological overtones. The use or incorporation of these features may be very dependent on the nature of the application.

While the importance of various human roles has been emphasized elsewhere, there is still much to consider about what software design features may aid the functioning of conference moderators, gatekeepers, group coordinators, etc. In most systems the conference moderators have special privileges for editing, deleting, and reorganizing material submitted to the discussion. In some applications it may be desirable for moderators to schedule the delivery of items.

It is also becoming quite evident from existing systems that individuals need multiple identities for the multiple roles they may play, even within a single group. By allowing the computer to sort incoming communications according to these identities, a person can deal with them more easily. The need for this sorting comes about in part from the increased pace of communication that takes place in this environment. In fact, people can easily overload themselves by seeking to get involved in every available conference. Approaches involving semiautomatic abstracting of a conference proceedings may help to alleviate this problem.

The designer of CC systems has to give as much weight to the behavior of his users and the social psychology of their particular communication process as he or she gives to the internals of the computer system.

Neisner (1964) was probably the first social scientist to pinpoint the importance of the user community in the successful operation of interactive systems. Englebart et al. (1976) appear to consider this factor fundamental in design work. A designer wants to foster a user community in which users aid one another and actively exchange information on ways to do new and useful things they have discovered. Usually successful user communities have been co-located because they evolved by people helping one another at terminals and exchanging recent printouts. A CC system makes it possible to foster this atmosphere for geographically dispersed groups. The design of such systems must incorporate ways for users to find one another according to their needs for exchanging experiences on both the mechanics of these systems and how they are used.

A Specific System

It may be useful to consider a specific system and have its designer explain at least the rationale, if not the justification, for the choices made. The Electronic Information Exchange System, or EIES, was developed at the New

ISBN 0-201-03140-X, 0-201-03141-8 (pbk.)

Jersey Institute of Technology. It represents the designer's effort to incorporate what appeared to be desirable features for a CC system, based on the original OEP work and a number of more recent systems. It also attempts to incorporate some of the general guidelines for immediate access systems discussed earlier, as well as to push the available technology toward some of the hypertext and adaptive text features.

User's Guide

The user is initially provided with a 30-page booklet explaining the 1-page user's guide, the front and back of which are presented in Figures 10-1 and 10-2. The 1-page user's guide is the single piece of paper to which the user should refer for prompting when using the system. It is hoped that users read the booklet, but if a user knows how to handle terminals, he or she will probably ignore it and start right out with a trial-and-error approach to following the user's guide.

The front of the guide presents the menus (boxes) the system provides and indicates which menu follows what prior choice. Entering a number leads the user to the second- or third-level (columns on the guide) interactions. Beyond the listed menus the user may be asked specific questions that, once answered, return him to the last menu he was in on this guide.

The *control features* at the lower right show the user how to move back up the tree structure or jump around from box to box by entering specific symbols, such as a plus sign, instead of the expected response to any question or menu choice.

On the menu itself, the *initial choice* determines what activity the user wishes to engage in, while the choice at the top of the second column determines the handling of the chosen class of text items and provides the same options for the user whether dealing with messages, conferences, notebooks, or parts of bulletins. These consistent menu choices begin to give the user the feel that a text item is a text item, regardless of the particular classification it takes on at any particular time. An attempt is made to keep choices in a regular order, so that, for instance, "get items" is always (1). The "get" function always provides the full text, whereas the "display" function provides titles only. The *review choice* provides a lateral function. For example, it will provide one line for each conference the user is in, indicate how many new items await the user, and provide the chance to get those items without entering a particular conference. Whenever the user signs on, she or he is told about new messages waiting and can choose to receive them.

The *explanation choice* is an on-line file (actually a public notebook) that contains information on using the various features of the system that the users can selectively retrieve.

USER'S GUIDE FOR
ELECTRONIC INFORMATION EXCHANGE SYSTEM

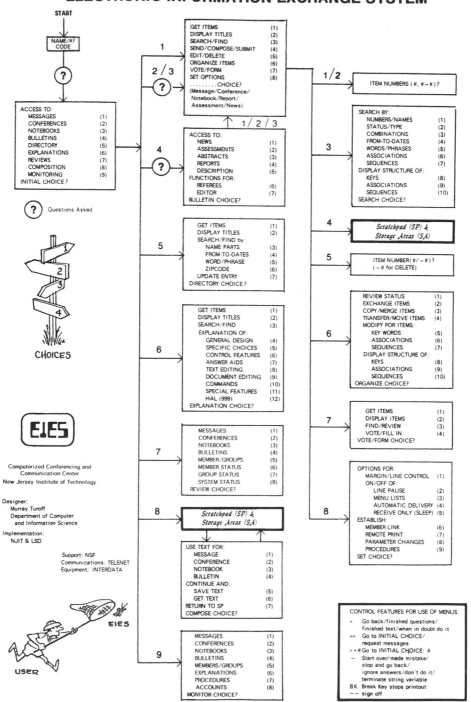

START

NAME/#? CODE

?

ACCESS TO:

MESSAGES	(1)
CONFERENCES	(2)
NOTEBOOKS	(3)
BULLETINS	(4)
DIRECTORY	(5)
EXPLANATIONS	(6)
REVIEWS	(7)
COMPOSITION	(8)
MONITORING	(9)
INITIAL CHOICE?	

? Questions Asked

CHOICES

EIES

Computerized Conferencing and
Communication Center
New Jersey Institute of Technology

Designer:
Murray Turoff
Department of Computer
and Information Science

Implementation:
NJIT & LSD

Support: NSF
Communications: TELENET
Equipment: INTERDATA

EIES
USER

1

GET ITEMS	(1)
DISPLAY TITLES	(2)
SEARCH/FIND	(3)
SEND/COMPOSE/SUBMIT	(4)
EDIT/DELETE	(5)
ORGANIZE ITEMS	(6)
VOTE/FORM	(7)
SET OPTIONS	(8)
.... CHOICE?	

(Message/Conference/
Notebook/Report/
Assessment/News)

2/3 **?**

1/2/3

4 **?**

ACCESS TO:

NEWS	(1)
ASSESSMENTS	(2)
ABSTRACTS	(3)
REPORTS	(4)
DESCRIPTION	(5)
FUNCTIONS FOR:	
REFEREES	(6)
EDITOR	(7)
BULLETIN CHOICE?	

5

GET ITEMS	(1)
DISPLAY TITLES	(2)
SEARCH/FIND by:	
NAME PARTS	(3)
FROM-TO-DATES	(4)
WORD/PHRASE	(5)
ZIPCODE	(6)
UPDATE ENTRY	(7)
DIRECTORY CHOICE?	

6

GET ITEMS	(1)
DISPLAY TITLES	(2)
SEARCH/FIND	(3)
EXPLANATION OF:	
GENERAL DESIGN	(4)
SPECIFIC CHOICES	(5)
CONTROL FEATURES	(6)
ANSWER AIDS	(7)
TEXT EDITING	(8)
DOCUMENT EDITING	(9)
COMMANDS	(10)
SPECIAL FEATURES	(11)
HAL (999)	(12)
EXPLANATION CHOICE?	

7

MESSAGES	(1)
CONFERENCES	(2)
NOTEBOOKS	(3)
BULLETINS	(4)
MEMBER/GROUPS	(5)
MEMBER STATUS	(6)
GROUP STATUS	(7)
SYSTEM STATUS	(8)
REVIEW CHOICE?	

8

*Scratchpad (SP) &
Storage Areas (SA)*

USE TEXT FOR:

MESSAGE	(1)
CONFERENCE	(2)
NOTEBOOK	(3)
BULLETIN	(4)
CONTINUE AND:	
SAVE TEXT	(5)
GET TEXT	(6)
RETURN TO SP	(7)
COMPOSE CHOICE?	

9

MESSAGES	(1)
CONFERENCES	(2)
NOTEBOOKS	(3)
BULLETINS	(4)
MEMBERS/GROUPS	(5)
EXPLANATIONS	(6)
PROCEDURES	(7)
ACCOUNTS	(8)
MONITOR CHOICE?	

1/2

ITEM NUMBERS (#, # – #)?

3

SEARCH BY:

NUMBERS/NAMES	(1)
STATUS/TYPE	(2)
COMBINATIONS	(3)
FROM-TO-DATES	(4)
WORDS/PHRASES	(5)
ASSOCIATIONS	(6)
SEQUENCES	(7)
DISPLAY STRUCTURE OF:	
KEYS	(8)
ASSOCIATIONS	(9)
SEQUENCES	(10)
SEARCH CHOICE?	

4

*Scratchpad (SP) &
Storage Areas (SA)*

5

ITEM NUMBER(#/ – #)?
(– # for DELETE)

6

REVIEW STATUS	(1)
EXCHANGE ITEMS	(2)
COPY/MERGE ITEMS	(3)
TRANSFER/MOVE ITEMS	(4)
MODIFY FOR ITEMS:	
KEY WORDS	(5)
ASSOCIATIONS	(6)
SEQUENCES	(7)
DISPLAY STRUCTURE OF:	
KEYS	(8)
ASSOCIATIONS	(9)
SEQUENCES	(10)
ORGANIZE CHOICE?	

7

GET ITEMS	(1)
DISPLAY ITEMS	(2)
FIND/REVIEW	(3)
VOTE/FILL IN	(4)
VOTE/FORM CHOICE?	

8

OPTIONS FOR:

MARGIN/LINE CONTROL	(1)
ON/OFF OF:	
LINE PAUSE	(2)
MENU LISTS	(3)
AUTOMATIC DELIVERY	(4)
RECEIVE ONLY (SLEEP)	(5)
ESTABLISH:	
MEMBER LINK	(6)
REMOTE PRINT	(7)
PARAMETER CHANGES	(8)
PROCEDURES	(9)
SET CHOICE?	

CONTROL FEATURES FOR USE OF MENUS:

+	Go back/finished questions/ finished text/when in doubt do it
++	Go to INITIAL CHOICE/ request messages
++#	Go to INITIAL CHOICE: #
–	Start over/made mistake/ stop and go back/ ignore answers/don't do it/ terminate string variable
BK	Break Key stops printout
– –	sign off

ISBN 0-201-03140-X, 0-201-03141-8 (pbk.)

Figure 10-1 User's guide for electronic information exchange system.

ISBN 0-201-03140-X, 0-201-03141-8 (pbk.)

ANSWER AND HELP AIDS

Interaction Aids:

cr	Carriage Return used to end input line/lists menu/ Assumes YES answer/ Makes no change in what is there already
CTRL and X	Line Delete (Cancel current line/Start over)
CTRL and H	Character Delete (May also be: backarrow/underscore/ backspace/terminal dependent)
?	short explanation provide
? #	Explains CHOICE: #
??	detailed explanation

To get Human help:

??? message	Send one line 'message' to console operator
??? #, message	Send to Member #
+ ON	Tells who is on now
+ LINK #	Request terminal Link to member #
+ CNM	Compose New Message

When all else fails phone 201-645-5503

Questions may be answered ahead utilizing:

;	separate answers
;	end multiple answers to one question
.	let this question be asked (+, . , Y)
, ,	equivalent carriage return
(y)	answer ACCEPT question YES

Line and Item Number Aids:

# - #	range of item or line numbers
#, # - #, #	any combination
# -	from # until end
- #	from first till #
-	all lines or items

CONTROL COMMANDS

(May be entered any time)

Text Control Commands for Output:

+ left #	set left margin at column #
+ left	set left margin at column 1
+ right #	set right margin at column #
+ right	set right margin at column 72
+ space	normal line spacing
+ space - 1	eliminates all blank lines
+ space #	# blank lines between text lines
+ trace #	copy the next # lines of output into the Scratchpad (up to 100).
+ trace # SA #	copy into SA # instead of SP

Set Option Commands:

+ SAM	Set Automatic Message delivery
+ SNAM	Set No Automatic Message delivery
+ SSM	Set Short Mode of interaction
+ SNSM	Set No Short Mode
+ SLP #	Set Line Pause every # lines (For pauses on output at CRT's)
+ SLP	Set Line Pause every 23 lines
+ SNLP	Set No Line Pause
+ SCM #	Set Conference Marker at item #
+ SAC code	Set Access Code (up to 12 characters)
+ SPEN name	Set PEN name (up to 24 characters)

Alerting Control Commands:

+ sleep	Terminal goes to sleep until new comment or message delivered
+ sleep mm:ss	Will print new items until time delay specified is reached
+ sleep MG#	Will sleep until member # gets on or off the system (break key negates sleep state)
+ alarm mm:ss	provides an alarm after time delay specified
+ alarm MG#	provides an alarm when member # gets on or off the system (alarm discontinued when user gets off)

CONTROLS

USER

TEXT COMPOSE/EDIT COMMANDS

Used in Scratchpad (SP) starting
with first character of line.
(# indicates line numbers/range)

+	text complete/finished
= #	go to line #
=	go to end of text
: #	print lines indicated
: -	print all lines as typed
: :	print all lines after editing
: : #	do for lines #
:/word/	print first line with 'word'
:/word/a	print all lines with 'word'
:/cat/ - /dog/	print from line with 'cat' to line with 'dog' in it
*	deletes line printed above
* #	delete lines indicated
* *	delete all of SP
/old/new/	replace 'old' with 'new' in line printed above
/old/new/ #	do for lines indicated
/old/new/a	all occurrences on line
/old/new/a #	for lines indicated
/old/new/a -	for entire scratchpad
↑	back up one line
↑ cat	replace letters printed above with letters 'cat'
↑ ***	delete letters above *
↑ ↑ cat ↑	insert 'cat' at first up arrow
!	blank this line
! #	blank lines indicated
!text!	fill this line with 'text'
!text! #	do for lines #
!text! (#)	place text in columns #
<	insert one new line here
< #	insert lines indicated here
< <	insert a series of new lines
>	end insert of a series of new lines
& #	copy lines #
: !	list blank lines
< !	insert blank line
* !	delete blank lines
: .	print lines with indirect edits
= 3; < 20; - 13	Use ; to do multiple edits

PENNAMES

SPECIAL MEMBERS:

name (nickname, number)
SYSTEM MONITOR (EIES, 100)
CONSOLE OPERATOR (HELP, 101)
OPERATIONS MANAGER (REQUEST, 102)
CENTER DIRECTOR (BOSS, 104)
USER CONSULTANTS: 110-119

PUBLIC CONFERENCES:

name	number
PRACTICE	1000
PROBLEMS	1001
SUGGESTIONS	1002
IMPACTS	1003
NOTICES	1004
WISDOM	1005
TERMINALS	1006

ISBN 0-201-03140-X, 0-201-03141-8 (pbk.)

ISBN 0-201-03140-X, 0-201-03141-8 (pbk.)

Figure 10-2 Data discussed on pp. 361 ff., Advanced Features Guide.

There are three ways in which users can organize the pages of text that make up conferences, messages, and notebooks:

Keys: Any text item can have a special line added to the title containing a set of key words or phrases. The system provides lists of the keys being used in a particular conference or notebook.

Associations: When an item is written (or later), the author can indicate another item with which it is associated. It is then possible to ask for all the items that have been written that are associated with a particular item.

Sequences: Since items are usually entered in sequence and numbered accordingly, the option is also provided to define a sequence of items independent of their initial ordering and to retrieve by that sequence.

These three organizational capabilities are actually quite powerful if used wisely, and are not too hard for new users to learn. The keys, in a sense, provide an indexing capability, with which any user of a high-level textbook or reference book is familiar. The association capability is most commonly used when someone makes a "proposal" in a discussion, and everyone who has a comment on that item directly associates the comment with the original item containing the proposal. It also provides a way of indicating an outline structure for a set of items, since associations can be made with items that are in turn associated with others. Many conferences develop a number of separate discussion threads that are intermixed according to the order in which they were written. The sequences allow the explicit definition of the separate discussion threads, or they can used to tie together a set of items into a single sequential report.

These are not the only possible organizational structures, nor are they the most flexible from a computer science viewpoint. They are more of an attempt to provide the user with easy-to-use structures that satisfy a majority of the requirements for structure in a text-oriented system.

Although authors of an item can define the relationships for the items they author, the moderator of a conference is privileged to modify these structures for any items in the conference. This means that the moderator can function to establish consistency in such things as what key words or phrases can be used.

The system also provides a text item to which the user can attach one or two voting scales chosen from nine predefined scales, such as desirability, feasibility, validity, pertinence, agreement, or likelihood. This allows all members of a conference to vote on those scales.

To write something, a user accesses a personal "scratch pad," which is a temporary storage space for any item being composed. On the back of the user guide are the *text edit commands*. Actually, only five specific options out

ISBN 0-201-03140-X, 0-201-03141-8 (pbk.)

of the whole list are taught to a new user, via an example, and this set provides the ability to do all necessary corrections. The other options merely make it easier to do more complicated corrections. All commands apply to the line the user is working with, unless the user adds specific line numbers or ranges to the end of the command. There is actually an option to define the range of a command by specifying beginning and ending substrings of the text across line boundaries, but it is not introduced to beginning users. The use of special symbols to trigger these edit functions is an attempt to segment them in the user's mind from the advanced editing features they may wish to learn later.

Although many editors specify the range application (lines or portion of text) before a command, we have done it the other way because the beginning user does not have to specify a line address to apply commands at the current line of text. The scratchpad saves what the user has done until such time as it is used or consciously deleted. There are 99 lines for composition available in the scratchpad and the system automatically renumbers lines to compensate for insertions or deletions of lines.

When the text is completed, the author can choose (1) to utilize a stored pen name to sign the item, (2) to remain completely anonymous; or (3) to sign the item with his or her real name.

The *answer and help aids* indicate how a user can get quick guidance with the single or double question mark, find out who is on line with +ON and send an urgent one-line message (???) for human help, or LINK to another user for assitance by letting that other user interact for her or him while observing the interaction. Through the +ON and the +CNM the user is introduced to the idea that there are commands that do the same things that can be done by the menus. The user is also reminded that a person can learn to anticipate answers by separating answers with commas or semicolons.

The *control commands* are provided to the initial user because they may be needed to operate the particular terminal in use. The "trace" function provides a way for the user to copy a transcript of his or her interaction into the scratchpad and send it to someone to illustrate any particular problem being encountered. The sleep and alarm commands are very convenient for users who are trying to do two things at once.

The user's guide also provides a list of special members or roles played by people on the system who are responsible for interacting with users, as well as the identities of the *public conferences* set aside to aid and help users.

Once the user has spent a little time practicing with the system, it is hoped that this one-page user's guide is all that is needed to interact with the system.

Because of the nature of the system it is very easy to hand an intermediary a concise list of things to enter that will not require the intermediary's understanding what is being done.

ISBN 0-201-03140-X, 0-201-03141-8 (pbk.)

Advanced Features Guide

The advanced features guide is for the user who wants to learn more; it offers numerous capabilities useful for particular circumstances. The front and back of this guide are illustrated in Figures 10-3 and 10-4.

The *Chinese command menu* is an invitation to the user to try commands. Obviously not every combination is allowed, but no harm is done if the user forms a command that does not exist. The *explanation* provides a complete list of all available commands selectively by verbs for any user who desires such a list. However, most users will focus only on those commands they commonly use. This approach of "compose your own" encourages the trial-and-error approach we desire users to adopt. It also provides about the shortest possible representation of the information for the benefit of the user.

A very important item for the user who ultimately wants to exploit the full power of the system is the form for addressing or specifying text items or even portions of text items. When in a conference, a participant has only to write the number of a conference comment in that particular conference in order to reference it (point to it or incorporate it in another item). If the user wants to reference an item located elsewhere, however, the full identification must be provided. It is the ability to address items anywhere in the system that gives the user the first hint that this system may be viewed as a single, large collection of text items, regardless of the current specific use of an item as a comment in a conference, a message, or a page in a notebook. This addressing of items is necessary to know how to use other advanced features.

In the *scratchpad* and *storage areas* the user is made aware that five images of the scratchpad are available for personal use and that text fragments that are being composed can be manipulated among the *scratchpad* and *storage areas*. These are working areas for items undergoing composition. It is also possible to pull into the scratchpad any item that a person has written or is priviledged to modify, to edit or merge it with other items or fragments, and to send it back out to replace the original. These capabilities of copy in, copy out, exchange, insert, add to, and replace provide the user with the equivalent of scissors, Scotch tape, and copying machines. This feature represents a key ingredient in providing computer power for the communication process.

Another item that turns out to be very useful in a communication environment is the *reminders*. A personal file of one-line reminders may be added to, modified, or deleted at any time. Upon logging in, the user is told how many reminders are in this file. A typical use for reminders is to record the identification number of a message that a participant is unable to respond to at the moment it is received. Since public conferences do not maintain a list of conferees, users may also note the identification number of the last item

ISBN 0-201-03140-X, 0-201-03141-8 (pbk.)

they have seen in a public conference. Of course, many users use reminders to remind them of things that have nothing to do with the system. This is to be expected and is a significant sign that the system has begun to be incorporated into their information environment.

The *document edit and control commands* represent another major step in providing powerful features. The "indirect edit" commands are actually placed in the text and executed when the completed item is read by someone. They allow for creating tables in a very easy manner, dynamically adjusting the text and the format. The "indirect control" commands allow such things as referencing, in the text of an item being composed, other items elsewhere in the system, so that when this item is printed out the other item will be incorporated at the point where the reference is made. It is also possible to actually transfer the reader to another text item with the "chain" command. In fact, a user can compose an item made up only of references to other items, and the one item might represent hundreds of pages of text. Also in this section is the capability to define questions and forms and specify the delivery of the resulting responses of readers back to the sender or to a conference or notebook.

The power of the foregoing features is considerably augmented when supplemented by the *adaptive text features*, which allow tests on the user response to determine what happens next. A person who learns these features as well has learned a programming language. The adaptive text features are plagiarized from the PILOT (Rubin, 1973; Luehrmann, 1977) computer-assisted instructional language, which has been successfully taught to teachers and students even at the junior high school level. However, allowing the user to advance as she or he wishes does not really make a conferee aware that a form of programming is being learned. This fact would probably have scared users off if it had been a starting axiom.

String variables allow a user to store an item of text titled by a letter of the alphabet and use that letter to supply the text item as input for the user's interaction with the system. This allows the user to form his or her own personal commands to carry out any sequence of operations frequently done, by typing only three keystrokes.

Numeric variables are in essence vectors that are analogous to string variables, with the exception of allowing numeric operations. Both string variables and numeric variables may be used with the adaptive text features to permanently record responses from those answering the "accept" questions.

All of these features, taken together, allow unlimited possibilities for users to structure and organize text in the system. They also allow for the potential impact of creating new styles of writing, in which the reader can be provided with control over the text, under guidance provided by the author.

ISBN 0-201-03140-X, 0-201-03141-8 (pbk.)

CHINESE COMMAND MENU

One may pick a verb abbreviation from column one and an object abbreviation from column two to attempt to form a valid command; no harm is done if not valid.

VERBS		OBJECTS	
A	Add	A	Associations
C	Compose/Send	B	Bulletin
CA	Copy & Add	D	Directory
CY	Copy	C	Conference
D	Display	CC	Conference Comment
E	Explain	F	Forms
EX	Exchange	G	Group
F	Find/Search	K	Keys
G	Get	M	Message
L	Let	MG	Member/Group
M	Modify/Edit	N	Notebook
O	Organize	NC	New Comment
P	Print	NM	New Message
R	Review	NP	Notebook Page
S	Set/Send	NV	Numeric Variable
SN	Set Negative	RM	Reminder
TA	Transfer & Add	S	Sequence
TR	Transfer	SA	Storage Area
V	Vote	SP	Scratchpad
		SV	String Variable
		SS	System Status
		T	Time and Date
		V	Votes

SAMPLE COMMANDS

+GC	name/#	Get Conference
+RMG	name/#	Review Member/Group
+G	item	Get any text item
+D	item	Display item title
+GM	#	Get Message #
+DM	#	Display Message #
+GMG	#/name	Get Member/Group #
+DMG	#/name	Display Members/Groups
+MM	#	Modify Message #
+MM	− #	Delete Message #

ADDRESSING OR SPECIFYING TEXT ITEMS

ADDRESSING OR SPECIFYING TEXT ITEMS

Building Blocks:

M#	Message #
C#	Conference #
CC#	Conference Comment #
N#	Notebook #
NP#	Notebook Page # (or P#)
SA#	Storage Area #
L#	Lines #
H	Item Heading Only
T	Item Text only

Alternatives:

NP#. #	Equivalent NP#L#
#, #−#, #	ranges allowed for text items or lines
#A	Items associated with item #
#FF	Items following in sequence from #

Any meaningful combination allowed:

M12384L21 -30
C72CC88L1 -30
M21456T

Sample Commands:

+G	item	Get item
+D	item	Display item
+CY	item<item	Replace item
+CY	item<<item	Insert or add
+EX	item><item	Exchange items

Other Text Items:

MA# Member # Address
MG# Member Group # Description
GM# Group # Membership List

INFORMATION OVERLOAD

 Advanced Features

Computerized Conferencing and Communication Center
New Jersey Institute of Technology

REMINDERS (RM)

+CRM reminder	Compose or Add
+ARM reminder	to end of list
+DRM/+GRM	Display or Get all reminders
+DRM #/+GRM #	Only ones specified
+FRM/ word/	Find reminder with 'word' in it
+MRM − #	Modify (delete −) reminder # and renumber others
+MRM # reminder	Replace reminder #
+ALARM RM# date	Place a warning date on this reminder

HARDWARE

SCRATCHPAD (SP) & STORAGE AREAS (SA)

All text composition is done in the scratchpad; each user has one, as well as five Storage Areas to save different items of text undergoing composition. These are identified as:

SA1   SA2   SA3   SA4   SA5

While in scratchpad text may be moved back and forth as in the following examples:

Exchange SP with SA:

&> <SA #	Exchange the contents of SP with SA #.
&< SA #	Same as above.
& SA #< > SA #	Exchange Storage Areas

Copy into and Replace (<) SP from SA:

&<SA #	Copy from SA # into SP beginning at the line the command is entered.
&<SA #L #−#	Copy only selected lines from SA #.

Copy In and Insert (<<) into SP from SA:

&<<SA #L#	Insert lines # from SA # without replacing anything in SP

Copy Out of SP and replace (>) in SA:

&>SA #	Copy out SP into SA # beginning at line 1
&#>SA#	Copy out lines # in SP into SA #
&>SA#L#	Begin at Line # in SA # for copying.

Copy Out and Insert (>>) into SA:

&>>SA #	Add SP to end of SA #
&>>SA#L#	Insert beginning at line #
&#>>SA#	Add indicated lines to end of SA #

General Forms of Above:

Copy out Replace (>) and Insert (>>)

&> item	Replace item with SP
&#> item	Replace item with some of SP
&>> item	Add SP to end of item
&#>> item	Add some of SP to end of item
&>> item L#	Insert SP at L# in item.

Copy in Replace (<) and Insert (<<)

&< item	Replace from this point
&<< item	Insert from this point

Example Item Types:

SA #	Storage Area #
SA#L#	Only certain lines
M#	Message #
C#CC#	A Conference Comment
N#NP#	A Notebook Page
NP#	Your Notebook
SP#	Scratchpad Lines #
C#CC#T	Text only
C#CC#H	Heading only

ISBN 0-201-03140-X, 0-201-03141-8 (pbk.)

ISBN 0-201-03140-X, 0-201-03141-8 (pbk.)

Figure 10-3 EIES Advanced features

DOCUMENT EDIT AND CONTROL COMMANDS

Used in text at beginning of line. Applies when
final item is printed (at output time). May be seen
in SP after editing (use : :).

Indirect Edit Commands:

. BLANK #	Print # blank lines here.
. BLOCK # – #	Block text following from column
. NOBLOCK	# to # until NOBLOCK encountered.
. CENTER text	Center 'text' within margins.
. EXACT	Prints text as literally entered,
. NOEXACT	ignore other indirect edits.
. FILL $	Replace symbol ($) in text following
. NOFILL	with blanks.
. INDENT $ #	Indent # space for each $ (any symbol)
. NOINDENT	encountered after column one (blank) in
	each line following until NOINDENT.
. LEFT #	Set left margin at column #.
. LEFT	Set to default of column 1.
. OVER # text	Begins 'text' over # spaces.
. PAGE	Generate a form feed character.
. RESET	Reset all margins and spacing to
	default settings.
. RIGHT #	Set right margin to column #.
. RIGHT	Set right margin to column 72.
. SPACE #	Space # Lines between text
. TABS $ #S, #L, #R, #C	Tab output following based upon
. NOTABS	occurrence of $ (any symbol) and space.
	Left, Right or Center justify within tabs.
. TEXT	Formats paragraphs by filling unused
. NOTEXT	blank areas on right of lines.
. !text! (#)	Indirect use of direct edit.
+ IMAGE	Convert what is in scratchpad to explicit
	copy of what edits produce (remove edits).
. RIGHT # ; . LEFT #	Semicolon used to separate commands
	on same line.

Indirect Control Commands:

. CHAIN item	Will start print of another text item
	and not return to this one.
. DELIVER #/C#/N#	Take all answers supplied to FORM,
	QUESTION and ACCEPT (with variables)
	and send to members, groups, a notebook
	or a conference.
. DISPLAY item, item	Printout titles only of items.
. EXEC command	Execution allowed of most EIES
	commands when the text is utilized.
. EXDIS command	Executes and displays command.
. FORM $	Same as FILL but reader asked to supply
. NOFORM	answers in place of $ symbols.
. GET item, item	Print out items indicated here and
	return to next line of this item.
. END	Return from a GET item if not end of text.
. INCLUDE text	Deliver 'text' with reply, do not print
	with text of FORM, QUESTION, etc.
. QUESTION text	Print 'text' and accept and store
	variable length answer.
+ INHIBIT	Inhibit execution of above commands while
+ NOINHIBIT	in Scratchpad.

ADAPTIVE TEXT FEATURES

(Abbreviations also shown)

. ACCEPT:	Allow reader to answer a question
. A:	asked just before.
. A: X $	Accept answer as a string variable.
. A: X #	Accept answer as a numeric variable.
. MATCH: S1, S2, . . .	Check if what reader put in
. M: S1, S2, . . .	matches any of S1 of S2 or . .
. TYPE: text	Type out 'text' but substitute wherever
. T: text	a string or numeric variable appears.
. T: X $	Type contents of string variable X.
. TY: text	Type 'text' if MATCH was Yes.
. TN: text	Type 'text' if MATCH was No.
. JUMP: line/item	Jump to line or item.
. J: line/item	Uncond'tional jump.
. JY: line/item	Jump if MATCH was Yes.
. JN: line/item	Jump if MATCH was No.
. REMARK: text	'text' not printed out, but
. R: text	stored for benefit of author.
. USE lines/item	Use indicated lines or item
. U: lines/item	and return to next line.
+ NOPILOT	Inhibit execution of above commands
+ PILOT	while in scratchpad.

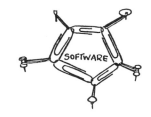

SOFTWARE

STRING VARIABLES (SV)

(letters A to Z)

+ SSV A = string	Set String Variable A equal to 'string'
+ GSV	Get all String Variables
+ GSV A	Get only SV A
+ DSV	Display letters used
+ FSV / string/	Find SV with 'string' in it
+ A$	Use SV A as input to EIES for an answer
+ MSV A/ old/ new/	Replace 'old' with 'new' in string variable A
+ MSV – A	Delete String variable A
+ SSV A = S1 , + Bs, S2	Incorporating another string
+ ASV A = string	Apend this string on end of current A string (128 characters allowed)

ORGANIZATION

NUMERIC VARIABLES (NV)

(letters A to Z)

Everything for String Variables applies to numeric
variables substituting NV for SV in above and:

. + SNV A = #, #, #	Set A equal to a series of numbers separated by commas
+ A #	Use NV A as input
+ DNV	Display letters used and how many numbers in each separated by commas
+ LNV A = B + C	Math operation on NV
+ LNV A (3) = A (3) + 1	For single element (Let 1 be added to 3rd element)

EVALUATION

ISBN 0-201-03140-X, 0-201-03141-8 (pbk.)

ISBN 0-201-03140-X, 0-201-03141-8 (pbk.)

Figure 10-4 Data discussed on pp. 368 ff., Advanced Features Guide.

Other Advanced Features

HAL ZILOG (HAL, 999)

Hal is a conference member who is really a Zilog microprocessor development unit with two floppy disks and his own telephone dialer. Hal is able to phone the conference system and receive and compose text items like any other user. The uses planned for Hal are

1. Providing various analysis and display graphic routines for use by an EIES members;
2. Obtaining data and output from data bases and models on other computer systems that might be of use to conference groups;
3. Interfacing to other computer-based communication and text-processing systems to move text items back and forth between systems.

In essence, a user calls from a public notebook, a stored form that she or he fills out and sends to Hal as a message. The particular form for a particular job constitutes the user-supplied information Hal needs. He then carries out the task and sends the result back in the manner requested by the user (e.g., as a message to the user or group, or as an entry in a notebook or conference).

Hal has been designed to be able to be easily taught to simulate a user at a terminal. Hal can therefore provide conference system users with any information they may need for their discussions that may be contained in other computer systems.

As a result, a CC system becomes potentially a centralized coordination node for a host of computer resources, in which individual members of a group familiar with a particular computer system become the transponders to make that resource available to the group as a whole.

INTERACT*: A Procedure Oriented Language

The adaptive text discussed in the "Advanced Features Guide" section represents one type of procedural capability in EIES which performs actions when items are being printed on a terminal. In addition, EIES supports the ability to create and execute procedures which may utilize all EIES resources by supplying EIES with commands as if they were entered from a terminal.

For the novice user this procedural capability provides a simple means for commonly entered commands and/or text segments to be made available for future use. The experienced user has available the features of a complete programming language designed to facilitate the development of specialized user interfaces and communication protocols. This string-processing-oriented

---

*This section was coauthored by James Whitescarver. See also Turoff et al., 1977.

ISBN 0-201-03140-X, 0-201-03141-8 (pbk.)

language[†] may be entered as part of the text items stored in EIES; it is aware of the EIES file structure as well as all variables that represent the communication processes taking place.

DESIGN GOAL OF INTERACT ON EIES

The procedural capacity of EIES is designed to meet the following objectives:

1. To provide an alternative means of supplying data to EIES by performing any action on EIES normally performed by the user from the terminal.
2. To provide the ability for generation of alternative user interfaces to the system.
   (a) Make it behave like other systems with similar capabilities.
   (b) Make it behave like special-purpose or limited-feature systems for special users.
   (c) Control the interaction of EIES with special terminals and other processors (e.g., floppy disks, cassettes, smart CRTs, Faxgram or Mailgram services).
3. To allow the implementation of new EIES features via procedures.
   (a) As a test-bed for new features that are proposed.
   (b) To allow the modification and extension of "Help" and "Explanation" procedures on line by nonprofessional programmers.
   (c) Permanent implementation of little-used features to reduce overall requirements for program memory and development cost.
4. Gaming and other multiterminal specialized interactions, such as (a) group interactive modeling, on-line classrooms, bridge games, simulation games, and other specialized communication structures.
5. Polling or other specialized message/item response.
6. CAI–PILOT support for computer-aided instruction applications.
7. Support of adaptive text in the form of embedded conditional text.

PRINCIPLES OF PROCEDURE EXECUTION

Normally the user communicates directly with EIES.

$$\text{EIES} \leftrightarrow \text{USER}$$

When the user calls a procedure, or a built-in procedure is called, the procedure can control the entire interaction process between itself, the user, and EIES.

$$\text{EIES} \leftrightarrow \text{PROC} \leftrightarrow \text{USER}$$

[†] A string-processing oriented language is one designed to facilitate the manipulation of symbols or sets of symbols, such as ordinary text. This particular language is called INTERACT.

ISBN 0-201-03140-X, 0-201-03141-8 (pbk.)

If when supplying an input to a procedure the user invokes another procedure, the user's input is supplied by the second procedure to the first.

$$\text{EIES} \leftrightarrow \text{PROC1} \leftrightarrow \text{PROC2} \leftrightarrow \text{USER}$$

PROCEDURE INVOCATION

A user may invoke a procedure by typing "+NAME" at any input in the system. "NAME" may be the key of a system procedure or an item on the user's "item index." Items may be executed directly by typing "+#ITEM," where "ITEM" is in standard format. For example, "+#NP10" would execute the lines in the user's notebook page 10. In addition, a one-line string of commands may be utilized in a string variable (see advanced user's guide).

EIES utilizes a very general event service scheduling mechanism to control the flow of information on the system. An important application of this mechanism is the scheduling of procedures. Procedures may be queued for execution at a later time by other procedures or by the system. The procedure will be executed when a certain set of conditions specified by the queue entry are met. Possible conditions include

1. State of any user (logged on, hit break key, entered conference, etc.).
2. State of any text item (viewed by certain user, modified, etc.).
3. A specific time delay has elapsed.
4. A user-defined event has occurred (set by a procedure).
5. State of any queue (number and/or type of entries, specific entry).
6. Any combination of the foregoing.

TOOLS FOR PROCEDURES

Interprocedural communication: Along with the standard communication methods (messages, conference comments, Bulletins), otherwise independent procedures may communicate by utilizing special event service queues to pass information between themselves.

Traps: Trap structures cause the queuing of event service routines for the period that the current procedure is active. Traps are useful in detecting special conditions and overriding EIES default action during the course of executing the current procedure.

Control structures: Structures for looping and programming alternative actions (If Then Else) provide the means for writing well-structured procedures without the use of labels for jumps.

Variables: Arithmetic vectors and scalars, text items, string variables, and a general-purpose stack are provided for the manipulation of local and/or global information.

Expressions: The expression evaluator provides numerous operators, and functions for manipulating vectors, scalars, strings, logical functions, and

ISBN 0-201-03140-X, 0-201-03141-8 (pbk.)

ISBN 0-201-03140-X, 0-201-03141-8 (pbk.)

functions for obtaining special information from EIES (e.g., time of day, members of group, name of member number X, state of events).

Pattern matching: EIES supports an extended version of the Dartmouth PILOT match statement. The internal pattern format, obtainable by a simple function call, may be used on EIES wherever a search string is used.

String replacement: An expression (vector, scalar, or string) may be embedded anywhere in a procedure by using the form: "@(expression)." Whenever such a form is encountered, it is replaced by the character representation of the evaluated expression before the statement is interpreted. When string replacements are nested, the innermost is performed first.

Structure simulation: A command is provided (+INCMD) to allow commands that follow a procedure invocation to be processed like control or information structures.

File structures: Commands usable by privileged procedures enable the generation and maintenance of file structures to allow the creation of new data structures, indexes, and file protocols.

Debugging aids: A view mode is available to allow the writer of procedures to trace any portion or all of a procedure execution.

Other: Commands exist for queue processing, event definition, logical item processing, command input/output, stack manipulation, and variable allocation.

As a result, the concept of procedures brings about the complete embedding of a programming language within the text items of EIES. For the user who has obtained or desires to obtain the degree of computer sophistication compatible with the use of procedures, EIES provides complete control of his or her communication practices and the ability to tailor a personal system. It also provides those of us concerned with the development of CC systems with an extremely powerful test-bed for experimenting with the human interface and group communication process.

Summary

The EIES interface is an example of the power the computer can provide to individuals and groups engaged in communications. It offers an extremely rich and diverse set of capabilities, both for the users and for experiments by designers and evaluators. However, it still represents only one example of the many communication structures that can be implemented on the computer for other objectives, such as educational, parliamentary, recreational, and management systems.

This particular area of utilizing computers for human communications is largely an untapped one, and it will take many years of experimentation before we can develop hard and fast truths about the nature of the human interface for these systems. Because of this, designers should design with the idea that changes in the interface will be the rule rather than the exception. Also, the intimate role that the user plays in these systems means that user expectations, desires, wishes, and reactions have to be given extremely serious consideration by designers.

For Discussion

1. Give some specific examples you have encountered where computers take on human-like aspects or roles in terms of the jobs they do and their relationship to humans.

2. If you have had direct contact with interactive systems, provide specific examples of some of their effects or characteristics, such as the bully effect or fishbowl effect.

3. Take one fanatic computer-type and one individual with a specialty in an area who knows nothing about computers. Have them try to hold a discussion on applying computers to that person's specialty, and have everyone else observe and try to determine the degree of actual communication that takes place.

4. What signals can you itemize for determining when a user is in one or another phase of development: uncertainty, insight, incorporation, or saturation?

5. Try to match various EIES features to the information domains or requirements of users. What EIES features support each domain: simple inquiry, consensus formulation, and report generation; discovery and analysis, planning and decision analysis; management and operations. What features appear to be missing for what domains?

6. Match EIES features to the design principles of forgiveness, escape, segmentation and generalizability, variety and flexibility, guidance, and leverage.

7. What are the possible differences in a paper book and one available through a computer terminal, i.e., How would you design an on-line book? If you find this too difficult to generalize, pick a specific type of book such as a cookbook.

8. Can you provide some other examples where analysis might be directly incorporated into the communication process to aid communications about a complex problem?

9. Design an on-line form of a Delphi exercise or a nominal group technique.

ISBN 0-201-03140-X, 0-201-03141-8 (pbk.)

10. For particular situations (of your choosing), how would you collect profile information on individuals to utilize for matching people who have things in common about which they might wish to communicate? In what situation would your approach be considered an invasion of privacy?

11. Describe some specific examples where graphical conferencing capabilities would be mandatory.

12. What changes or improvements would you make to the EIES design?

13. In what situations would you define other types of text items besides those in EIES: comments, messages, and pages.

14. What sort of organizational structure or capabilities would you use for a specialized group such as a legislature?

15. Try composing some common types of letters as an adaptive text item in which the text provided to the reader at various points is conditional on the reader's answers. Consider advertisements, love letters, letters home, etc.

16. Can you describe some specific situations where information from data bases or models is very necessary to a discussion?

17. What sort of personality, if any, would you give Hal Zilog?

18. What sort of person is likely to wish to or desire to use all of the EIES capabilities?

19. In what discussion situations would someone who knows certain EIES capabilities have a significant advantage or edge over someone who does not?

20. What sort of communication-oriented games would you design to these systems, or what games would you add communications to?

21. Suggest situations where experienced users of CC systems would wish to write very specialized or unique programs to do their personal tailoring of their communication activity or structure. Consider individuals who might have a very special role to play in some group communication process.

ISBN 0-201-03140-X, 0-201-03141-8 (pbk.)

APPENDIX

Hardware and Software Technology

Introduction

This fairly technical Appendix is written for those who are familiar with the hardware and technology of computers. However, some of the flavor of the internal problems of design will be evident to the reader not intimate with computers.

To date the technology of computerized conferencing systems is better described as having been demonstrated as feasible rather than perfected. Only about ten conference systems have been implemented. The effort currently required to implement a system represents one to six person-years by system-level programmers. This considerable investment is necessitated by having to impose conferencing requirements on software and hardware systems not originally conceived of or designed for this type of application.

To better convey the nature and extent of this effort, we review here some of the specific implementations that have been done and the problems they presented. Based on this review, we make some observations about what developments with respect to the technology of implementation are likely to occur in the future and why they appear to be desirable.

Review

The very first system implemented at the Office of Emergency Preparedness was written in a version of the BASIC language called XBASIC and developed by Language and Systems Development Incorporated for use on Univac 1100 series computers. The features available as a result of the unique combination of the characteristics of the operating system on the 1108 (EXEC VIII) and those of the XBASIC were what allowed the creation of the original Delphi conference system and the EMISARI system that evolved from it. The operating system for the 1108 was designed to support multiprocessing operations so that N processors could share the work load under a single operating system. Therefore, features were designed into both the hardware and software that proved to be very useful for a conferencing application. In terms of the hardware, the instruction set was well designed for reentrant code, so that more than one processor or more than one user (in the conferencing case) can utilize the same program in the computer core.

ISBN 0-201-03140-X, 0-201-03141-8 (pbk.)

The hardware also allowed memory protect* for segments of core memory and dynamic assignment of the protected areas, as well as the ability to transfer variable-length strings or records among core locations without using a loop of load and store instructions and consequently extra hardware registers. As a result of the requirements for a multiprocessor environment, the operating system had an excellent file structure with the ability to change dynamically the read and/or write privileges on small segments of any file. Added to these featues were the following unique XBASIC features.

- The ability to incorporate operating system commands directly in the XBASIC code so that all of the file manipulation capabilities of the EXEC VIII were available to any XBASIC program.
- One of the most powerful sets of string functions available in any language of the time. String manipulation is a key task of a conferencing function.
- Functions that provide operating system status on the success or failure of an executive-level operation at the XBASIC language level.
- A compiler for XBASIC separate from the compile and go system which compiled fully reentrant code with all error checking stripped out, so that the conference program could do its own error checking on user input and ensure that the user was never thrown out of a conference program.
- A versatile "chain" feature that allowed the segmentation of a conference package to small software modules that could share common data.

The overall EMISARI package represented some 7000 lines of XBASIC code but no one module exceeded 500 lines. The modularity allowed both its use and development to proceed simultaneously with as many as five programmers working on the system at the same time by utilizing design specifications of the input and output of program modules. This ability to develop features and incorporate them easily into an operating conference system was the key to allowing the system to evolve according to user requirements developed out of user experience. This is a requirement that is very necessary in organizationally oriented systems, and has not received due attention in the area of management information systems, in particular.

Practically none of the implementations that followed the original OEP work had anywhere near the "semi-ideal" circumstances represented by the 1108 and XBASIC combination. The hardware and software problems were tougher, which seems on the surface a bit inconsistent with what one normally pictures as technical progress in the computer field. It must be recalled that the design of EXEC VIII on the 1108 was somewhat ahead of its time, and UNIVAC, almost three years late in getting a working software

---

*Internally a computer is performing many independent jobs in one common blackboard of memory. "Memory protect" is the problem of preventing one job from writing on the area of a computer's blackboard reserved for another job.

ISBN 0-201-03140-X, 0-201-03141-8 (pbk.)

package, never met the original design specifications. Most of the second generation of operating systems never attempted to deal with a multiprocessor environment; in fact, the major concern, from a time-sharing viewpoint, was the protection of files and data belonging to a single user. Therefore, it was extremely difficult with some operating systems to have simultaneous users utilize the same common file. For the standard operating system on the IBM 360 series it was, in fact, strictly forbidden, so that the few attempts at conference systems by delayed merger of files on IBM systems turned out to be largely unworkable, except as one-time experiments. There was also a distinct trend toward isolating the operating system from application languages such as FORTRAN, so that there was no danger of a user's program causing interference for the operating system. This is an obvious approach to general-purpose time-sharing systems that must service a wide variety of user applications.

The basic conflict that results from the need for security and operational reliability precludes allowing the users of a higher-level language some of the inherent power of the computer hardware and the operating system features. The need for files and programs servicing different users and processing applications to be able to communicate and exchange data is only now beginning to be recognized in the area of operating system design. How to satisfy that need, from the point of view of hardware and operating system software, is not yet clear. One approach that has evolved is referred to in the literature as "transaction oriented" processing systems for large-scale computer systems. However, smaller computers dedicated to and optimized for conferencing system environments may be a more desirable approach.

Some of the other efforts that have been made since the OEP project illustrate the problems. At the 1975 National Computer Conference, Dr. Turoff chaired a panel where a number of designers and implementers were asked to present some of the difficulties they faced and what they felt were desirable alternatives. The following is abstracted from the notes prepared for that panel by some of the participants.

Minicomputer Implementations

Two of the efforts reported at that panel were minicomputer implementations. The first was the system Bell Northern Research created on a PDP 11/45. This system was designed by Hilary Williamson, who described the internal design as follows.

The system handles four distinct types of tasks:

1. User tasks, which communicate with users and monitor their response.

ISBN 0-201-03140-X, 0-201-03141-8 (pbk.)

2. Command tasks, which are initiated by user tasks when a user request needs to be serviced.
3. Command analysis tasks, which schedule service that is provided on a round-robin basis.
4. File access tasks for retrieving and updating files.

All the tasks communicate through a global shared-data area and by means of intertask communication primitives. The system is event driven and task swapping is forced by priority management and by frequent execution of a small task that does nothing but exit.

Data sharing for the global area uses a concurrency control algorithm. For files, this sharing is done by passing all requests through a single task. These functions are minimal functions that cannot be interrupted. (Note that this capability of noninterruption is not usually available at the application level in general-purpose time-sharing systems and is one approach to getting around the problem of two users updating the same item—either a text item or a vote total, for example—at the same time.)

Command analysis is performed on the basis of a classification of all possible parameter types. Message text is structured as a linked list of "superlines." A further logical structure of superline "units" with associated attributes is superimposed on this. Other message data structures include order lists, reflecting message order in time; key phrase and label tables; and a linked list structure of "associated" messages.

The system was implemented in FORTRAN but required assembly-level routines to handle strings and bit masks. Both the disk access time and memory core limitations (256,000 bytes) constrain the system to approximately eight to twelve simultaneous users. This type of system is largely I/O bound, which can be offset by larger internal core buffers when sufficient memory is available. By conglomerating sets of small files, the costly need to continually open and close files was minimized. This is an area where modifications to make the operating system more suitable for this application would have been desirable. Memory limitations combined with the awkwardness of the swapping mechanism caused problems. These were overcome to some degree by trading off the extent of overlaying and reentrancy of tasks. A further problem in the area of optimization of task swapping was that the most obvious candidates for swapping, user tasks, could not be checkpointed. Potentially, a deadlock situation might occur as the user tasks manipulate priorities of other tasks. Though the method of passing file requests through a single task speeded up the implementation, it is probably not the most efficient way of doing it. The use of semaphores might be a better alternative. Writing more of the system in assembly language would have led to a higher proportion of pure code to impure code and less core requirement. Other

ISBN 0-201-03140-X, 0-201-03 41-8 (pbk.)

difficulties could have been minimized by facilities for file sharing and facilities to trap interrupt conditions at the FORTRAN level. A separate disk for task swapping/overlays would also have been desirable. In general, Williamson does not feel that a minicomputer, or at least the PDP 11/45, is an ideal machine. Certainly an operating system with time-sharing capabilities is essential.

The second minicomputer implementation described was for EIES on an Interdata 7/32. This was done by Tom Hall, a systems designer for the Language and Systems Development Corporation.

The host language was FORTRAN IV. This choice was essentially forced, since it was the only higher-level language available at the time. The main logic of the system was a set of programs corresponding to the user interaction steps with several sets of "black box" subroutines to do the nitty-gritty things. These were

1. I/O package, which was modularized for one programmer to program; all others were told to use the subroutines according to calling sequence specifications only. The lowest levels of this package are the only ones dealing with the operating system and disk organization details, allowing for improvements in this area without disturbing the rest of the systems. Also, a higher-level I/O package handles all types of input a user can make and is parameterized to catch user input errors and provide error messages and correction opportunities to the user before data is passed back to the main program.

2. The telecommunications package, which insulates the main system from the interrupt-level operations naturally involved in terminal handling. This package provides an interface from FORTRAN coding to the system features available for communications work (i.e., Interdata's ITAM telecommunications package). The setoff modularity of this package allows for conversion to various network disciplines as this becomes necessary to support system growth.

3. The service subroutine package to do all the character manipulation and the data structuring that needs to be done but is so awkward in FORTRAN. The subroutines in this package sit between the FORTRAN coding in the main logic and the low-level subroutines in the two packages above. They provide such services as disk paging, control of index sequential file structures, and encoding/decoding of lines to and from the terminals.

The main logic of the system is designed and coded as if there were only a single user signed on (i.e., as the system would have been coded on a time-sharing system). Communication between users is via common-access data files, although disk block paging implies that the most-used parts of

ISBN 0-201-03140-X, 0-201-03141-8 (pbk.)

these files are really in core. Only the terminal handlers and a scheduling set of routines deal with multiple terminals and they shield the main logic from this fact. The scheduler works by switching from one user to another at predictable program points, namely real or virtual I/O calls, so that the main logic does not worry too much about being purely reentrant at all times. The scheduler also works with a memory paging system to switch from one user's data area to another utilizing hardware memory access control registers on the Interdata machine. The main-line FORTRAN program is essentially unaware that the user is changing out from under it at I/O calls but proceeds to act on the data provided which reflects a different user.

The system is organized as a tree structure, formed by the user interface design, consisting of a sequence of menu choices. A set of commands to use instead of menu choices is superimposed on this structure by a command translator and a procedural language, which can dictate the use of the features of the conference system.

Although called a "multitask" system, the Interdata OS/MT could not support a multiuser environment, having only essentially fixed memory partitions, and no memory swapping or true time-sharing system features. Neither did it have any true reentrant programming disciplines allowing for one copy of a program with multiple data regions, one for each user, plus common regions. Also, the system did not allow memory allocation to switch between data areas without moving data, even though the hardware had memory paging registers ideally suited for doing this. The system loader/ linkage editor did not support the segmentation of data areas into blocks that could be switched around using the paging registers, having no way to allocate data areas to virtual space that was not actually available at load time.

FORTRAN itself posed problems. Its code was nowhere near reentrant in nature, even if the operating system would have supported it. The generated code's subroutine calling/receiving sequences were especially inadequate, both for reentrancy and for an indirect recursion capability which was desired.

The difficulties with reentrancy and time sharing forced the system itself to have a scheduling routine and a data area management routine, both working with the terminal routines, to divide time and memory among users. Given the problems with the FORTRAN-generated code and the loader's inability to segment data areas well enough, the scheduling routines had to have considerable help before they could do anything. Part of this help came from establishing coding conventions, so that the main logic would use "common" blocks for certain classes of data items and not for others, to distinguish between strictly local variables, variables common to the whole system, and variables specific to the current user's level of recursion. To resolve the calling sequence problems and to collect these classes of data into areas to be manipulated with paging registers, it was necessary to take the generated code produced by the FORTRAN and modify it with an intermediate processor. Fortunately, the FORTRAN compiler was of the simplistic type that generates symbolic assembly code and calls in the assembler. Therefore, inserting a processor in between was possible. This "fudging" processor had to do an unfortunately large amount of work, but by doing so made everything else possible, in spite of the operating system.

Executive modifications were also necessary for some of the foregoing. The EIES implementors were trying to convince the system to do things for which it had not been well designed. Most of these executive changes dealt with I/O control and file-handling procedures.

From such an effort a number of observations can be made about operating system features and language features that would have been desirable.

Operating System Features

1. Communications routines should be more naturally integrated into the overall system, including interface from higher-level languages, so that general programmers can write multiuser systems without a wizard systems programmer to handle communications problems.

2. Better disk management facilities. Index-sequential files would be useful if fully integrated into the overall system. The system also has to have flexibility in file buffering, utilizing user areas for index tables if desired, so that the number of open files is not limited to the fixed number of buffers generated in the operating system.

3. A reentrant programming discipline, coordinated throughout the system, including the FORTRAN compiler, assembler, loader, and run-time libraries. Even without a true multiprogramming, time-sharing system, this would make it possible to write programs to be used by several users at once.

4. Better ways of using the raw computer power without interference from the operating system. For example, it should not have been necessary to modify the operating system to make use of the memory access registers, for which purpose they were already ideally suited.

Language Features

1. Character data handling has always been a problem area in FORTRAN, and unfortunately in many other otherwise high-level languages. A natural way to deal with string data is urgently needed for conference systems as well as other human-interacting systems. We had already dealt with this problem and defined subroutine packages for the essen-

ISBN 0-201-03140-X, 0-201-03141-8 (pbk.)

tial functions, but the code produced using subroutine packages is nowhere near what it should be in terms of readability and natural structuring. It manages to be structured but awkward.

2. The lack of natural structuring in FORTRAN programs was also evident in coding the system. The tree structure of choices and decisions evident in the user interface description does not translate well to a language without block structures or even a very good "if" statement. The structure required had to be coded into tables to drive the programs. There is a loss of clarity in the coding-versus-the-user-interaction relationship resulting from this approach.

Hall's overall opinion is that they had to do entirely too much working around things rather than working with things. He suspects that their experience in getting the system to do what they wanted in spite of itself might indicate that our coding could be implemented on other systems, but at the same price of learning how to fudge things where necessary.

As programmers, working around things and coding in FORTRAN with all its known inadequacies did not really bother them too much. Some simplicities in the system turned into benefits. For example, the FORTRAN compiler that produced assembly code enabled them to overcome the reentrance problem in a very powerful way. In general, programmers expect these sorts of difficulties and can get around them.

Hall's final summary of the situation was

> Attempts to remedy such problems might help in some cases, but might just as well pose new problems in others. I personally am less fond of systems that do almost what you want but not quite than I am of systems that don't make pretensions about doing everything. I also distrust attempts to patina "new" features over an inadequate system to make it more useful, since the underlying weaknesses still exist. For example, adding structuring to FORTRAN programming is not my favorite subject.

These previous two attempts represent the first major efforts to create fairly sophisticated conference systems in dedicated minicomputer environments. Of course, the Interdata 7/32 configured with half a million bytes of core and 300-megabyte disks begins to look less like the classic concept of a minicomputer. There are proponents who believe that the future of these systems is as a part of large-scale general time-sharing systems, and others who feel that dedicated specialized computers are the answer. Depending on the scope and objective of the conferencing system, both schools may be right. Straightforward messaging and simple notebook systems can easily be accommodated in general-purpose environments. However, a comprehensive package intended to provide a total communications and word-processing system places very

ISBN 0-201-03140-X, 0-201-03141-8 (pbk.)

different demands on hardware and software resources than do computational applications. A look at some other implementations will shed further light on some of the problems involved.

Specialized Implementations

At least two implementations represent attempts to superimpose a conference system on large-scale multiterminal systems of a specialized nature. One such case is CONFER, designed by George Carter on the PLATO system at the University of Illinois. This system was written entirely in TUTOR. It is not surprising that a powerful computer-assisted instructional language allows the creation of a conferencing system, since the features must be available to pass inputs from students back to teachers when appropriate. PLATO has had for many years simple simultaneous message passing among on-line users. However, Carter's system was the first that attempted a stored discussion type of conferencing system. The system represented 30 man-months of programming, which is also not surprising in light of other experiences with TUTOR.

The system called CONFER allowed limited graphical-type interactions through the use of a set of codes for users to specify diagrams. The primary problem that resulted was the existence of the limitations set by the objectives of the PLATO system, which imposed severe constraints on the conference package. Because PLATO's goal was the support of 1000 simultaneous terminals, the amount of in-core work space per terminal was severely constrained. Since a conference system provides more unpredictable file-retrieving requirements than a CAI lesson, the response time for CONFER was marginal because of increased need for disk accessing under conferencing. The per-terminal limit of 2000 to 3000 machine instructions per second was also found to be insufficient. In this implementation we have the classic case of a major system optimized for a particular application providing poor operational characteristics when conferencing is forced into the original environment. While TUTOR provides a good language capability for describing the individual's interaction, it has no descriptive capability for the communication and specialized file aspects of conferencing, which means the programming effort was just about as bad as utilizing FORTRAN.

The "Memo from Turner" system designed by Bert Liffman represented an attempt to create a conference system in a dedicated APL environment on the I. P. Sharp–Timesharing system. Both I. P. Sharp and Scientific Timesharing have made major modifications to APL to accommodate file structures more suitable for data base applications. The conference package followed fairly standard design practices, such as a single routine to control and monitor all user inputs and functional subroutines to carry out command

ISBN 0-201-03140-X, 0-201-03141-8 (pbk.)

features, and performed fairly well. There were, however, certain noticeable problems. APL is a great personal computing language but there are so many different ways that programmers can accomplish the same thing that consistency at the coding level is difficult to maintain. A relatively minor change, if made by a programmer different from the original one for that routine, can require a major programming effort. Constraints on work spaces and other resources meant to optimize more typical computational applications placed limitations on the design and performance of the conference package. The inefficient and rather abstract method that APL uses to deal with strings meant that any reasonable amount of string manipulation or searching was far too expensive according to the normal charging mechanisms. String characters are encoded as integers and APL primitive arithmetic operations must be used. The symbolism of APL may be beautiful for mathematics, but it is intolerable for describing either string manipulation, communications, or file manipulation. The biggest problem, however, was that the conference system could not be isolated from the operation of the APL system environment. Users could suspend the program by hitting the wrong key and be thrown back into APL. The continuous file access requirements of a conference design forces use of periodic file update techniques, so that some simultaneity for the user was removed. Even with attempts to minimize string processing and file accesses, operation of the system was still in the neighborhood of $30 to $40 an hour if standard APL charging algorithms were used.

Other nasty problems resulted from such things as the very large output buffers utilized to hold material for all the terminals. If a crash or interruption occurred before the n messages or items waiting for a user were actually delivered (printed out), the conference package had no way of telling that they had not been delivered. In other words, certain system status conditions were not available to the APL program.

In both these specialized implementations, it is obvious that any large-scale use of conferencing would require language changes to either TUTOR or APL, and quite possibly front-end processors that could more efficiently handle some of the word-processing and terminal I/O operations. A more expensive alternative to the latter one is the use of intelligent terminals. However, any general-use conferencing system must be able to deal effectively with dumb terminals in order to maximize the potential user population.

Observations

Aside from the Univac EXEC VIII on the 1100 series Univac machines, the only other second-generation machine that seemed amenable to implementing conference systems was the time-sharing operating system on the PDP 10.

ISBN 0-201-03140-X, 0-201-03141-8 (pbk.)

This was the target machine for the efforts of both the Institute for the Future and Language and Systems Development Corporation. This is not surprising, in that the PDP 10 was the mainstay of research efforts in computer science for that hardware generation. However, even on the PDP 10, the problem of two people possibly updating the same item at the same time is resolved by careful scheduling of disk accesses to reduce the chance that an error is made to an extremely small probability. The ultimate solution is the test-and-set hardware commands on some of the new machines, which can test and conditionally change a status bit in one uninterruptable machine operation.

As a result of the implementation efforts that have taken place, the requirements for both hardware and software in implementing conference systems are becoming more evident. A review of these requirements seems the appropriate conclusion to this Appendix.

Review of Technical Requirements

At the lowest level, the particular hardware and software features for handling I/O operations and interrupts can greatly influence the efficiency of these systems, and in turn the degree of capability that the designer can offer to the user.

The designers of computer hardware and operating systems have to date paid more attention to improvements to enhance computational and arithmetic capabilities. Neither in hardware nor in most software systems do we typically find the same macros for the manipulation of variable-length records and their transfer operations (as typified by, for example, single and double precision floating point capabilities).

As a result of unsuitable file structures at the system level, CC systems are often far more complex than they need to be. There is really no need today for not making the dynamics of file access and control more of a hardware macro operation. File structures must be more overlapping rather than exclusive structures. In other words, the same element may be a member of one or more files.

The current trend toward insulating the higher-level language from the operating systems needs to be reversed, so that the conference program in that language can control the operating systems. This means the status and control of interrupts and I/O have to be incorporated at the higher level. The only computer application that has a certain technical similarity is a process control system, where a host of different sensors and control devices must interface to one application package. In a sense the participants in a conference system present the same problems as the sensors and control devices. Most conference packages today have to make frequent transitions into the domain of the operating system command language. All user I/O (break key,

ISBN 0-201-03140-X, 0-201-03141-8 (pbk.)

ISBN 0-201-03140-X, 0-201-03141-8 (pbk.)

etc.) has to be removed from the operating system to the conference system so there is no way for the user to get at the operating system.

Problem areas that are currently open ended are those that have to do with distributed processing involving a number of conferencing systems able to work in some sort of cooperative network. The ARPA message system approach is to make a separate copy of any multiply addressed message and put it in each recipient's private storage area (i.e., a message to 20 people is going to need 20 times the storage of the original message). As yet no real work has been done on distributed systems for this application.

Reliability of systems is still not easily attained, even with dual processor systems. Security and integrity of data requirements are as severe as in any other computer system.

While the foregoing issues are important to the better operation of conferencing systems, there is one final issue that is crucial to the long-term development of the technology. At the moment, most of the systems that have been implemented represent many person-years of programming effort. There is no suitable higher-level language that allows a concise and appropriate specification of a human communication process. All current efforts at using FORTRAN, APL, BASIC, etc. are like describing a picture of the Mona Lisa in words only.

Even in the professional areas that deal with human communications there is no such language. Any paper on a group communication experiment or even an edition of *Robert's Rules of Order* speaks to the fact that we must use general English at present to describe a communication process. It is part of the challenge of computer and information sciences to find suitable languages to translate in an efficient manner a real-world process into terms the computer can deal with. Currently we are beginning to understand the basic functions that characterize conferencing systems and it can be expected that a computer language will evolve that can allow these systems to be created in a few person-months of effort. We now know that we must incorporate into a higher-level language the following types of features.

1. Communicating Entities: Data types representing persons, roles, and groups that engage in communications and have certain attributes and privileges. Associative specification techniques such as those utilized in SIMSCRIPT may be appropriate. Entities may include files.
2. Communication Items: The items passed along a communication channel that may be specified by virtual or associated composites or fragments of other items.
3. Communication Nets: Actual graph-type specifications among entities.
4. Concurrent Processes: Being able to specify processes that are concurrent in nature and the triggers for inhibiting or invoking such processes. Dynamic forms of decision tables may be appropriate for this function.

Ideally, there should be developed higher-level language tools enabling the designers of human communication processes, such as management or social scientists, to specify their own structures in such a language, much as an engineer would use FORTRAN to specify a numerical computation. The existence of such a language would go a long way toward making computer power more available to the social science experimentation area and would allow for more experimentation with this new communication medium than is now possible (see Chapter 9).

The functions that go into a conferencing system are quite extensive and embrace what a computer does normally, and then some:

1. Composition: This can go from simple text editing to form fill-ins, spelling correction, speed writing and document editing, and even type-setting. All of word processing is included here.
2. Filing: Just about any way a participant would like to file and index an item or collections of items could be appropriate for a particular conferencing system.
3. Addressing: The computer allows the extension of the addressing function to be conditional and adaptive, so that address by subject or by what is happening is a possible requirement.
4. Coordination: Processing of communications on a group basis, such as routing for approvals, votes, gathering of data in a structured manner, etc.
5. Alerting: Informing of what has taken place, informing on a conditional or adaptive manner, suspension of actions.
6. Delivery: May also be conditional on other actions or adaptive by grouping of items.
7. Information Retrieval: The whole gauntlet of search-type capabilities.
8. Personalization: Specialized interfaces tailored to the user, user profiles for selective gathering of material.
9. Computational: Use of computer analysis aids in particular for subjective judgment analysis by individuals and groups.

In addition, conference systems seem to mandate incorporating unique aids, such as user calendars or reminders. What we end up with is reducing to computer-understandable specifications a rather complete agenda of current and potential human communication and information-processing tasks. The conferencing systems of the next generation are only hinted at by those we have seen to date, and will truly represent integrated communication and information systems.

ISBN 0-201-03140-X, 0-201-03141-8 (pbk.)

THE BOSWASH TIMES

The Computerized News Summary Service of the Megalopolis

December 26, 1985

ISBN 0-201-03140-X, 0-201-03141-8 (pbk.)

Who Controls Space?

Bergamo, Italy, Dec. 25, 1985. Mr. Roberto Colleoni, a mild-mannered, distinguished Italian gentleman, seems out of place as a modern pirate of the airways. However, Mr. Colleoni is president of the Swiss-based company that has launched and is operating the "Volk" communication satellites. These satellites are not authorized by any known national or international bodies. With their specialized $200 transmitter–antenna system it is possible for anyone to beam or receive digital transmissions around the world.

We asked Mr. Colleoni how his company makes money from this operation. He replied that they monitored the broadcasts and merely sent out bills after the fact to those doing advertising or communications via the systems. Since rates are so much lower than any of the "legal" systems, the Sicilian subsidiary has had no problem in collection. "We, of course, only collect from those who can afford to pay. Furthermore, this is not our only business or service. Our worldwide accounting and bookkeeping service based in Sicily also benefits greatly from the availability of this communication service."

Numerous protests have been made in the United Nations and by various European P.T.&T.'s (Postal, Telephone and Telegraph agencies) about this operation, and requests have been made to the United States to shoot down the satellites. The Soviet Union, of course, has an active jammer operation, but this has not proven effective outside of major metropolitan areas.

Mr. Colleoni stated that this technology of low-cost people-to-people communication has been available for some time but not provided by the international carriers because of their wish to maintain their highly profitable rate structures. Most governments at this point are uncertain what actions to take because of the rapidly growing popularity of this worldwide digital version of CB communications (CBC). The CBC'ers, as they are referred to, now have active clubs around the world.

When asked where the initial money came from to finance this operation, Mr. Colleoni was extremely evasive.

However, it is well known that numerous multinational corporations have been very disgruntled at the regulatory and price barriers in existence for international communications. It is thought in some circles that the influence of the multinationals has caused hesitation on

393

the part of governments to take action against the Volk system. Mr. Colleoni emphasizes that space is international, just as the high seas are, and that any action against the satellite system would be illegal.

Informatronics Declares Bankruptcy

Wilmington, Dec. 25, 1985. The financial world was shocked today when one of the glamor stocks of the early eighties faded entirely from the Big Board. Trading was halted when this Delaware-based high-technology company declared bankruptcy.

Informatronics, Inc. was known as the company with the most up-to-date and complex computer–tele-communications equipment. All of its executives had audio components on their large-screen computer terminals, which enabled them to input and retrieve communications in spoken form as well as in writing. They could display a TV-like picture image of individuals with whom they were communicating, in addition to sending the usual written communications through their computer networks.

Analysts close to the company say that in recent months, the internal politics of personal animosities and coalition building in a struggle over the replacement for the soon-to-retire Chairman of the Board became so bitter and time consuming that most middle- and higher-level personnel began spending all of their time communicating with peers in the company. As a result, sales and income dropped to the point where the company could not pay the bills on its lease-purchase equipment and wide-band communications lines.

ISBN 0-201-03140-X, 0-201-03141-8 (pbk.)

394

ISBN 0-201-03140-X, 0-201-03141-8 (pbk.)

395

Technical Details

- Our informational environment designer will work with you and your associates to create the optimum environment from our rich selection of options, which include wall, ceiling, and floor covering displays. Less than 1/4 inch thick, this integrated flat display technology may be utilized to cover all the surfaces of any room or to tailor any shaped chamber, such as our popular spherical cubicles. Standard is 16 color variations, 32-level intensity scale as well as 128 dots per inch resolution. For slightly extra our panels can be sensitized to any of 256 unique signals on any square inch spot, triggered by our patented laser pencil signaler. Our laser pencil can be adjusted to provide complete painting and brush control. No more cleaning crayons off regular walls—let your children paint the whole chamber when you are not using it for other purposes.

- Wireless voice activation and input devices allow full control of the chamber as you move around it. A very popular option for auditorium size configurations.

- A unique electrostatic device is available to allow the incorporation of texture on the display surface. This provides the virtual impressions of hard, soft, smooth, rough, jagged, or clammy, as well as tingling and mildly shocking sensations.

- Special electrosensitive transparent sheets (of any size) can be placed upon the surface of any display area and pulled off in one minute to produce a hard copy of the display for use in any of a number of standard mass printing devices for hard-copy publishing. Also suitable for production of micro forms.

- Our holograph projectors produce three-dimensional displays within the chamber capable of instantaneous sizing up to 256 times smaller or larger than original display size. Dance remotely with life-size partners or have a complete dance company perform on your table.

- A complete quadraphonic or octophonic speaker system with a library of sounds and music to accompany any information and communications.

- Completely computer-controlled climate environment regulates temperature, humidity, static electricity, wind motion, oxygen concentration, pressure, and smells to create the ultimate in information–communication environments.

- Our "Robby" simulation computers can be set up to remember and imitate any selection of chamber conditions you dictate throughout a communication process. It can also be set to learn and infer the types of environments you or your family members prefer to accompany various informational and communication activities.

- A large selection of physical motion augmentators, from simple headsets, gloves, and shoes to complete bodysuits, can be

ISBN 0-201-03140-X, 0-201-03141-8 (pbk.)

adapted to any type of control you desire over the chamber. For example, a slight nod of the head in any direction can be set to rotate the complete display 180 degrees. Walking or swimming motions can be utilized literally to find your way through a three-dimensional display of your information and communications. Hand motions can stretch, elongate, condense, zoom, and otherwise warp and manipulate the displays. Create your own personalized dances to the tune of your communications.

• Our special biocontrol and stimulus unit is capable of tuning control and feedback directly from brain wave and other physiological and neurological body activity. Coupled with a multichannel synthesizer, a group of people can experience collective emotions or engage in various emotional transfer experiences between individuals.

One session in a total Information–Communication Chamber will convince you that you will want one tailored for your home or office. Rentals or leases also available in major urban areas.

ISBN 0-201-03140-X, 0-201-03141-8 (pbk.)

CHAPTER 11

Technology, Economics, and Utility

Introduction

There is no better way to describe Western civilization's relationship with technology than as a long-term "love affair". Technology has proven a captivating and exciting mistress, seeming to provide us with an endless variety of new delights. Perhaps no area of technology has more personified this image in recent years than computers. The rate at which technology for electronics and computers has advanced far surpasses that of anything heretofore—machines, automobiles, airplanes, etc. As a result, we have been conditioned to expect that, no matter what delights we experience today, something new and more wondrous will be proffered tomorrow.

With the recent emergence of the home computer, this love affair is about to extend beyond the computer subculture to involve the greater mass of the society. We can expect the computer to replace the automobile as the love object, status symbol, ego extension, and pseudo companion of future generations of Americans.

Although there are many reasons to rejoice at this prospect, significant problems also arise when we become preoccupied with technology as an end in itself. Blind love, oblivious to all faults, can have its undesirable consequences. Computerized conferencing represents an excellent case in point. Computerized conferencing systems (CCSs) can be, and have been, implemented utilizing today's technology in terms of commercially available and off-the-shelf equipment. To those seduced by the promise of the technology that is on the laboratory bench or is possible with significant added costs, today's generation of CCSs may seem like a drab mistress. However, we consider this a warped view, more characteristic of an affair with the technology than of reasoned consideration of the consequences and impacts of this form of communication.

Starr Roxanne Hiltz and Murray Turoff, The Network Nation: Human Communication via Computer

ISBN 0-201-03140-X, 0-201-03141-8 (pbk.)

398

From another perspective, computerized conferencing is but one component of what Joseph Coates calls "the telecommunications tinker toy"
(Coates, 1977a, pp. 196–197).

> At the core of telecommunications are those now relatively prosaic devices, the
> telephone, radio, television, cable, computers, and calculators. Their ancillary,
> derivative and support devices are of course much more numerous. Together
> they comprise the "tinker toy kit" from which applications are assembled (Table
> 1). Those devices surely do not mark the limits of exploitable basic knowledge.
> Bob Forward has pointed out that only a small portion of the phenomena
> known in nature are used in the present communications media . . . .

Table 1
Components of the Telecommunications Tinker Toy.

Radio	Computers
Telegraph	Computer utilities
Telephone	Illiac-IV
Touchtone	ARPANET
Facsimile	Plato
TV	Calculators
Cable, one and two way	Data banks
Porta packs	Punched cards
Tapes	Integrated microcircuits
Frame grabbers	Photography
Cassettes, audio and visual	Microfilm–microfiche
Videodiscs	Optical scanners
Microwave	Sensors
Lasers	Satellites
Fiber optics	Intelsat
Picturephone	ATS-6
Electronic scratch pad	ERTS
Credit cards	Master/slave manipulators
	Automation devices

ISBN 0-201-03140-X, 0-201-03141-8 (pbk.)

Without the computer, these components would be so many loose pieces of
technology lying around the house or office. It is the computer that can
integrate the pieces into a powerful information and communication network.
That is ultimately the direction in which we see the technology developing:
toward many specialized telecommunications services used for specific purposes, but integrated and based on networks of computers and in-home micro-
processors, along with various information and communication media.

That computer-based communication systems represent a synthesis of
communications with information systems constitutes the fundamental technological advance, and this advance is already well under way. For instance,

Cohen (Cohen et al., 1977, pp. 1, 5) has written

> The technology and utilization of information systems and communication
> systems have come to a point where the two are beginning to overlap. For
> example the digital computer, a core ingredient for large scale information
> systems, is becoming an equally important factor in the switched telecommuni-
> cations network. From the other perspective, we note that networks of com-
> puters and the construction of large time-shared information systems are depen-
> dent upon the availability of communication networks. Thus, it is not unreason-
> able to assume that one day the two systems may become a fully-integrated,
> comprehensive information–communication system which would be reached
> through a single multi-purpose apparatus available to everyone . . . .
>
> The long range prospect is one of a completely digital transmission and
> computer switched telecommunications network that will be far more reliable
> and more economical than the present electro-mechanical network.

Thus, the key technological advances have to do with decreasing the cost
and increasing the reliability of existing technology, more than with making
new breakthroughs. These key changes in the economics of computer-
mediated communications–information systems will occur within the decade.

To substantiate this perspective, we will review the technological components required today for CCSs and examine the potential advances in
technology usually assumed to be necessary for future systems. All these
advances will, in fact, offer certain benefits for selected situations. However,
the major impacts on a societal basis will largely result from widespread
availability of straightforward technology at the lowest possible cost. There
will always be those who must have their Cadillac, but it is the Model T that
will make it all possible.

Technology

Until recently computer technology has been a seller's market, largely
restricted to the commercial or organizational marketplace. With high technology in a seller's market, the often unconscious goal on the part of the
industry is to utilize advances in technology to maximize the performance of
new units without reducing price. The marketing of computers and terminals
is characterized by this approach. However, the strategy for the mass public
market would be one of using technical advance to minimize price. This is the
reduction the industry must pursue. The first major example of the mass
market approach in this field was "the little things that count"—calculators.
We can now expect to enter an era where the other elements of computer
technology follow the pattern set by calculators.

ISBN 0-201-03140-X, 0-201-03141-8 (pbk.)

The basic technological components of existing CCSs are

1. A computer terminal able to handle upper- and lowercase characters at a transmission rate of 30 characters per second;
2. A phone line capable of providing communication from the terminal with reliable transmission of digital data at the rate the terminal can transmit;
3. A computer system providing the CC software and the centralized file of communication transcripts;
4. A digital data network to provide low-cost transmission of digital data when long-distance phone charges would otherwise have to be borne.

Decreases in the cost of these minimum components, rather than major new technological breakthroughs, will have the largest impact on the speed and comprehensiveness with which such systems enter the mass market.

Most of what we have talked about in this book, in terms of the potential benefits of computerized conferencing, involves the design and structure of the CCS itself and is a function of the software providing the technology. Developing the software is a one-time cost that will substantially decline in the future, as the requirements for this application area are included in the capabilities of standard system and support software packages. There are, however, numerous other advances that can be applied to the hardware components of CCSs. The ones commonly discussed are sophisticated terminals and new communications network technologies.

Sophisticated Graphic or Intelligent Terminals

Enhancing a terminal amounts to putting a small computer and memory into it and adding local exchangeable memory units, such as cassettes, or small random-access storage devices, such as floppy disks. There are also display and input devices capable of handling diagrams and pictures, or of optical character recognition. An example of a sophisticated terminal could be a complete word-processing system allowing a person his or her own personal file system and the ability to prepare material without being tied into a central computer. There are many specific situations where this type of terminal is advantageous for a particular application area. In a business it would facilitate the capturing of all text material in electronic form, thereby making it easier to extract pertinent material for use in a conferencing system. Local text editing can minimize communication costs if these are a significant factor, as when international communications are involved, or if the regulatory situation maintains communications costs in excess of what technology should make possible. In areas like construction or design, the addition of graphical capabilities may be absolutely necessary.

ISBN 0-201-03140-X, 0-201-03141-8 (pbk.)

Rather straightforward cost–benefit analyses can be done to show the tradeoff between doing some processing in the terminal and doing it all via the central computer, with the proper accounting of communication cost factors as well. The ideal situation for saving is where a single organizational responsibility exists that can specify the terminal, communications, and central computer characteristics, such as a conferencing system owned in total by one company and used for its own purposes. In any other systems, the central computer system must be able to service the "dumbest" terminal available in order to reach the widest possible market of users, and the overhead for this range will be reflected in the cost structure to all users. As a result, the tradeoff for an individual user in terms of intelligence in the terminal versus increased time needed for communication becomes largely a function of the communication costs.

Today a "dumb" visual terminal costs approximately $1000 and a dumb printing terminal about $1500. We expect these prices to fall to the $300 to $600 range by 1980. Since the primary societal impact of computerized conferencing will be felt when there are sufficient terminals around so that all the individuals in a communicating group can access a terminal, the widespread availability of terminals is the key element. It is very likely that mass marketing of dumb terminals will occur in the mid-eighties, and that they will be sold primarily for home computers and other computer services that will become available to the public—games, recreation, education, advertising, transaction services, etc. With sufficient terminals in place, public-access computerized conferencing will become popular. Some home computers may also be tailored by small groups of individuals as their own personal store-and-forward message system for themselves and their close friends. Some of this is already taking place among computer hobbyists.

Among the options that will be available for the $3000 terminals within 10 to 15 years are the following.

(a) Terminals equipped for high-speed transmission and reception and the use of floppy disks, bubble memories, or other local storage devices. This will greatly decrease costs and time for transmitting large amounts of text and/or data from one terminal to another and for retrieving and searching from computer-based files.
(b) Voice input and output, and video input–output options, to allow persons to send and receive mixed-media communications.
(c) A selection of print styles, color, and easily created vertical and horizontal lines to produce visually striking written output.

All of these frills, however, will be costly, in terms of both hardware and transmission. Nor do they necessarily represent a *communications* advantage, as will be discussed further in what follows. The market for more sophisti-

ISBN 0-201-03140-X, 0-201-03141-8 (pbk.)

cated terminals will always be a strong and sensitive function of the costs of communications between the terminal and the computer.

To summarize, future development of CCSs will be a function of the costs and distribution of the simplest possible terminal providing a full character set for alphanumerics, a reasonable page size and quality of output, and at least 30 characters per second (cps) printing capabilities. In 1978, at least one company was planning to market, within a year, a portable, high-quality, hard-copy 30-cps terminal, with a three-color printing option, for $600. Although there will be a tremendous organizational market for more sophisticated devices and there will always be individuals and organizations willing to pay for the Cadillacs and Edsels of the terminal market, the important factor for the growth of CCS's will be costs and sales of the cheapest units. A future marketplace can of course be imagined where the salesman tries to entice the customer with all the accessories that could be added to the basic unit, and the prestige to be derived from owning the sports model rather than the basic sedan.

The one exception will probably be the home or personal computer capable of acting as both a word-processing (or text-processing) machine and a terminal. Units of this sort, selling at around $1000 dollars in the early eighties, will immediately find a market in the home office environment and replace or substitute for the high-quality typewriter. The home word-processing system is likely to be the major marketable item for personal computing beyond video games. Since digital communication networks and centralized computer services for individuals will come into their own in the late eighties, these word-processing units will proliferate at a time when paper still provides the basic medium of communication among individuals.

Communications Networks: Telephones, Cables, Satellites, Fiber Optics, etc.

A normal telephone line can be made to print 120 characters per second as opposed to 30, although this capacity requires more expensive electronics in the terminals and there are certain locations in the country where the basic phone plant would have to be upgraded. In many foreign countries even 30 characters a second over average phone lines is too demanding. If coaxial cables (like those for cable TV) can be used, the speed can go up to thousands of characters a second. With sophisticated terminals it is possible, given the corresponding communications capability, to transmit and receive large volumes of information in very short times and later utilize the terminal to peruse the material.

The question arises whether this high speed is really necessary in the CC environment. Individuals who have access to high-speed communication by direct hookup to the computers or digital networks claim it allows them to

ISBN 0-201-03140-X, 0-201-03141-8 (pbk.)

skim the material quickly. An alternative approach to high-speed skimming is to develop in the software of CCS's more sophisticated filters that the individual can tailor to his or her own requirements to allow the computer to pick out what is significant. High-speed skimming has a number of characteristics that could tend to defeat the atmosphere and objectives of a CCS:

High-speed transmission represents a time compression and psychologically reduces the significance of the text to the reader in terms of the subconscious awareness of the effort of the reader relative to the effort the writer put into it.

It reduces the assurance to the writer that his or her written material will be viewed as significant and lowers the psychological pressure to deliberate or reflect carefully on what is written. In other words, it lowers the value of what is communicated.

High-speed skimming takes the feedback out of the communication process, whereas filters would allow feedback to writers as to whether their material is being selected or viewed.

High-speed reception increases the potential for users to get themselves into an information overload situation.

High-speed skimming may be very appropriate for standardized data bases and elaborate document files; however, it could very well represent the wrong design objective if blindly applied to CCS's. We require a much better understanding of the psychological processes involved in this area before high speed can be accepted as a desirable goal. It would be all too easy, as is being done with some message systems now, to reduce the CCS design to a telegram- or memo-type system producing reams of useless output for the record only. The way to prevent this is to ensure that there is sufficient time to read each piece or segment of communication that is delivered. Only in this way does the group pressure exist to keep each communication segment relevant. This is one reason why many current systems do not permit an individual permanently to refuse printout of a particular item of communication that has been addressed to him or her.

Most of the other touted advances, such as fiber optics and satellites, are viewed in this context only as promising lower costs for the transmission of digital data. As for its effect on the conference system, whether data are transmitted via microwaves, radio waves, light waves, or pony express, (if the horses are fast enough) is a matter of indifference. With respect to communications, realizing a cost reduction is largely a policy or regulatory problem. Cost reduction will require the acknowledgment of a public need for digital communications as opposed to current business-use-only assumptions.

Few TV cable systems are being designed with sufficient network versatility for highly variable two-way transmission. This lack of flexibility results in

ISBN 0-201-03140-X, 0-201-03141-8 (pbk.)

tremendous sunk costs, and the communities involved will benefit little as regards future computer potentials, such as computerized conferencing. Hence, a mass market for computerized conferencing will depend for decades on the public's being able to utilize the home phone to reach either the computer or a digital data network. Those who desire high speed will require their own private lines and will have to be willing to shell out the added costs of lines and equipment. At best, the digitalization of the phone voice network is expected to take many decades because of the sunk investment that must be turned over by the phone companies. Only after that is accomplished may higher-speed digital communication become inexpensively available to most homes in the country. Whether that goal is on the phone system's hidden agenda is unclear. It has not been presented as an objective to the Federal Communications Commission, nor have objectives been set on public transmission capabilities and costs of digital data.

For the foreseeable future public use will have to piggyback on the phone system and we will have to hope that phone rates do not increase to the point of wiping out the potential for home use of computerized conferencing. Currently the use of 800 numbers to enter a digital data network costs about $15 per hour and represents the upper limit on local phone costs. The lowering of that cost should be a primary aim of advances in communications technology. To bring it down to the charge for a local phone call, without any economic differential to the conferee based on the geographical distribution of participants, would be the single key factor in bringing about the Network Nation. Certainly, if as a society we choose to accept as a value the minimizing of energy expenditures, then the logical extension of this value is to remove distance-based charges for the transmission of electronic forms of information. The technology under evolution in this area makes distance charges unjustifiable, and their continued existence is an expression of the inability of policy and regulation to keep pace with technology. The more meaningful parameter of cost is volume of data transmitted, since a digitized communication system allows sharing of communication channels. The growth of digital data newtorks or value-added networks and the degree of flexibility these networks will offer for the rerouting of communications among many separate computer communication systems will ultimately determine the scope of public use.

The technology needed will continue to evolve because of the market for basic message systems as part of office automation systems, and computerized conferencing will benefit from the spillover effect. Reliability and security will be noticeably improved. Frankly, limited message systems based on replacing TWX and internal mails will be far easier to sell to businessmen, who are unlikely to understand initially the full potentials offered by conference systems.

ISBN 0-201-03140-X, 0-201-03141-8 (pbk.)

However, many European countries (where all forms of communications are monopolized by one government organization) now place very high and artificial rates on digital data networks. The reason for this is the fear that the electronic message systems that organizations could place on these nets would wipe out the rather sizable profits in current TWX-like offerings. Nevertheless, enough people have been bitten by the love bug of technology to ensure that a reasonable number of full conference systems will emerge; some noticeably excessive uses of this and associated teleconferencing technologies are also likely.

Digital Broadcasting

A significant potential for networking is the use of broadcast or radio digital systems, provided the regulatory mess in this area can be overcome. Consider that a single FM radio channel could broadcast the equivalent of thousands of digital bands at the 30- to 120-cps transmission rate. A person could use a telephone and a terminal to leave items with the digital broadcasting service, to be entered into this steady stream of broadcast digital signals. An individual's smart terminal (personal computer) could be programmed to pick out and store items of particular interest to that person, such as, has IBM gone down four points? Has anyone offered item X for sale? Has anyone made comments on subject Y or news item Z? The smart terminal would scan the broadcasts and accumulate only those items its owner was interested in, to play back later on. Such a system is technically feasible today and would obviously revolutionize local community forms of communication and advertising. In this case the personal home computer becomes a personal listener and "gatekeeper" for information.

Computers: Macro, Mini, Micro, and Multiple

Computer hardware will evolve more features tailored for this application area. Whatever the size of the computer used, we see a tendency toward multiple-processor systems all doing essentially the same job, so that any breakdown in one processor can allow the job to be off-loaded to others. Because of the need for redundancy, the size of any one processor is likely to be in the medium to mini range. Since this type of application utilizes the resources of a computer in a different distribution of requirements than the more typical computational job, we are also likely to see machines dedicated to this type of operation. This and the fact that many organizations will want to own their own computers to deal with their sensitive communications will reinforce the trend toward using smaller dedicated machines. Today a person can set up a conferencing system with about a half-million-dollar investment in resources to service 1000 or more daily users. A system to service 10 to 25

ISBN 0-201-03140-X, 0-201-03141-8 (pbk.)

friends would cost about $20,000. Over the next decade these prices will go down to a fifth or a tenth of these costs. This means any entrepreneur could find it easy to offer a specialized system if allowed to by regulation and policy.

Codes for Security and Privacy

A recent breakthrough in coding techniques is the "trapdoor" code, where the code used to encrypt a text item differs from the one needed to decrypt it. This means people or services can broadcast to all a code to be used in sending them confidential material. This heightened degree of security could make computerized conferencing more secure than the mails, and thus remove any inhibitions about what is communicated, because a conferee would not have to fear a security leak in this part of the communication process. At one extreme, services could be offered for collecting and issuing X-rated material or dialogue. Since only those having subscribed to the service would have the electronic module able to decode the material, there would be no meaningful grounds for applying obscenity laws to this type of service, since the actual transmission in encrypted form would not be obscene. At the other extreme, the double protection of the trapdoor encryption could also be used to ensure security, as in the transmission of important scientific data, rather than mere privacy.

Graphics

Our own work with computerized conferencing and a number of other efforts dealing with text editing have led to observations that the way people deal with text in a computer system can be radically different from their treatment of other written forms, such as letters, books, and articles. The CC experience is unique in that it is quite common, within the context of a conference, to have a large group of people essentially working on one document. Even the word "document" takes on a somewhat different meaning or form within this context. Individual writing styles are observed to change and evolve with the use of this medium. There are aspects of these observations that indicate there are present limitations placed on the evolution of CCS's by the lack of graphics-oriented terminals.

A CCS, as opposed to a message system, is designed to foster human group communication about a topic. As a result, textual comments follow a process of building on one another and each may represent a particular aspect or "fragment" of a topic or subtopic.

Perhaps the best analogy to what is taking place in these conferences is a series of discussion threads with many nodes in common, so that a representation might appear as a complex web representing the relationships among these discussion threads. The webs would be as variable in nature as if they had been created by drunken spiders. The categorical classification system allows our spider to build in a many-dimensional space. In addition, a conferee has the ability to classify text items as to intent or motivation: point to be made, summary, analogy, information giving, information requesting, positive feedback, negative feedback, footnote, side issue, etc. The more complex the topic, the richer the diversity of "organization" and "classification" of the elements of the discussion that is needed. With all this taking place within the computer-mediated discussion, the user is confronted with an interface to it that is more constrained than what he or she can normally do with a tabletop, sheets of paper, scissors, and tape. Because of the terminal, the user becomes the fly in the web and may become entrapped in a phenomenon we have observed of "information overload," which is a result of both the amount of contributions by the group and the inability to view them in total because of the limited window, bringing about a form of mental distress.

In order to provide the user with a window appropriate to a CCS or to any system where users are dealing with a large amount of textual material from the point of view of composing, reflecting, considering, and analyzing topics of concern, a number of unique requirements emerge.

The screen size desirable should probably accommodate more than 10,000 characters in order to allow the user to manipulate and observe the relationships among many fragments of text at one viewing. This drive to larger screen sizes for text is developing, to a small extent, in the word-processing industry. Hovever, the process of creative use of text, as opposed to error correction and page layout, makes more demands on screen size.

Graphics are needed to display relationships and create qualitative content-oriented models of the perspectives relating different fragments of text. Such use of graphics would develop a whole new way of describing or writing about things. One of the few disciplines that now makes use of this style is architecture. However, the usual graphics system strives for preciseness. When dealing with the sort of models evolving in a CCS, the user has to be able to express the tentativeness of a relationship, the importance of a comment, the strength of a belief, and a host of other emotional contents that would normally be conveyed nonverbally in face-to-face discussions. In other words, one important role of graphics is to replace the losses that result in computerized conferencing from the absence of nonverbal communications.

In addition, we believe graphics will allow the addition of dimensions of communication content not really possible with the nonverbal mechanisms. In terms of the technology of graphics, the user should be able to easily draw and compose things with fuzzy lines, bold lines, broad brush strokes, faint or seemingly crude and handdrawn lines, curves, outlines, etc. We need graphic

ISBN 0-201-03140-X, 0-201-03141-8 (pbk.)

systems that deal with visual images of degrees of preciseness, emotion, desirability, importance, etc. This is a very different orientation from that which underlies most graphics efforts to date. In addition, the lack of high-quality output for finished graphic products is still a major technological shortcoming, although it is hoped that the work in areas such as ink jet* will reduce the costs of obtaining such output.

Multimedia Technology

It is becoming increasingly possible to mix or integrate communication media at a reasonable cost. The ideal system to some technologists appears to be the ability to talk to someone while watching him or her on a TV screen and transferring written material digitally. This media-rich communication can be further augmented by having the spoken word digitized (the typewriter you can speak to) so that high-status users need not demean themselves by using a keyboard. The image of a high-ranking executive sitting in front of such a complete array of communication alternatives in a kind of corporate command and control center is very appealing to an individual who already has a slightly bloated ego.

A significant number of companies have invested large amounts in video conferencing systems that largely gather dust but represent too much of an investment to rip out. We see little real benefit in such multimedia setups, since they defeat the time dispersion benefit of computerized conferencing and the psychological atmosphere encouraging reflective thought that CCS's generate. Multimedia communication systems foster formality, impose status relationships on the participants, or accentuate such relationships that already exist. The situation with respect to multimedia is reminiscent of the psychology behind the early concepts of management information systems, when these systems were often sold to chief executives on the basis of giving them more complete control of "everything" right at their fingertips. Those executives who already believe that they have imbeciles working for them and find it impossible to delegate authority will find multimedia systems as attractive as the early concepts of MIS.

As for talking to your terminal, this technology could be useful for some handicapped individuals or for those with very strong aversions to print media. In some factory situations, too, voice input is important for workers whose hands are occupied in inspection or other similar activities. During the next few decades, however, it will probably be cheaper for a person to spend an hour a day for three weeks becoming a casual typist than to purchase such

---

*The ink jet sprays droplets of ink electronically onto the paper and is not limited to standard characters on keys. Thus it could produce graphics or mathematical notations.

ISBN 0-201-03140-X, 0-201-03141-8 (pbk.)

a device. At some point in time use of computer terminals will become so widespread that it will be an embarrassment not to be able to utilize one. An atmosphere can be imagined in which the management person unable to operate a terminal is regarded as behind the times—hardly an impression leading to advancement.

We can expect many companies and even individuals to become overenamored of the technology and oversold on technological flourishes. For the serious designer and the user interested in improvement in communications with other humans, the technology should become a black box and anything that adds unnecessary frosting to the contents or detracts from the cognitive meaning of the communications themselves should be considered a liability rather than a benefit. Also, no technological frill is going to determine, for those involved in the communication process, how much or about what they are going to wish to communicate.

Economics

Over the long term, the costs of this form of communication relative to those of other forms will be the driving force for widespread adoption. To illustrate, we will first develop a few simple general models that have been previously discussed in the literature (Turoff, 1972b, 1976). Then we will examine specific comparative costs for EIES as opposed to other media. The interesting economic comparisons to make are the costs of computerized conferencing relative to spoken communication, such as face-to-face meetings and the telephone, and to written communication using the mails or facsimile.

Computerized Conferencing Relative to Spoken Communications

In a face-to-face meeting or telephone conversation, the speed of communications is based on the average rate (r_v) and is a function of the relationship

$$r_v = \frac{W_v}{T_v} \tag{11-1}$$

where W_v is the number of words spoken in time T_v.

In a CC session of time T_c all the participants may be typing at an average rate for the group r_t or reading at an average rate r_r at time intervals of their choosing. However, we shall assume that the average time a person spends typing is t_t and the average time reading is t_r. Therefore we assume on the average for any individual that

$$T_c = t_t + t_r. \tag{11-2}$$

ISBN 0-201-03140-X, 0-201-03141-8 (pbk.)

Furthermore, we assume that in the interval of total time all the words put in by the group are also read by the group, to provide the equation

$$W_c = Nr_t t_t = r_r t_r. \qquad (11\text{-}3)$$

Note that a single individual could spend T_c total time in many different sessions. Also, since no one has to wait for someone else's typing (i.e., material is being contributed asynchronously), we have a factor of N, the number of people in the conference, in the foregoing equation. Utilizing Eqs. (11-2) and (11-3) a little algebra, we have the effective communication rate of words per unit of time:

$$r_c = \frac{W_c}{T_c} = \frac{r_r}{1 + (r_r / Nr_t)}. \qquad (11\text{-}4)$$

Note that as the size of the conference group increases, the throughput rate r_c approaches the reading rate or the terminal print rate, whichever is lower. However, if a person does something else while the terminal is printing, the higher throughput of the reading rate could be realized. Potentially, computerized conferencing allows more words to pass among a group in a given length of time or the same number of words in less time than a spoken conversation involving the same number of people.

Table 11-1 exhibits the effective throughput rate in words per second based on:

Terminal output rates of 6 words per second and 24 words per second, to reflect the current available rates over regular telephone lines;

Typing speeds of 0.2, 0.5, and 1.0 word per second, to reflect experienced hunt-and-peck, casual, and professional typing, respectively.

For a person sitting at a terminal there is no real advantage for the throughput rate to be faster than his or her reading rate. Higher throughput rates, however, are cost-effective when people take the printed material out to be read later (off line), as is common when secretaries or other intermediaries operate the terminal.

These throughput rates should be compared with spoken communication rates, which generally fall between 1 and 2 words per second, with any real give-and-take discussion being closer to 1 word per second. This provides the comparison in Table 11-2 of how many people are needed to make computerized conferencing a faster means of communication than spoken communication.

The numbers needed are reflective of small group sizes and there are conditions when even two-party CC communication is faster, as when two secretaries act for their employers in the exchange of messages.

This whole comparison, by the way, places no economic value on the ability of the communicating individuals to engage in the communication

ISBN 0-201-03140-X, 0-201-03141-8 (pbk.)

ISBN 0-201-03140-X, 0-201-03141-8 (pbk.)

Table 11-1

Throughput of Computerized Conferencing[a]

Print:	6			24		
Typing:	0.2	0.5	1	0.2	0.5	1
$N = 2$	0.38	0.86	1.50	0.39	0.96	1.85
3	0.55	1.20	2.00	0.58	1.41	2.67
4	0.71	1.50	2.40	0.77	1.85	3.43
5	0.86	1.76	2.70	0.96	2.26	4.14
6	1.00	2.00	3.00	1.14	2.67	4.80
7	1.14	2.21	3.23	1.32	3.05	5.42
8	1.26	2.40	3.43	1.50	3.43	6.00
9	1.38	2.57	3.60	1.67	3.79	6.55
10	1.50	2.72	3.75	1.84	4.14	7.06
12	1.71	3.00	4.00	2.18	4.80	8.00
14	1.91	3.23	4.20	2.51	5.42	8.84
16	2.09	3.43	4.36	2.82	6.00	9.60
18	2.25	3.60	4.50	3.13	6.55	10.29
20	2.40	3.75	4.62	3.42	7.06	10.91
22	2.54	3.88	4.71	3.72	7.54	11.48
24	2.67	4.00	4.80	4.00	8.00	12.00
26	2.79	4.11	4.88	4.27	8.43	12.48
28	2.90	4.20	4.94	4.54	8.84	12.94
30	3.00	4.28	5.0	4.80	9.23	13.33

[a] Rates are in words per second; 10 people with an average typing speed of 0.5 word per second utilizing terminals printing at 6 words per second would be exchanging communications at an effective rate of 2.72 words per second.

Table 11-2

Number of Participants Necessary for Computerized Conferencing to Be Faster than Spoken Communication[a]

Print:	6			24		
Typing:	0.2	0.5	1	0.2	0.5	1
Speaking:						
0.5	3	2	2	3	2	2
1.0	6	3	2	6	3	2
1.5	10	4	2	8	4	2
2.0	16	6	3	12	5	3
2.5	22	9	5	14	6	3
3.0	30	12	6	16	7	4

[a] All rates are in words per second. If a group of people had an average group discussion rate of 1.5 words per second and an average typing speed of 0.5 word per second, then when utilizing 6 word per second terminal output, only four members of the group would be needed to make CC faster than a spoken discussion by the group of four.

process at different times, or on loss of time in spoken communications reflecting the attempts to reach someone by phone or to get to a face-to-face group meeting. Therefore, these models are extremely conservative with respect to the potential economic benefits of computerized conferencing.

In terms of effective typing speed, some loss results from the mechanics of the interaction or the time spent by the user instructing the computer what he or she wishes to do. Based on very crude statistics this loss appears, for both EIES and the Planet CC systems, to be between 0.1 and 0.2 word per second. The minimization of this loss is a significant goal for designers of these systems.

Besides the basic throughput, the other factor is the cost. In principle a CCS can save people time in the process of communicating. The value of their time saved can then be applied against the cost of the system. The cost of communicating orally is

$$T_v NV \tag{11-5}$$

where V is the average value per unit of time of the time the N people are involved in an oral discussion lasting time T_v. For a CCS the cost is given by

$$T_c N(C_c + V) \tag{11-6}$$

where the factor C_c is the cost of the conferencing system per unit of time. To be conservative we shall ignore travel or time costs to get to a face-to-face meeting or telephone costs for a conference call. In essence we are comparing a CCS with a face-to-face discussion where everyone has only to walk down the hall to a meeting room. We will look at the tradeoff situation where we consider the two costs equal:

$$T_v NV = T_c N(C_c + V) \tag{11-7}$$

or

$$V = \frac{T_c C_c}{T_v - T_c} = \frac{C_c}{(T_v / T_c) - 1}. \tag{11-8}$$

Since

$$\frac{W_c T_v}{W_v T_c} = \frac{r_r}{r_v(1 + r_r / Nr_t)}, \tag{11-9}$$

we may reduce Eq. (11-8) to

$$V = \frac{C_c}{r_r / r_v - (1 + r_r / Nr_t)}\left(1 + \frac{r_r}{Nr_t}\right) \tag{11-10}$$

by assuming that the same number of words are to be exchanged in each case.

ISBN 0-201-03140-X, 0-201-03141-8 (pbk.)

ISBN 0-201-03140-X, 0-201-03141-8 (pbk.)

Table 11-3

Minimum Number of Conferees for Savings to Occur in Use of CC Relative to Oral Communication.[a]

	Reading rate:	6			24		
Typing rate	Speaking rate: 1	1.5	2.0	1	1.5	2.0
0.2	7	11	16	6	9	11
0.5	3	5	7	3	4	5
1.0	2	3	4	2	2	3

[a] Rates are in words per second.

Table 11-4a
Value Coefficients (V/C_c) [a]
Print speed $r_r = 6$ words per second

Typing:		0.2			0.5			1.0	
Speaking:									
N	1.0	1.5	2.0	1.0	1.5	2.0	1.0	1.5	2.0
2	×	×	×	×	×	×	2.0	×	×
3	×	×	×	4.9	×	×	1.0	3.0	×
4	×	×	×	2.0	×	×	0.7	1.7	5.0
5	×	×	×	1.3	5.6	×	0.6	1.2	2.7
6	×	×	×	1.0	3.0	×	0.5	1.0	2.0
7	7.3	×	×	0.8	2.1	×	0.4	0.9	1.6
8	3.7	×	×	0.7	1.7	9.5	0.4	0.8	1.4
9	2.6	×	×	0.6	1.4	5.0	0.4	0.7	1.2
10	2.0	×	×	0.6	1.2	2.7	0.4	0.7	1.1
12	1.4	5.3	×	0.5	1.0	2.3	0.3	0.6	1.0
14	1.1	3.7	×	0.5	0.9	1.6	0.3	0.6	0.9
16	0.9	2.5	22.2	0.4	0.8	1.4	0.3	0.5	0.8
18	0.8	2.0	7.8	0.4	0.7	1.2	0.3	0.5	0.8
20	0.7	1.7	4.9	0.4	0.7	1.1	0.3	0.5	0.8
25	0.6	1.2	2.7	0.3	0.6	1.0	0.3	0.4	0.7
30	0.5	1.0	2.0	0.3	0.5	0.9	0.2	0.4	0.7
∞	0.2	0.3	0.5	0.2	0.3	0.5	0.2	0.3	0.5

[a] All rates are in words per second; the print speed r_r is 6 words per second. At an average typing speed of 0.5 word per second and a group size of 10 able to talk at an average rate of 1.5 words per second, the value coefficient is 1.2. Therefore for a conference system costing $10 per hour a savings results if the value of the time of the people involved is equal to or more than $12 per hour. (This assumes they will do something useful with the time they save.)
× means a CCS would be more expensive than face-to-face communication.

Table 11-4b
Value Coefficients (V/C_c) [a]

Typing:	0.2			0.5			1.0		
Speaking: N	1.0	1.5	2.0	1.0	1.5	2.0	1.0	1.5	2.0
2	×	×	×	×	×	×	1.2	4.3	×
3	×	×	×	2.4	×	×	0.6	1.3	3.0
4	×	×	×	1.2	4.3	×	0.4	0.8	1.4
5	×	×	×	0.8	2.0	7.6	0.3	0.6	0.9
6	7.0	×	×	0.6	1.3	3.0	0.3	0.5	0.7
7	3.0	×	×	0.5	1.0	1.9	0.2	0.4	0.6
8	2.0	×	×	0.4	0.8	1.4	0.2	0.3	0.5
9	1.5	7.9	×	0.4	0.7	1.1	0.2	0.3	0.4
10	1.2	4.3	×	0.3	0.6	0.9	0.2	0.3	0.4
12	0.9	2.9	10.9	0.3	0.5	0.7	0.1	0.2	0.3
14	0.7	1.5	3.9	0.2	0.4	0.6	0.1	0.2	0.3
16	0.5	1.1	2.4	0.2	0.3	0.5	0.1	0.2	0.3
18	0.5	0.9	1.8	0.2	0.3	0.4	0.1	0.2	0.2
20	0.4	0.8	1.4	0.2	0.3	0.4	0.1	0.2	0.2
25	0.3	0.6	0.9	0.1	0.2	0.3	0.09	0.1	0.2
30	0.3	0.5	0.7	0.1	0.2	0.3	0.08	0.1	0.2
∞	0.04	0.07	0.09	0.04	0.07	0.09	0.04	0.07	0.09

[a] All rates are in words per second; the print speed r_r is 24 words per second. Using the conditions of the example for Table 11-4a, the value is 0.6. Therefore, for a CCS costing $10 per hour, the value of the people has to be only $6.00 or more per hour in order for a savings to occur.
× means a CCS would be more expensive than face-to-face communication.

We note that the denominator in Eq. (11-4) can be zero or negative for given values of r_r, r_v, and r_t as a function of N. This implies that there is a minimum number of people necessary for a net savings in cost to be possible in utilizing CC instead of oral communication. This minimum is itemized in Table 11-3. It represents the number of people necessary to make the denominator positive.

Whether or not a savings occurs can be determined from the table of value coefficients V/C_c (Tables 11-4a, b). Multiplying these numbers by the per-hour cost of CC tells you the minimum value in dollars per hour of an employee to make CC cheaper than face-to-face meetings, even if all of the individuals are in the same building. The tradeoff is much more in favor of CC if we include a loss for travel time, transportation expenses, phone costs, etc. (see Turoff, 1972a). For a simple comparison, consider technology costing approximately $10 per hour (Tables 11-4a, b).

Costs of commercial conference and message systems in the late seventies varied anywhere from $20 to $40 per hour whereas EIES demonstrated costs

ISBN 0-201-03140-X, 0-201-03141-8 (pbk.)

Table 11-5
Number of People Needed to Produce a Savings.[a]

A. When a person's time is worth the computer cost:							
	Print		6			24	
Typing rate:	Speaking:	1	1.5	2.0	1	1.5	2.0
0.2		15	30	60	11	18	24
0.5		6	12	25	5	7	10
1.0		3	6	12	3	4	5

B. When a person's time is worth twice the computer cost:							
	Print		6			24	
Typing rate:	Speaking:	1	1.5	2.0	1	1.5	2.0
0.2		10	18	30	8	14	18
0.5		4	8	14	4	5	7
1.0		2	4	6	2	3	4

[a]All rates are in words per minute. We see that CC costs of $10 to $20 per hour and group sizes of 10 to 20 people make CC economically viable for positions from middle management up in today's organizations.

of $5 to $10 per hour on a nonprofit basis, or for a company owning its own dedicated minicomputer-based system. Since these costs include the use of nationwide data networks, we can compare this amount with the cost of an hour-long telephone call, which is not factored into Table 11-3. We would expect costs to drop to the $1 to $5 range over the next five years if it were purely a matter of efficiencies of new technology. For EIES in particular, however, the central computer cost has gone down to the point where it equals the digital communication network costs. That cost is more a function of regulation than of technology, and therefore it is harder to say if technological efficiencies will be realized in terms of reduced costs or higher profits.

We illustrate one possible use of Tables 11-1 to 11-4 by creating a table that yields the number of people required in order to realize a savings when the value of a person's time is equal to and then twice the cost of the CC service (Table 11-5).

Computerized Conferencing Relative to Mail

It is also quite useful to make a straightforward comparison of computerized conferencing to mail. The model for this comparison must distinguish between private messages (sent from one person to one person) and group messages (sent from one person to N persons making up the group).

ISBN 0-201-03140-X, 0-201-03141-8 (pbk.)

The first equation expresses the conservation of total time T for an individual:

$$T = t_g + t_p + t_r \qquad (11\text{-}11)$$

where t_g is the time spent writing group messages, t_p the time spent writing private messages, and t_r the time spent receiving messages. For the entire group, the theory of the "conservation of words" (i.e., what is sent is also received) gives us

$$Nt_g r_t + t_p r_t = r_r t_r \qquad (11\text{-}12)$$

where r_t and r_r are the rates for transmitting and receiving respectively. In actual practice private messages are sent to more than one individual, a situation that would favor computerized conferencing if factored into Eq. (11-12). However, we are content to make a conservative case for this technology.

If we divide Eq. (11-12) by the average size of a message or letter(s), we have, for the number of messages received,

$$NM_g + M_p = M_r \qquad (11\text{-}13)$$

where M_g is the number of items of group messages or letters written and M_p the number of items sent or mailed to any one individual. The cost of doing this by CC is $C_c T$ and by mail is $C_{po} M_r$ where C_{po} is the postage cost per item. We now ask when these are equal.

$$C_{po} M_r = C_c T$$

$$\frac{C_c}{C_{po}} = \frac{M_r}{T} = \frac{r_r t_r}{S(t_g + t_p + t_r)}$$

$$= \frac{r_r}{S[1 + (t_g + t_p)/t_r]}. \qquad (11\text{-}14)$$

Using

$$t_r = \frac{r_t(Nt_g + t_p)}{r_r}, \qquad (11\text{-}15)$$

from the conservation of words and a little algebra we have

$$\frac{C_c}{C_{po}} = \frac{r_r}{S\left[1 + \dfrac{r_t(1+y)}{r_t(N+y)}\right]} \qquad (11\text{-}16)$$

where $y = t_p/t_g$ and indicates the ratio of the number of private messages to group messages written by an average individual. Since r_r must be expressed

Table 11-6
C_c / C_{po} **for Infinite** N

Letter size	Print speed 6	(words per second) 24
100 words	216	864
200 words	108	432
400 words	54	216

Table 11-7
Allowable CC Cost in Dollars per Hour

Letter size	Print speed 6	(words per second) 24
100 words	28.08	112.32
200 words	14.04	56.16
400 words	7.02	28.08

on a per-hour rate to be compatible with C_c, we will include a factor of 3600 seconds in the hour. For the limit case of infinite N we have Table 11-6.

If the factors in Table 11-6 are multiplied by the cost of a postage stamp, we know how much we can afford to pay for a CCS to be just as cheap as the mails. A CCS or message system costing less than this would save money. For the limit case of infinite N and 13 cents postage we have the results shown in Table 11-7.

Table 11-8

C_c / C_{po} **for** $y = 0$**, All Group Messages**[a]

Print speed:		6			24	
Typing rate:	0.2	0.5	1.0	0.2	0.5	1.0
$N = 2$	14	31	54	14	35	66
4	26	54	87	28	67	123
6	36	72	108	42	96	172
8	96	86	123	54	123	217
10	54	99	135	67	149	254
20	87	135	166	123	254	196
30	108	154	180	173	332	480
40	123	166	189	217	394	540
50	135	174	193	256	441	585

[a] Rates are in words per minute.

ISBN 0-201-03140-X, 0-201-03141-8 (pbk.)

Table 11-9

C_c/C_{po} for $y = 1$, All Private Messages[a]

Print speed:	6			24		
Typing rate:	0.2	0.5	1.0	0.2	0.5	1.0
$N = 2$	10	24	43	11	26	51
4	17	37	54	18	43	82
6	23	49	80	25	59	110
8	28	59	93	31	74	136
10	33	68	103	38	89	161
20	56	101	138	71	155	263
30	74	122	156	99	211	339
40	88	137	167	127	258	398
50	99	147	175	152	300	446

[a]Rates are in words per minute.

Tables 11-8 and 11-9 reflect the tradeoff for finite group size. We have used 100 words per letter, so the results for other cases can be obtained by dividing by 2 or 4 entries.

To illustrate the use of these tables, in Table 11-10 we compute the allowed CC cost for a letter 200 words long and a typing rate of 0.5 word per second. These appear to be representative average operational values for conference systems. Postage is assumed to be 13 cents.

We can observe from Table 11-10 that with computerized conferencing now emerging in a $5 to $10 per hour rate, the technology is cost competitive

Table 11-10

Conference Allowed Costs in Dollars per Hour for $S = 200$ and $r_t = 0.5$[a]

Print speed:	6		24	
y:	0	1	0	1
$N = 2$	2.02	1.56	2.28	1.69
4	3.51	2.41	4.36	2.80
6	4.68	3.19	6.24	3.84
8	5.46	3.84	8.00	4.81
10	6.44	4.42	9.69	5.79
20	8.78	6.57	16.51	10.08
30	10.01	7.93	21.58	13.72
40	10.80	8.91	25.61	16.77
50	11.30	9.56	28.67	19.50

[a]Rates are in words per minute.

ISBN 0-201-03140-X, 0-201-03141-8 (pbk.)

with the cost of postage for from 10 to 50 people who desire to communicate as a group. The cost tradeoff does not reflect delays with mail nor the automatic confirmation of delivered items, which could be valued by using the cost of postage for a registered or certified letter requiring a receipt rather than a first-class letter. If the U.S. Postal Service is right in its prediction that postage rates will double in five years even more rapid substitution of electronic forms of communication for the mails can be expected. Furthermore this analysis is ultraconservative because the time needed to type the letter on line is used in the CC model but no comparable cost is considered as part of postage cost. The recent emergence of electronic funds transfer (EFT) systems becomes very understandable from an economic viewpoint when a check or bill is considered to be a letter of less than 20 words.

In practically any economic way of examining this technology we find the potential for competition and substitution for other communication forms. However, the economic questions merely indicate that tremendous opportunity exists in this area; they do not address the more interesting issues associated with computerized conferencing. Psychological and sociological factors will drive the use of these systems; economic factors only create the opportunity.

Table 11-11
EIES Cost Analysis

Yearly hours ($\times 1000$)	Telenet charges ($\times \$1000$)	Center cost ($\times \$1000$)	Total cost ($\times \$1000$)	Cost per hour ($\$$)	Cost per item ($\$$)
10	35.0	150.0	185	18.50	1.08[a]
15	52.5	162.5	215	14.33	0.84
20	70.0	175.0	245	12.25	0.71
25	87.5	187.5	275	11.00	0.64
30	105.0	200.0	305	10.17	0.59
35	122.5	212.5	335	9.57	0.56
40	140.0	225.0	365	9.13	0.53
45	157.5	237.5	395	8.78	0.52
50	175.0	250.0	425	8.50	0.50
55	192.5	262.5	455	8.27	0.48
60	210.0	275.0	485	8.08	0.47[b]
65	227.5	287.5	515	7.92	0.46
70	245.0	300.0	545	7.79	0.45
75	262.5	312.5	575	7.67	0.45

[a]Actual cost for the pilot period.
[b]Budgeted cost for second year of operation with approximately 600 users.

ISBN 0-201-03140-X, 0-201-03141-8 (pbk.)

A Specific Example of Costs and Performance for Computerized Conferencing

From October 1, 1976, to October 1, 1977, EIES had 10,000 hours of use; 40,000 items of text were composed and 123,000 items were delivered. Over 2,000,000 lines of text were delivered to about 150 users. These data represent the largest single field trial of computerized conferencing to that time for which published statistics are available. The cumulative use of EMISARI since 1971 may have been larger, but no systematic collection of data has ever been published.

The cost of EIES is a function of the level of use made of it. The operational design point is 60,000 user-hours per year and reflects a cost of $8.08 per hour or 47 cents per item received. The latter figure is taken from the use during the experimental year, which averaged 3.5 minutes invested per user per item received, or 17.1 items received per hour of interaction. The variability in cost as a function of use is illustrated in Table 11-11.

Having an actual cost level allows us to make a number of specific comparisons between EIES and other methods. First, however, we must summarize some comparative data:

1. From the paper "The Evolution of Office Information Systems" by J. Christopher Burns (1977, p. 62) we borrow the following values

Cost of a page of Facsimilie	$1.97
Teletype rate	$2.42 per 66 words
Cost of internal memorandum	$4.55
Cost of a letter	$6.41

2. New Jersey Bell Telephone charges for a Newark to Washington D.C. phone call (3 minutes, prime time):

Station to station	$1.00
Person to person	$3.00

3. From a text-processing study on 1000 professionals and 300 secretaries by Exxon, reported by Len Keating at the American Management Association meeting on the automated office of the future (December 5–7, 1977):

Cost of a professional person-minute	$.30
Cost of a secretarial person-minute	$.15
Effective throughput of a secretary	16 words/minute
Professional handwriting speed	15 words/minute

With these data we can make some comparisons with the common non-computer alternatives to EIES.

ISBN 0-201-03140-X, 0-201-03141-8 (pbk.)

FAST WRITTEN FORMS

For a 221-word item (the average size of EIES items during the test period), we have the following costs:

Facsimile	$1.92
Teletype	$8.10
Mailgram	$3.96
(EIES)	($.45–1.08)

U.S. MAIL

The secretarial cost of preparing a letter is $2.07. We ignore professional time involved in initial drafting or dictation and checking, since this would be expended on EIES anyway and at 15 words per minute for handwriting the costs would seem to be equivalent, based on the average for the test operation. However, more experienced users were demonstrating 19 or more words a minute and the comparison could be made more favorable by factoring in this rate. Since average circulation on EIES was three people on a per-item basis, we must divide the $2.07 for typing by 3 to get base costs of 69 cents. The variable cost per item delivered is either 13 cents or 73 cents if a confirmation (as exists on EIES) is made. The confirmation of delivery of a message or the status reporting of how much everyone has read in a conference is an important part of the psychology of communication on EIES and has been noted by observation to be a triggering mechanism in creating new communications. In addition, a charge per copy of the letter to all three recipients must be included at 5 cents per copy with one copy remaining with the sender. This results in a cost range for the U.S. mail of an item sent to three people:

$$\text{Low cost} = .69 + (.13 + .05) = \$.87 \text{ per item received;}$$

$$\text{High cost} = .69 + (.73 + .05) = \$1.47 \text{ per item received.}$$

As we see, even the cost of mail is more expensive once EIES usage builds to 15,000 hours per year. As we have stated before, this technology is today cost equivalent to the U.S. Postal Service. When all the associated costs of filing, storage, etc. are factored in, true costs of mail are much higher and more reflective of the $4 to $6 range found in the literature. Exxon in looking at their typing of professional pages found a true total cost per page of text in the area of $20. Even without the inconveniences of mail and the impracticality of holding discussions through the mail, it would prove to be too expensive a mechanism to compete with EIES. Only if all the professionals involved were assumed to send xerographic copies of long handwritten material would mail be economically competitive.

TELEPHONE

At a speaking rate of 1.5 words a second, 2.5 minutes are needed to deliver 221 words (the average EIES item) over the phone. However, this is an

ISBN 0-201-03140-X, 0-201-03141-8 (pbk.)

investment of 5 minutes of professional time (two people are involved) as opposed to 3.5 minutes on EIES per item received. This adds 1.5 minutes of indirect cost, or 45 cents, to the basic 3-minute call. The cost of a station-to-station call is low because it is assumed that the party being phoned is in at the time the call is placed. We assume one and a half calls are made, in the average, to reach the other party. The person-to-person call would be a more realistic option for comparison with EIES and we take person-to-person rates as the upper limit and ignore lost professional time in placing calls that did not reach the other party. This results in:

$$\text{Low cost} = 1.00 + .50 + .45 = \$1.95 \text{ per item received;}$$
$$\text{High cost} = 3.00 + .45 = \$3.68 \text{ per item received.}$$

In theory we should multiply these costs by 3 to account for the circulation of an item on EIES, but the costs are already far in excess of EIES. Furthermore, it is very probable that to communicate the same material, many more words would be needed in a telephone call. However, the latter point is still a conjecture without sufficient experimental backup to measure or estimate such effects.

We have used Newark to Washington D.C. as a typical long-distance rate. If other locales are used, and the factor of 3 is included for circulation, the basic observation will not change.

We doubt the viability of the telephone for the types of discussions that take place over EIES due to the lack of written material or common file ability provided by the phone. Even if this were not the case the phone would be out of the running on economic terms.

FACE-TO-FACE MEETINGS

Since the average circulation was three items received for each sent on EIES during the test period, we will look at a face-to-face meeting of four people where three had to travel to the location of the fourth at a travel cost of $100 (equivalent to Newark to Washington D.C.) and expenses of $50 per day. We assume they meet for a full 8 hours per day at a talking rate of 1.5 words per second, so that 43,200 words are exchanged in a day. This is equivalent to 195 EIES text items. These assumptions result in the following comparison as a function of the length of the meeting in days.

Cost per Item with Direct Costs Only

Days of meeting	1	2	3	4	5
Items exchanged	195	391	586	782	977
Cost/item	$2.30	$1.53	$1.28	$1.15	$1.07

ISBN 0-201-03140-X, 0-201-03141-8 (pbk.)

As we see, the meeting would have to run for five days before it became cost equivalent to EIES at the lowest use level. This comparison is not completely fair, however, since a person on the terminal at our current rate of 3.5 minutes per item will receive only 137 text items in an eight-hour period. Therefore, each person would have to invest 203 minutes every day to receive the additional 58 items over EIES. At 30 cents a professional minute, this $61 per day per person must be taken off the face-to-face meeting as an indirect savings. Then again, the individuals waste travel time in getting to the meeting and for our simple case we shall assume six hours total round-trip travel time, which is representative of a Newark, N.J., to Washington D.C. trip. If we now add this indirect cost back as well and estimate the cost per item received in the face-to-face example as a relative cost to EIES, we have

Cost per Item with Indirect Costs or Savings Included

Days of meeting	1	2	3	4	5
Cost/item	$2.71	$1.12	$.58	$.32	$.16

As we see, at least a three-day meeting is required for a face-to-face conference to become cost competitive with the EIES test operation, and a four-day meeting is required to become cost competitive with the designed operational levels. This little exercise also assumes it is possible to break up the EIES exchanges into four-person subgroup meetings and neglects the value of the written form. In addition, the trip used is somewhat optimistic with respect to costs of travel. Finally, it should be pointed out that long meetings (three days or more) are seldom practical or necessary; on the contrary, the meeting that lasts less then eight hours is probably most frequent, and the shorter the meeting, the greater the time and cost per item for the face-to-face condition.

We have been working with a meeting among four persons; in actuality most conferences involve a larger number of participants. At higher numbers of participants, throughput becomes important, as described earlier in this chapter.

Ultimately, we do expect computerized conferencing to substitute for a significant percentage of one- to three-day meetings. The estimates we have just exhibited illustrate that the economics are in favor of this proposition.

ELECTRONIC MAIL

One alternative to a dedicated multipurpose CCS is to purchase an electronic mail service. In "The Future of Computer Communications" (Cerf and Curran, 1976), we find the following cost estimates based on 1976 commercial message services: a 1000-character message sent 1 to 1 will cost $3.25, 1 to 5 will cost $1.11 or $5.55 for all delivered.

ISBN 0-201-03140-X, 0-201-03141-8 (pbk.)

Since EIES has a 1 to 3 circulation average for the test operation and an average item size of 1105 characters, the interpolated cost is $1.53 for a commercial message service, where $2.95 is for composition and 64 cents to deliver each of the three copies. This $1.53 is significantly above the $1.08 figure for the EIES test period and does not reflect any storage costs typical of commercial message systems.

In a paper by David Brown (1976) reviewing both analyses and experimentation conducted on the Hermes system of Bolt, Beranek and Newman (using both Tymnet and Telenet), I. P. Sharp's message system, and Scientific Timesharing's message system, the following conclusion is made: "We have reason to believe that no unsubsidized commercially available electronic mail service can currently be used for an average of less than $15.00 per hour."

The foregoing cost analysis does explain, however, why industry in the last few years has taken an active interest in electronic mail. There is a growing realization that letters, mail, and travel are not as inexpensive as they sometimes appear. However, that interest or awareness is still confined to "message systems" and the rather limited view that what is being talked about is a cheaper TWX or Teletype service. The concept of utilizing the computer to structure and facilitate group communications is still rather foreign in the commercial applications environments and we suspect will remain so for a number of years into the future. It is very likely to take a good deal more research and development of a knowledge base on the impacts of such systems on things like the quality of communications before we see commercial availability of CCSs.

Measures of Performance

Underlying the cost per item or per hour of a CCS are measures of performance such as the circulation rate and the throughput rate. We will report some observed data for these measures during the EIES test period and the field trials of Planet–Forum.

The circulation ratio must be looked at specifically for different types of items (e.g., messages, comments, pages) as well as for text fragments such as words or text lines. For the EIES experience, circulation has been larger per text line than per item, meaning that the larger items are sent to more persons. Individuals seem to be motivated to circulate larger items to a greater number of others, since they represent a greater investment of effort. The circulation ratio is also crucial to the estimation of the time invested per word received.

$$\text{Time/word} = 1/(\text{circulation} \times \text{input rate}) + 1/\text{output rate}. \quad (11\text{-}17)$$

Increasing either the circulation or the input rate offers the greatest potential gain for throughput in these systems, since 80% or more of the time on line is usually spent composing.

ISBN 0-201-03140-X, 0-201-03141-8 (pbk.)

The relative proportion of items sent and received is significant. For the EIES test period, while 75% of the items composed were messages, over 50% of those received were conference comments. Because messages averaged 173 words and comments 296 words, the difference is even more marked when examined on a per-word basis. The differences in circulation and item size exhibited on EIES indicate that different kinds of uses are made of messages and comments.

The effective input rate for formula (11-17) is arrived at by estimating the time used for output from the amount of material delivered and the output rate. This time is subtracted from the total time, which is then divided into the amount of material composed in order to provide a words-per-minute input rate estimate that includes the time needed to carry out the mechanics of operating the system as well as the actual typing time. For the EIES test period, this ranged from 6 words per minute for beginners to 25 words per minute for experienced users, with an average of 15 words per minute overall, which is about equivalent to handwriting.

The amount of material a person deals with per session and the length of a single interaction are also interesting parameters. These can be utilized to estimate a session rate (words per minute), which reflects how easy or hard it is for a person to deal with textual material within the context of the human interface design.

The Planet and Forum systems of the Institute for the Future represent systems intended to provide conferencing capabilities. They also provide the ability to send private messages addressed to only one person. Under their research activities (through 1977) they accumulated 4687 hours of use over an 18-month period, a good portion of that operational on commercial time-sharing systems such as Tymshare. A recent report (Vallee et al., 1977) summarizes their experiences with a group of 141 geologists who utilized 1100 hours of time over a 15-month period. The following data are taken from that report, with the exception of those marked with an asterisk, which were obtained via personal communication.

Hours	1,140
Sessions	10,839
Messages	4,825
Circulation of Messages	1.00
Comments	3,613
Mean circulation of comments	8.61*
Average cost per hour on Tymshare	$16.45
Average size of message	47 words*
Average size of comment	63 words*

ISBN 0-201-03140-X, 0-201-03141-8 (pbk.)

These data allow us to construct Table 11-12:

Table 11-12

	Planet–Forum	EIES sample[a]
Items received/session	3.3	5.4
Items sent/session	.8	1.8
Total transactions/session	4.1	7.2
Session length (minutes)	6.3	18.1
Session rate (words/minute)	41.	88.
Item size (words)	63.	221.
Circulation	4.25	2.93
Effective input rate (words/minute)	8.6	24.
Time/word received (seconds)	1.8	1.0

[a]EIES sample excludes use of the system by programmers and direct support staff of the EIES operation.

In Table 11-12 the session rate is a figure derived from the number of transactions per session times the average item size divided by the session length. For face-to-face or spoken conversation, the session rate is about 90 to 120 words per minute.

That the EIES and IFTF approaches to computerized conferencing start from two very different philosophical bases is expressed in this passage from a recent IFTF report (Johansen, DeGrasse, and Wilson, 1977, p. 4);

> There are differences of opinion, however, over what comprises "computer conferencing." In the New Jersey Institute of Technology system, for instance, computerized conferencing is combined with other computer resources, such as a journal system, a text editor, and even a kind of management information system. While such a system provides more computer power, it does so at the expense of the simplicity of operation we felt was necessary for an initial exploration of the utility of small group communication through computers. PLANET is a simple system which enables social scientists to explore the potentials of computer conferencing without requiring that they control for the effects of peripheral elements involved in more complex computer services. Our approach has been to base our assessments of computer conferencing on this basic system for group communication through computers.

The philosophy of design that underlies EIES has always been that the objective of computerized conferencing is to utilize the computer to tailor communication structures and to build as an integral part of such communication structures any computer aids or functions that would act to facilitate

ISBN 0-201-03140-X, 0-201-03141-8 (pbk.)

the communication process. Therefore, EIES, as a system designed for long-term use by scientists, is designed as a rich and complex system to meet what are felt to be diverse needs. It is actually very useful that the two major efforts in this area to date have approached the endeavor from two very different directions. The state of the art is such that a diversity of views and directions should be taken. It is quite clear from the results to date that there are very distinctive differences in results. Out of diversity often emerges knowledge.

Terminals

The foregoing comparative analyses do not reflect one significant factor— the cost and distribution of terminals. In one sense this omission is justifiable in that terminals will be bought initially for other applications and represent a multi-use device. A $2000 terminal, amortized over a five-year period on a basis of 2000 hours of use with $200 a year maintenance, would add about 30 cents an hour to the cost of a conference system. Another way to look at it is that a rental of $70 a month would cost about the same as coast-to-coast travel to only two meetings. From this standpoint, the cost of the terminal is a fairly trivial matter. However, if terminals have to be justified solely for this application, the current cost of terminals for all participants is such that it is a larger capital expense than the conference system itself. Therefore the declining cost of terminals will be a significant factor in the penetration of these systems to areas where interactive computers are not currently in use. Public schools, small businesses, local governments, independent professionals, and homes are examples where lower costs for terminals would produce a sizable market potential.

An equally important factor is that the utility of a communication system is somewhat dependent on the size of the population that can utilize it. In the early days of the telephone the system was not very useful if the people a person wanted to call did not have phones. Thus it behooves us to look closely at the current and possible future distribution of terminals.

In 1975 there were 659,000 terminals in use and in 1978 it was estimated by International Data Corporation that there would be 1,275,000 terminals. This indicates a current doubling time on the number of terminals of three years if we believe this technology is in an exponential rate of growth phase. Fitting this situation to an exponential growth curve gives

$$\text{millions of terminals} = 0.659e^{t(0.19)} \qquad (11\text{-}18)$$

where t is measured in years and $t = 0$ is 1975. We can compare this with the number of phones, which now appears to be in a linear growth phase:

$$\text{millions of phones} = 150 + 6t. \qquad (11\text{-}19)$$

ISBN 0-201-03140-X, 0-201-03141-8 (pbk.)

Since each terminal can be connected via a phone, it is interesting to make the following observations, based on the extrapolated curves:

1. By 1985 as many terminals will be sold as there will be phones installed, or approximately 6,000,000 units.
2. By 1991, 10% of all phones in place will have terminals to go with them.
3. By 1999 50% of all phones will have terminals.
4. By 2001 terminals will be available for over 90% of all phones.

However, although exponential growth is considered a reasonable model for the early distribution phase of a new technology, the logistic substitution curve is considered a more reliable estimator over the long term. Fitting the two data points above provides

$$y = (1 + 226.8e^{-0.184t})^{-1} \tag{11-20}$$

where y is the ratio of terminals to phones and we are assuming a phone is being replaced by a phone–terminal combination.

This curve provides the following projections:

1. By 1993 10% of the phones will have terminals.
2. By 2005 50% will have terminals.
3. By 2017 over 90% will have terminals.

It is obvious there are not enough data points to place any statistical confidence in these extrapolations. Usually planners do not feel confident that a technological substitution will take place until at least 10% substitution level is reached, and as of 1978 only a 0.76% level of substitution prevailed.

However, it is quite reasonable to expect that the whole area of computer technology will mushroom over the decade at least, so that very rapid growth in terminals will take place. We suspect that the error in reaching a 10% level is less than five years and that it will occur between 1990 and 1995. A 50% level between 2000 and 2010 seems reasonable. Although we hesitate to project 90%, it seems sure to occur before 2040, barring any major disasters to our economy and computer industry.

The rationale for this projection is that the eighties will be the decade when most businesses and organizations will have to have terminals. When we approach the 10% limit, the business market will begin to saturate but major production facilities will be in place and manufacturers will be anxious to keep them fully utilized. The result of this will be a major push for a home market in the late eighties, the nineties, and on into the new century. That push for the home market is what will bring us levels of 50% and more in the number of phones having terminals in place. This is not a daydream when we look at what has happened for other technology areas, such as the calculator. In other words, even without sufficient statistics, there exists a rationale based

ISBN 0-201-03140-X, 0-201-03141-8 (pbk.)

upon current activities in the computer industry to support a prediction of rapid growth in the availability of terminals, and such a rationale supports the use of exponential growth and logistic substitution models rather than linear extrapolation.

It is really a foregone conclusion or a self-fulfilling prophecy that the turn of the century will see the real emergence of the information-based society. The changes already taking place on a small scale are shaping the stuctures and characteristics of the society that will prove to be the dominant structures of the Network Nation.

Utility

The number of words passed in a communications process is not a measure of the actual cognitive transmission of information. We do not have sufficient knowledge to quantify the relative efficiency of the transmission of information in audio form versus written form. We suspect that the more complex the situation under discussion, the more beneficial is the written form. We also suspect fewer words are needed in written form because what is transmitted receives a greater investment of prethought. The suspected impacts on group processes have been discussed elsewhere, but once again they are not adequately supported by experimentation to allow quantification for any cost–benefit analyses.

For any particular application opportunity costs can be factored in. In other words, what is the value to every member of the group of being able to choose his or her own time for participating in the communication process, independent of the others in the group? Depending on the activity level of communications of the members in the group, we can make a statistical estimate of the lost communication opportunities due to lack of a CCS. An example would be the added free time salesmen have to communicate with customers if all internal communications are conducted via computerized conferencing.

There are a host of considerations pertaining to the fundamental utility of a CCS that can have impacts far in excess of the economic considerations of mere word throughput. Although these are discussed individually throughout the book, it is useful here to itemize those whose measurement and quantification via suitable experimentation would vastly improve the ability to do meaningful cost–benefit analyses.

1. *Quality of decision.* Given a nonroutine situation that involves a complex set of factors and options, each of which may be best understood by different individuals, computerized conferencing is likely to lead to (a) a wider variety

ISBN 0-201-03140-X, 0-201-03141-8 (pbk.)

of options raised; (b) a more thorough consideration of the relevant data, advantages, and disadvantages of each option; (c) a greater likelihood of the group's adopting a good decision.

2. *Degree of group commitment to the results of a discussion.* It is our hypothesis that the greater freedom to explore alternatives can lead to a greater group commitment to any group decision resulting from the CC environment. Even where the decision arrived at would have been the same in the face-to-face environment, the aftereffect on the individual of the verbal environment is often: "Did I really agree to that?" In any organizational context the degree of commitment can mean the difference in ultimate success or failure of any actions implied by the group deliberations.

3. *Impact of group size on participation and morale.* The flow of material in a CCS seems to be optimum at higher numbers of conferees than is the case for a face-to-face or audio-only discussion. The optimum group size is probably over 20, as compared to 5 to 15 for face-to-face conferences. While this optimal size is bound to be a strong function of both the objective and type of discussion, as well as of the characteristics of the participants and the moderator, it may very well be that for any fixed set of these parameters the range of optimum group size for computerized conferencing is completely different from that for face-to-face conferencing. Since numerous studies have shown that active participation in a discussion is related to satisfaction of participants, the result is that CC should improve morale in medium-sized groups.

4. *Equal participation.* Observation and a few controlled experiments seem to confirm that a greater degree of equality of participation exists in computerized conferencing than in face-to-face discussions. The desirability or utility of this equality is a function of the objective of the meeting.

5. *Retention.* It is hypothesized that a higher individual retention rate for information is possible in the CC environment than in oral discussions. This is probably a complicated function of characteristics of written material and fatigue factors resulting from any prolonged verbal discussion.

6. *Conciseness.* It is hypothesized that material will be presented in a more concise manner in the written form than in the oral form, and that as a result more complex material can be conveyed or exchanged among the group than is otherwise possible.

7. *Computerized aids.* How do we measure or quantify the impact or benefit of aids such as voting, anonymity, pen names, special routing of messages, data formatting, and polling capabilities? All of these will have distinct benefits for particular applications.

8. *Hard copy and responsibility or accountability.* Hard-copy or written records in electronic form mean accountability and minimization of misquoting or imperfect recall of results, as often occurs in an oral discussion. For

ISBN 0-201-03140-X, 0-201-03141-8 (pbk.)

many organizational applications this can result in crucial benefits. Quantification of costs is actually straightforward since the same capability can be had by tape-recording and transcribing all group discussions. Benefits are more difficult to quantify.

9. *Self-activation*. The examination of opportunity costs and benefits really results from the fact that the protocols and procedures are under the complete control of the individual and not imposed by the group. Quantification, however, is probably possible only for a very specific situation.

10. *Electronic compatibility and/or transferability*. A benefit of having the group discussion in electronic form is that it can be manipulated for the preparation of documents and reports, and easily merged with information from data bases and models.

11. *Value of information*. If we could account for the value of information in the same way we account for dollars, there would probably be little problem in showing the economic benefits of computerized conferencing on a true utility basis. We do know from the work with computerized data bases that the timeliness and selectivity provided by the computer are the two major ingredients in the value-added nature of a data base over that of a paper file containing the same information. We propose that the computer in the CC manifestation has the same value-adding effect applied to the information contained in human communications. However, it must be possible to ascribe value to elements of human communication before the proposition of value-adding can be quantified.

Although any cost–benefit study must deal with estimates of the foregoing factors for particular situations, it is unlikely that these factors will be amenable to accurate quantification for some time to come. Many of the factors have not been measured for oral communication modes. Perhaps this is symptomatic of our society's love of technology to the point of ignoring the real issues, which are the impacts in terms of costs and benefits to the end user.

Although it is hoped that we will gain a better understanding of the relative effectiveness of alternative communication processes, improved awareness is not going to be really important for the future use of these systems on an overall basis. A cursory analysis of our use of audio or written communication processes tends to support the position that the current mechanisms are utilized very inefficiently and that, for example, more committee meetings are wastes of time than are not. Very likely, much material sent over CCSs would also be a waste. However, computerized conferencing penalizes the individual receiver less than audio modes of conferencing do, and in the long run this savings may be the greatest overall economic gain. The receiver can choose to ignore what represents a waste of time for him or her. However, it would be

ISBN 0-201-03140-X, 0-201-03141-8 (pbk.)

ISBN 0-201-03140-X, 0-201-03141-8 (pbk.)

hard to visualize justifying such a system in organizations by proving it reduces the cost to the organization of wasteful communications, since this requires the admission that they exist. The management position would have to be that wasteful communications should be eliminated. From a pragmatic viewpoint this goal is an idealized or unreal one. From the economic point of view communications are a wasteful or inefficient process; in the social sense, some waste and redundancy appears necessary to humankind's sanity and society's functioning.

The real benefit of understanding the human communication process is an indirect one. This understanding will allow the design of better systems that can minimize adverse effects and enhance good ones for particular applications. The economics and associated trends in costs are already such that there is little doubt that this technology will become widely available and penetrate most major organizations and industries. This is a technology that may alleviate such conditions as increasing communication burdens on individuals in all organizations, rising travel costs, increasingly complex problems requiring consideration by larger groups of individuals, pressures for more flexible working conditions, and rising people costs. There is, as a result, every form of nutrient available to feed and produce a flourishing growth of this communication medium. The greatest inhibiting factor seems to be that we are talking about utilizing a computer for human communications.

Attitudes Toward Computers

Potentially more important than improvements in existing technology and costs in determining the rate at which the Network Nation will be formed are shifts in public attitudes toward computers and computerized services. Fear, embarrassment, and deep-seated distrust characterize most people whose sole contact with computers is through computerized bills and paychecks. A "blame the computer" syndrome is so widespread that the ACM has defensively set up an ombudsman-type committee to try to counter the tendency of service personnel and the mass media to use the computer as a scapegoat for all mistakes and lack of courteous and timely attention to customer requests. Consumers are told that "the computer" made mistakes and they are difficult to rectify; or that the computer "will not allow" the consumer's preference. In reality, of course, the blame lies with programmers or data-entry clerks who make mistakes, and with computer systems that were not designed with consumer convenience and protection of privacy in mind. In terms of CCSs the existing negative attitudes call for an emphasis on "black boxing" the computer and making its presence as unobtrusive as possible. (Hardly anyone perceives that the telephone network is highly computerized and getting more so every day.)

Countering these attitudes is the spread in education and experience with computers. A few hours of successful on-line experience with a well-designed interactive system tends to shift attitudes toward viewing the computer as a fascinating toy and potential helper for all sorts of tasks. Among middle and upper management, a certain one-upmanship and glamor attaches to the ability to demonstrate familiarity with the latest in computer technology and systems. During the next decade, as more and more young people have direct experience with computers in school and on the job, and as in-home micro-computers become common, the public attitudes toward computers are likely to shift toward viewing them as flexible and powerful tools that can be used by the ordinary person to help perform their everyday tasks.

Attitudes toward computers will influence the speed with which this technology is accepted and utilized. Current trends toward assembly lining white collar work via computer technology provide the alternative scenario of a widespread reaction against computer technology that would significantly counteract the economic factors discussed in this chapter.

Summary

The costs of computer-mediated communication are already competitive with those of face-to-face meetings, telephone conversations, and the mail. Drops in the cost of computers and the emergence of commercially available hardware and operating systems that are optimized for this function will bring down the cost further. We expect that the number of computer terminals will grow exponentially during at least the next decade from the more than 1.25 million that are now in place, and that their cost will continue to drop dramatically. In addition, we can expect some advances in technology, such as voice input or output, which would be useful for specific kinds of applications. But we do not need any technological breakthroughs to build the Network Nation: a continuation of current economic trends to increase the market penetration of existing technology is all that is necessary.

At this point in time the real utilities cannot be completely quantified, and the value of economic models lies in debunking attitudes like, "Isn't a 10-cent telephone call or 15-cent letter cheaper than using an expensive computer?" Yes, they are cheaper if the true costs to the end user and the organization are ignored. Communication is too pervasive a component of our societal and organizational processes to be left to suboptimizing considerations and/or purely cost-oriented models.

Thus, the costs of human communication via computer will drop to the point where it is clearly more economical than telephone or face-to-face

ISBN 0-201-03140-X, 0-201-03141-8 (pbk.)

meetings. Whether it will become a dominant communications form will depend on

1. the quality or comparative outcome of this communication process;
2. whether potential users will overcome their negative attitudes toward computer technology, and invest the time and resources needed to try the new technology; and
3. whether regulation of the technology imposes artificially high costs or restricts the development of a variety of available services.

This last factor is the subject of the next chapter.

For Discussion

1. What new technological wonder would you most like to add to your own personal telecommunications tinker toy? Can you justify its benefit on the basis of economic savings or increased effectiveness for specific communications purposes? Why do you really want it?

2. It would seem to be just a matter of time until increases in the cost of postage will create strong economic pressures to replace hand-carried mail with "electronic mail." What do you predict will be the interplay between postage costs and noneconomic factors in motivating use of the computer to replace different kinds of mail: advertising, bills, greeting cards, personal letters, etc.?

3. What specific applications of computerized conferencing will be most likely for white collar organizational employees having terminals in their offices; and by middle-class people with terminals in their homes?

4. What psychological and sociological factors would you consider most significant to incorporate into a more comprehensive cost–benefit analysis of computerized conferencing? What specific indicators could you use to measure these factors? How would you place a relative value or benefit on them, or how would you combine them into a higher-order measure? What experiments would be necessary to gather the required data? How feasible are those experiments?

5. Consider the following communications situations: (a) learning; (b) a complex pragmatic or applied problem; (c) situations involving hostility or intense emotions; (d) information gathering from heterogeneous or homogeneous groups; (e) bargaining, bid and barter, or negotiation.

    How are the requirements of these different kinds of communications situations likely to be related to the relative effectiveness and time needed to complete the communication via face-to-face meetings, telephone, computerized group, etc?

ISBN 0-201-03140-X, 0-201-03141-8 (pbk.)

6. How would you assign a monetary value to the self-activating nature, time dispersion of participation, and availability of a written transcript of discussion offered by computerized conferencing situations? How general can you be, or is the value always situation dependent?

7. At what travel costs will the number of one-day business trips along the Boswash corridor (Boston–Hartford–New York–Philadelphia–Baltimore–Washington D.C) drop substantially? What fraction of business trips do you feel are enjoyed by the travelers? Would that fraction differ for one-day hops as compared to longer trips? How would you set up a benefit–cost tradeoff analysis comparing one-day business trips to use of computerized conferencing?

8. For what situations do you see the value of being able to talk to a terminal instead of typing? How much more would a person be willing to pay for this privilege over other alternatives and how does that compare to the investment costs of time and typing lessons or the hiring of a typist?

9. When it is cheap to convert printed words to digital form (say 500 current U.S. dollars), what do you see as the most significant impact on the use of computerized conferencing?

10. Encryption has been suggested as a method to increase security on computer-based communication systems. An example would be one-way codes, which are ciphers a person can reveal to others that will automatically put their messages to them into code. However, a separate key is needed to decode the items written and usually is retained only by the receiver to ensure the security of what is transmitted. In what CC applications would this be useful? How much per message would you be willing to pay for it?

11. What sort of norms might evolve about the use of personal simulators capable of writing for a person ("electronic ghosts")? What countermeasures might be introduced to inhibit or detect the use of electronic ghosts and in what situations are such countermeasures likely to be employed?

12. Provide some examples of situations where the degree of group commitment to the results of a discussion can mean attaining or not attaining the agreed upon objective.

13. Describe some tasks for a group where equal participation in discussion appears to be a crucial factor in obtaining a better result. In what situations would you want unequal participation? Might this also be encouraged by a set of rules or communications structures in a CC system?

14. Describe a family stiuation where the decision is likely to be made that

ISBN 0-201-03140-X, 0-201-03141-8 (pbk.)

two terminals are needed in the home rather than just one (e.g., like the consideration of needing a second car).

15. What sort of computerized conferencing is likely to result in a demand for pay terminals (like pay phones) located in public places? When do you see this occurring?

16. In what situations would graphical conferencing be necessary? At what cost level would it appear to be feasible for the situations you suggest?

17. At what cost levels would computerized conferencing allow a market for different segments of the society and different applications? How low must the cost be to be affordable for teenagers; college students taking courses; adult education; housewives; stamp collectors; etc?

18. Since students do not receive a salary, how would you justify, in a cost–benefit analysis, the value of introducing this technology into the educational process?

19. Devices are being offered today that claim to be able to detect, from a person's voice, if he or she is lying or under emotional stress. What impact is this technology likely to have on the spread of computerized conferencing? What communication activities are likely to shy away from spoken communications because of the widespread use of such devices?

ISBN 0-201-03140-X, 0-201-03141-8 (pbk.)

THE BOSWASH TIMES

The Computerized News Summary Service of the Megalopolis

Information Wars Warm Up

Washington, D.C., August 5, 1989. The snow has finally hit the fan here in this city of compromises. All attempts by the President to head off what has been called by some the "Armageddon of Communications" have failed. The various interagency committees and special commissions have come to a complete standstill and what seems to be emerging is a rash of court cases and attempts to rush new legislation through Congress. In many of these situations various federal agencies and regulatory bodies sit on completely different sides of the fence. Among the actions in preparation for the courts appear to be the following.

The U.S. Postal Service is filing to restrict the offering of any electronic mail system by any private corporation, and to launch an injunction against both the FCC and the Federal Reserve Board to invalidate all regulations they have promulgated governing messaging capabilities in systems subject to their jurisdiction. Needless to say, the FCC and Federal Reserve Board are filing countersuits, with support from the banking, telephone, and data processing industries.

The Justice Department is filing against an as yet unnamed major corporation, which is charged with attempting, through unfair pricing, to capture the computerized conferencing market.

The National Newspaper Association and a number of other publishing organizations are filing against the Copyright Office with the claim that recent interpretations of the copyright law are intended to destroy their industry and go beyond original congressional intent. The Consumer Protection Agency will finally issue its draft guidelines on the collection and dissemination of mailing lists gathered from various historical data on computer communication systems. The CPA's action is likely to bring this multibillion dollar industry to a standstill.

The IRS is pushing for legislation in Congress, after losing a recent court case, to have the income tax laws apply to transactions using computer "credits" instead of dollars. Congress seems sympathetic to this issue, but may impose restrictions limiting this extension to commercial transactions only, and excluding the now thousands of local public systems for trading goods and services among private citizens.

One senator was heard to mutter that "the years of procrastination have caught up with us. What we

ISBN 0-201-03140-X, 0-201-03141-8 (pbk.)

have now is a battle over billions and billions of dollars of ongoing communication services."

On the New York Stock Exchange the values of the stocks involved are behaving like yo-yos, as investors vacillate about who will come out on top in the current confused mess.

Common Cause has launched the most dramatic court case, and seems to have triggered the breakdown in any organized executive response. It is asking the courts to wipe out the jurisdiction of all current laws and regulations governing computerized communications. Its suit contends that there are a great many paradoxes and catch-22's involved in trying to apply these laws and regulations to new technology and request that they all be nullified pending congressional action. This would essentially remove the current conflicting controls of the FCC, the Federal Reserve Board, and the U.S. Postal Service in one fell swoop.

Senator McFarlen introduced into Congress today a bill to bring about "Electronic Pollution Control." A number of industries and professional societies have been pushing hard for regulation to erase trivia from the nation's memory banks. The accusation is that the computer communications industry has devoted too much memory capacity to public use and not enough to less profitable corporate and professional use.

"The public is saturating the nation's memory bank with utter trivia—old electronic love letters, their children's first homework assignments," laments Senator McFarlen; "Computers are being used like family scrapbooks."

There has of course been considerable lobbying by scientific, professional, and business organizations to introduce limitations on public use of what is considered by them a precious national resource that must be preserved for "significant" information. That not all professionals feel this way was made plain by the keynote debate at the annual meeting of the American Association for the Advancement of Science. At that session leading scientists from the natural sciences debated a group of social scientists on the question whether the growing takeover of the nation's computer memories by the public represents an increase or decrease of entropy for the society as a whole.

Investments

August 6, 1989. This year marks the first in which the amount of money spent by the average U.S. family on home computer equipment and services exceeds that spent on transportation. The home computer market is now bigger than the automobile industry. One cannot help recall the disbelief that existed in investment circles when IBM bought out *Good Housekeeping* in 1979. Today marks the confirmation of that decision as a wise one.

ISBN 0-201-03140-X, 0-201-03141-8 (pbk.)

439

CHAPTER 12

Policy and Regulation

General Problems of Policy Formulation and Evaluation

A common fallacy relating to potentially revolutionary technology, such as computerized conferencing (CC), electronic funds transfer (EFT), and similar computer communications systems, is to plan, formulate policy, and make decisions based on the premise that this technology is only a more efficient (faster and/or cheaper) way of accomplishing what we do now. Thus in recent years the term "electronic mail" has become a popular label for some of the technology we deal with in this book. People have a natural tendency, when confronted with something new, to pigeonhole it as something they feel they already understand. Thus the automobile was first known as the "horse-less carriage" and initially viewed as a more efficient horse. The telephone was perceived as a substitute for or improvement of the telegraph.

When people perceive a new technology only as a substitute for or more efficient version of existing technology, this perception may constitute a self-fulfilling prophecy; that is, the limited perception of the technology may become the reality. This possibility becomes greater as the degree of regulation and policy that governs the use of the technology increases. It is particularly likely that this may occur in the area of computers and communications, where a multitude of differing regulatory, legislative, policy formulation, and decision-making bodies can influence the future of this technology and its application or utility for society.

Our approach to this area will be, first, to examine the general problems of policy formulation and regulation that impact on any new technology, and then to present the specific issues that concern this particular technology. There are a number of elements that underly the difficulties of dealing with technological progress, and only by clarifying these can a reasonable basis be established for some of the positions held by the authors.

Starr Roxanne Hiltz and Murray Turoff, The Network Nation: Human Communication via Computer

440

ISBN 0-201-03140-X, 0-201-03141-8 (pbk.)

Incremental Approaches or "Muddling Through"

The natural tendency of governmental bodies is to seek slight changes in or modifications of existing law, or to devise slightly different interpretations of policy or precedents in order to deal with a new situation. This phenomenon is built into the mechanisms of bureaucratic organizations. Where this approach collapses is where the rate of change of the situation (or new technology) is faster than the speed with which policy and legislative changes can be brought about. The introduction of electronic funds transfer is an excellent example of a situation in which governmental and regulatory bodies at both the federal and state level have been caught flat-footed. For example, the problem whether an EFT terminal constituted a branch bank in terms of current rules and regulations was one that sent the lawyers into spins.

Approaches to policy formulation for new technology are not likely to be ground-breaking if conducted by an abundance of lawyers in governmental bodies with an inherent tendency to advance with their eyes to the rear.

The governmental approach to preventing deleterious effects is, therefore, to think in terms of only slight modifications of the status quo and to extend to the new technologies the kinds of regulations that served to control the old ones. Sometimes this procedure is fairly satisfactory: the automobile license can be seen as an extension of the brand on a horse with some effectiveness for record keeping; however, it probably has less effect in preventing "rustling." The myriad regulations that evolved in the effort to control excessive speed, pollution, and unsafe features, to compensate accident victims through insurance, etc., were devised only as reactions to abuses that caused considerable suffering or harm to people, resources, and the environment. In this reactive posture, society never did come to grips with the facts that it has ended up spending a large portion of its income on automobile-related items (cars, highways, roads, gas, insurance, financing, traffic controllers, garages, parking lots, etc.); that the automobile would reshape our cities into urban sprawls without decent mass transit systems; or that productivity would be greatly diminished as people wasted more and more time behind the wheel of an automobile. It seems increasingly clear that incremental approaches to our problems are no longer the answer.

What Will Be, Will Be

It is interesting to observe people trying to deal with the future, whether it be an individual dealing with his future, a corporate executive concerned with his company's future, or a government official trying to cope with society's future. There is a basic conflict that underlies most such planning activities; seldom clarified or made explicit, it generates policy and decision options that are in conflict. On one side are those who believe that a person or society has

ISBN 0-201-03140-X, 0-201-03141-8 (pbk.)

very little control over the direction that future events will take. On the other side are those who believe that major changes can be consciously made or shaped by the individual or society. In planning or forecasting these perspectives are formally known as the *exploratory* and *normative* approaches. The exploratory approach in simple terms utilizes past history to extrapolate current trends into the future, whereas the second approach describes the future as the planner desires it to be, and then works back to determine the possible paths that can be taken to reach that future. This book is a normative book in which we have characterized desirable futures. In this chapter we are going to have to deal with some of the extrapolative trends that may drive us to less desirable futures.

The area of computer communications has reached the point where not the technology, but questions of policy, law, and regulation, will determine the degree of benefit that society will derive. If we extrapolate current trends, such things as public use may be *artificially* delayed by decades; more will be said on this point later in this chapter.

Technology Assessment

Technology assessment has been touted as the key to understanding the consequences of new technology. Although the words themselves have a scientific sound, this endeavor is at best the softest of the social sciences. In a sense, this area has been imposed on science and technology rather than evolving from it. Its standards of achievement have been established by government funding sources rather than by the scientific process. As a result it has come to be viewed as a theory, study, or thought process in which a group of competent people can somehow examine the future. While theory is half of the scientific process, the other half is experimentation, which is conspicuously missing from the class of efforts labeled technology assessment. To be completely fair, there are many cases where experimentation is impractical. We cannot decide to install nuclear power plants as an experiment, to be ripped out if some undesirable consequences develop. There are, however, a significant number of areas where technology can be experimented with—where the future can be tried out on a limited experimental basis—and computer communications is one of these. Experimentation within the scope of technology assessment is not demonstrating the feasibility of a system; rather, it is akin to a social experiment, such as providing a sample of people with income maintenance and evaluating whether such an arrangement has more beneficial effects than welfare payments. Paper-and-pencil studies of computer communications technology are not going to provide the insights necessary for laws and policy that will produce good results over the long term. However, paper-and-pencil studies seem to be the cheapest and easiest kind for government agencies to fund in investigations of

ISBN 0-201-03140-X, 0-201-03141-8 (pbk.)

new technology. As a result, the decisions being made today and in the near future may be the wrong ones.

The catch-22 of this particular technology is that people generally cannot comprehend this form of communication until they have experienced it. It is not possible to ask a scientist, businessman, teenager, social worker, or housewife how each would use computerized conferencing and expect valid answers until those individuals have experimented with it enough to understand what it is.

Values: Prejudging Future Options According to Current Preferences

Probably the single greatest problem in getting people to conceptualize future possibilities is the implicit assumption that values will remain the same. We seem able to hypothesize future consequences of technology, but then we discover policy formulation based on those predicted consequences uses today's values in appraising their desirability. For example, today many elderly couples from very conservative middle-class backgrounds live together "in sin" rather than have their social security payments reduced by getting married. To those who drafted the Social Security law this value change fostered by the law in this particular community would probably have been quickly discounted if anyone had thought of it. It is impossible to even hypothesize changes of this type unless those doing so start from an explicit effort to look at potential value changes.

Since human communications are the mechanism by which values are transmitted, any significant change in the technology of that communication is likely to allow or even generate value changes. Certainly, the beginning value changes will have to do with how we relate to other people and the striking availability of commentary about things most people will not talk about except perhaps to very close relatives. In computerized conferencing, both the ability to use pen names and the comfortable, psychologically open feeling that no one can reproach you for holding a very different set of values will allow much freer discussion among large groups about value-laden issues and values themselves. Therefore, if we really want to try to understand the future of these systems in order to generate appropriate policies, the study of values will have to be explicit in any such efforts.

Discounting

Discounting, formally, is a procedure in accounting and economics in which a dollar today, in terms of either costs or benefits, is more significant than a dollar in the future. Its folklore equivalent is "A bird in the hand is worth two in the bush." It is both an explicit and implicit factor in the way we approach policy and decisions and is terribly ingrained in our outlook.

ISBN 0-201-03140-X, 0-201-03141-8 (pbk.)

Given any set of options or alternatives, it tends to consciously or unconsciously drive us to the choice that maximizes today's return and pushes the costs as far into the future as possible. When formally taught in business courses it is often presented as a recipe, with little clarification of the axioms underlying it or the circumstances in which it may not be valid. One fundamental axiom behind this way of thinking about alternatives is that costs or benefits are independent of time scales or cycles: a given cost or benefit is assumed to be realizable in a given year with no history or memory effects in the process that lead to cumulative impacts. We know, however, that certain phenomena violate this assumption. Low-level leakage of radioactive wastes is a good example of costs accumulating that cannot be discounted formally. Even a minimum discount factor of 1% would bury the fact that a deadly dose of radioactivity from a particular leakage would occur in 200 years.

This discounting attitude partly explains our inability as a society to come to grips with problems such as energy. Another example of a cumulative impact phenomenon is our educational system, where the education provided to an individual in the first few years of schooling can impact on the education obtained in the last few years. However, from a planning standpoint we treat these learning periods as separable, and decisions are usually made on cost–benefit analyses of a particular grade or program for that grade. Only recently are people becoming aware of the need for life cycle cost–benefit analysis of social systems or programs such as education.

Computerized conferencing viewed as a mechanism for generating human social systems can only be evaluated properly on a long-term basis. The significant impacts, both benefits and costs, are more likely to be accumulative with time than separable year to year. One cannot evaluate in terms of what happens immediately if CC is introduced into a high school. The results of examining this situation would be very different from the assessment of its availability from elementary school through high school and college, and into the job environment.

Prescriptive versus Descriptive Systems

This is the classic philosophical problem of the distinction between the concepts of "is" and "ought to be." We have classically tended to view science as a process of defining what "is." Usually the approach to introducing any computer system is to define what "is" and to design a system based on that knowledge. However, computer systems integrated into a community of human beings represents at least a nonlinear feedback system where new laws of behavior evolve, and the phenomenon of prescribing behavior occurs. For example, today we are building systems to computerize patient medical

ISBN 0-201-03140-X, 0-201-03141-8 (pbk.)

records in hospitals, which seems like a pure descriptive use of a computer. However, those assessing hospital performances will now find it easy to correlate performance data on individual doctors—essentially just a reorganization of data in a computer system. Such correlation could influence insurance companies to set malpractice insurance rates based on individual doctors rather than a specialty class. As a result, we must ask what will be the attitude of doctors toward accepting high-risk cases, which today are commonly taken on by more experienced doctors or specialists. What we see here is the transition of a descriptive data base into a system context in which the system has prescriptive impacts on its elements.

The intimate involvement of people in a computerized conferencing system (CCS) means that as a system its major consequences are likely to be prescriptive in nature. For that reason it is important that we make a significant effort to understand this phenomenon. Certainly, the major impacts of automobiles have been the prescriptive ones. In dealing with human systems we are dealing in prescriptive effects, and the descriptive approach of the "hard" physical sciences, (e.g., physics and chemistry) is inadequate. We are dealing, in essence, with adaptive systems able to respond to their environment in new ways.

Regulatory Mechanisms

All too often the regulatory and policy mechanisms of government have been subverted by the industries they exist to control. Although this takeover has not usually been intended by the formulators of these mechanisms or the laws setting up such agencies, many factors lead to this corporate domination when the regulation involves a rapidly changing area. First is the fallacy that the scope of regulatory responsibility can be determined either purely on the basis of the technology, such as radio communication, or on the scope of the application, such as banking. This leads to the regulatory mechanism's inability to deal with gray areas or overlapping impacts of regulation across established regulatory agencies.

For any regulatory area involving high technology the regulatory body must have impartial expertise available that represents people at the forefront of the technology. Such people cannot be retained in an organization that does not incorporate a mission that provides them meaningful research and development activities. Those regulatory bodies that incorporate research laboratories (e.g., the Federal Aviation Agency) have seemed to understand better the technology's penetrating or impacting on the formulation of policies and regulation.

Another problem in resolving public policy issues is that decision making is disaggregated among at least three levels of government and numerous

ISBN 0-201-03140-X, 0-201-03141-8 (pbk.)

agencies at each level. "While no one person, agency, or institution is in charge or has a clear field or the authority to accomplish things, often dozens, if not scores, of units of governemnt have the power to intervene, to slow down, or to stop action by others" (Coates, 1977c, p. 11).

The final outstanding problem is the hearing process, which assumes that the views of all interested parties will be aired. The two related fallacies here are the beliefs that all interested parties can afford the travel, time, and effort to prepare for such hearings and that all interested parties know that they should be interested. More than any others, these misconceptions tend to cede the major influence on regulatory mechanisms to the industry they regulate. Regulatory bodies do not usually have the funds nor the human resources to really examine or question who should be the concerned or interested parties and ensure that these interests are represented in the hearing process. To solve these problems, extensive assessments should precede or parallel the hearing or inquiry processes.

Computerized Communication–Information Systems: Specific Policy Issues

There does not appear to be a problem about those involved in the computer communications industry and concerned government bodies recognizing that there are major issues to be dealt with as a result of the technological revolution taking place in these areas. The following extracts from a *New York Times* article (Birnham, 1977) testify to that recognition. Wrapped up in the issues this article refers to are billions of dollars of revenues a year for those organizations legally able to provide services under whatever policy, legal, and regulatory situations exist. However, these issues, which represent only the more visible and immediately obvious ones, mask many of the more fundamental or crucial issues that may well govern the course of our society as we move into an era where information and communications are dominant factors in shaping the society.

Nation Facing Crucial Decisions Over Policies on Communications
by David Burnham

Washington, July 7—Fundamental changes in the way that Americans communicate are forcing the Federal Government to make policy decisions that could alter the fortunes of the nation's largest communication organizations and the lives of most citizens.

The pending decisions confronting the Carter Administration, Congress, and the Federal Communications System involve a complex balancing of such questions as the competing rights of large corporations, the possible demise of the United States Postal Service, delicate diplomatic negotiations over the use of

ISBN 0-201-03140-X, 0-201-03141-8 (pbk.)

the air waves by individual nations and the importance of privacy to individual citizens.

The communications changes, made possible by the increasing use of several technologies, are expected to transform the way that many Americans conduct their lives . . . . According to interviews with scores of communication experts and reports by dozens of different agencies, commissions and consultants, some of the most difficult pending questions awaiting resolution are the following . . . .

*Should Congress follow the lead of the House Communications Subcommittee and attempt a total revision of the Communications Act of 1934, the basic framework for regulation of all communications in the United States and a document that many experts contend has been largely outmoded and is now slowing the application of the most advanced technologies?

*Should the Federal Communications Commission continue its policy of gradually subjecting the American Telephone and Telegraph to increasing competition?

*Should Congress grant the United States Postal Service a legal monopoly on electronic mail similar to the one it enjoys with conventional mail . . . ?

*Should there be a single clearing house within the Federal Government to serve as a highly efficient central switching point for the steadily increasing volume of financial transactions completed by electronics and computers, or would such a clearing house pose a threat to individual liberty?

The situation in the United States is very different from that in most other countries in the world. In much of even Western Europe the postal and telephone systems are run by a single government agency and it has been accepted practice that governments can exercise a high degree of "censorship" on various forms of communication, even newspapers. Because of this, the problems we are beginning to confront may not arise elsewhere for many years. It also means that we have the opportunity at the moment of having the leading edge in the development and application of this technology. Conversely, we may lose this edge if we formulate the wrong policies. The immediate problem is whether or not the real issues can be brought into the limelight of public debate and whether matters can be resolved in terms of the true public interest and welfare. The public interest is not likely to be well represented by the rather limited perspectives offered: those of IBM versus AT&T; the Postal Service's problems and fear of competition; etc. We can begin to see the emergence of the real issues by building on a set of specific ones.

Who or What Is the Market?

The presumption that seems to underly current regulatory inquiries, congressional hearings, tariff filings, and economic investigations by interested organizations is that this technology is essentially for business or commercial

ISBN 0-201-03140-X, 0-201-03141-8 (pbk.)

use. There seems to be no realization of potentials for public use among those in positions of power who set policy, make laws, and approve tariffs. In particular, the value-added networks emerging now are being allowed to set rate structures that will inhibit private citizens from participating in this technology. Common practices that do this are monthly minimum costs for service or subscription fees and discounts for volume users. The latter practice is being reexamined in the energy area in terms of the conflict between the "right" of citizens to have energy at a reasonable price, and encouraging efficient utilization for large-volume consumers.

It is our belief that we are entering an era in which the ability of an individual to function as a citizen of that society will depend on adequate access to computerized information and communication systems. Imagine today a person trying to function without being able to use a telephone. We believe that within 10 to 20 years computer terminals will be as necessary as telephones are today.

It is surprising that an object lesson has not been learned from citizen's band radio (CB) in terms of the pent-up demands for communications. When CC is compared to CB, the potential for citizen use increases manyfold over what has taken place with CB radio. The argument is often made that there is no money to be made in serving public use, but only commercial use. This shows a lack of basic economic knowledge in that approximately one eighth of all U.S. households (roughly 7,000,000 families) have greater gross income than 85% of all U.S. proprietorships (roughly 8.5 million firms), and greater gross income than one-third of all U.S. partnerships and corporations (roughly 1,000,000 firms). In other words, the 7,000,000 U.S. households with adjusted gross income of more than $25,000 per annum have greater cash flow than about two thirds of all U.S. businesses. The issue is not business versus residential use, but rather the more basic one of wealthy versus poor users, and the degree of need for the government to aid, in some manner, small business and low-income disadvantaged groups. This might justify either subsidies or a governmental system via the post office to ensure access by all segments of the society. (Of course, we do not mean that the post office should merely try to computerize the mail, but rather that it could develop and offer a general-purpose communication and information system.)

Is Computerized Conferencing a New Form of Telephone System?

Is this technology subject to regulation by the Federal Communications Commission? Should telephone companies be allowed to offer the service? Should computer companies be allowed to offer the service? Should either have exclusive rights to offer it? Are certain forms or services under computerized conferencing analogous to broadcast radio or TV, or more analo-

ISBN 0-201-03140-X, 0-201-03141-8 (pbk.)

gous to newspapers? Under the laws governing the operation of the FCC, the power over renewal of licenses gives a considerable potential for censorship of what is broadcast over radio or TV, whereas newspapers are not subject to this control. Is it in the best interests of the public to have one standard nationwide system with one interface, like the telephone system, or to have a diversity of different systems tailored to different purposes? Telephone companies are likely to opt for a standard system, whereas computer companies are likely to develop a more diverse set of systems.

While CCS's often use the telephone to connect between the terminal and the computer, or a value-added digital data network, the telephone system itself is only a transmission mechanism. Cable TV is another option for transmission if it has the right options built into its network. However, cable TV is strictly regulated on a local community basis, where there is usually little understanding of future potentials on the part of those involved in choosing the local cable TV option. Another communications option is provided by satellites and rooftop antennas, thus skipping the necessity for any "hard wiring" of the communications link.

The telephone companies, spearheaded by AT&T, have made a number of moves in recent years that have negative consequences for the development of this technology. Through the promotion of legislation, AT&T is attempting to overturn the Carterphone decision, which permits non-AT&T attachments, such as computer terminals, to telephone lines. The AT&T-sponsored legislation (formally called the "Consumer Communication Reform Act of 1977," or informally, the "Bell Bill," and sometimes the "Monopoly Preservation Act") would place the power for setting attachment standards with each state telephone regulatory commission. With telephone company lobbying, it is very likely that the result would be different attachment standards in each state, which, of course, did not apply to the nonforeign attachments provided by the phone companies. This would make it very difficult for any company except AT&T to have a nationwide market, and would probably reduce considerably the hundreds of companies now engaged in the manufacture of computer terminals and other attachments. The ability of telephone companies to deal independently with federal and state regulation has in some cases allowed them to circumvent federal regulatory mechanisms, and represents a major flaw in the regulatory process.

Recently the telephone companies have begun a push to eliminate unlimited phone use tolls for residential services. The companies' position is that this will allow cheaper rates for poor communities, subsidized by higher middle-class rates, a rationale that follows, in principle, their pattern of residential rates being subsidized by business rates. The latter claim is, however, open to question by those who have tried to decipher AT&T accounting practices.

ISBN 0-201-03140-X, 0-201-03141-8 (pbk.)

When a person gets on a computer terminal she or he usually spends from a half hour to hours at one sitting. Unlimited call service makes this very cheap from the home, as many computer professionals who have terminals in their homes can testify. Phone company planners are not so dumb as to fail to recognize that computer applications in general are driving up, or will drive up, the average holding time far above the current less-than-three-minutes in most areas. In fact, it is very likely that the breakdown in phone capacity in New York City a number of years ago was the result of an unforeseen amount of data processing communications cannibalizing existing capacity.

It may, in fact, be the poorer or disadvantaged segments of society who have the most to gain from the continuation of unlimited local telephone service rates, when computerized information and communication services become available to them. Furthermore, as the telephone company moves to digitize its network, it would be more appropriate to assess charges based upon the volume of data transmitted rather than the amount of time a line is in use. This alternative approach would be extremely beneficial to computerized services of all sorts.

Is Computerized Conferencing a New Form of Mail?

Obviously, a CCS can be designed to look as if it is an "electronic mail" system and Scientific Timesharing's early Mailbox system is one excellent example. In that system a person could send registered, certified, and special delivery as well as regular mail. It serves, through its design and wording of features, to give the impression that the Postal Service is functioning through a computer. What a user fails to realize is that it represents only one example of a specific communication structure and set of protocols.

The Postal Service and those concerned about it in government have come to realize that there is severe trouble ahead for that organization. As an increasingly large volume of financial transactions is handled through various general and specialized computer networks, the volume of first-class mail will dramatically decrease. Since somewhere between 60% and 80% of first-class mail is business mail, and since this represents the prime money maker, the substitution process will create a severe deterioration of Postal Service revenues. Currently it is estimated that it costs only about 7¢ to process 13¢ first-class letters. One proposal by the post office, to increase first-class postage for business users, will only serve to speed up the substitution process now under way. It is quite clear that the individuals in policy positions concocting such ideas do not grasp what is actually taking place.

Of far more concern are those in Congress and elsewhere clammering for a Postal Service monopoly on something called "electronic mail." Such legislation would present the interesting situation of an organization that has exhibited no competence in this new area, no significant experimentation or

ISBN 0-201-03140-X, 0-201-03141-8 (pbk.)

development research work, being given, with the stroke of a pen, carte blanche to provide a new technology in a noncompetitive environment. In many ways this possibility is more frightening than either AT&T's or IBM's gaining such a monopoly. If the Postal Service is to get into this area, as it probably should, it should do so on a competitive basis. In fact, there are arguments that validate a form of common service provided by the government. In principle, such a service could be very competitive economically, if the Postal Service could overcome its real problems of management and the conflicts inherent in attempting to be a private company, a government agency, and a Congressional pet all at the same time.

What's in a Name?

The previous discussions bring out the problem of names. In the literature, all of the following designations have been applied to what we have been talking about in this book, or various aspects of it

> Teleprocessing (TP)
> Electronic mail (EM)
> Message systems (MS)
> Teleconferencing (TC)
> Office automation (OA)
> Telecommunications (TC)
> Computer communications (CC)
> Computerized conferencing (CC)
> Computer-based conferencing (CBC)
> Computer-mediated interaction (CMI)
> Electronic information exchange (EIE)
> Computer-assisted teleconferencing (CAT)
> Computer-mediated teleconferencing (CMT)

In addition to these attempts at a general name, there are also the many names tacked onto specific systems: EMISARI; Partyline; Discussion; Conference; Forum; Planet; EIES; Confab; Mailbox; Confer; Hermes.

The choice of a name represents an implicit statement of a value or predecision on policy. Obviously anyone using the prefix "tele-" is at least unconsciously providing those who may not completely understand the ramifications of this technology with the impression that it belongs in the pigeonhole in which we place telephones. The terms "electronic mail," "message systems" (akin to telegrams), and Mailbox would all tend to place this technology in the pigeonhole for the postal system.

Unfortunately people who grasp these terms to try to understand the situation from a limited base of knowledge may be legislators, lawyers, executives, managers, bureaucrats, regulators—many of them ultimately responsible for setting or influencing policies, decisions, and regulations.

ISBN 0-201-03140-X, 0-201-03141-8 (pbk.)

Other complications arise from terms like "office automation," which imply that this technology is intended only for business and organizational use. The word "computer" has become disturbing for many individuals and groups in society, as akin to surveillance, invasions of privacy, errors, and inflexibility. The only remaining labels, "electronic information exchange" or "electronic exchange," while all right for professionals, are a little too abstract. Our own bias is toward computerized conferencing because it does imply both computers and group communication. Furthermore, now that computer technology is entering the public or home market, we feel that the negative connotations of the word alone will begin to disappear.

What we face is a situation in which the major explicit debate at policy levels is where to pigeonhole the technology. It should not be surprising, but even many of the "experts" in this technology are taken in by their own propaganda or preconception of where this technology lies. This is the sort of debate that lawyers relish. People are so preoccupied with definition and the resulting jurisdiction of control and regulation that the real issue is being sidestepped. That issue is what is truly in the "public interest" and for the "public good," and it cannot be understood by trying to find a narrow, traditional label under which to include this technology.

There is a severe threat that the policy regulation, and other decision processes now laying the groundwork for this field will unintentionally close off the options that could have the greatest potential benefits for society, and inhibit the investigations that must take place to realize and justify those benefits.

Any successful attempt at this time to make computerized conferencing fit into one of the existing frameworks for regulation is likely to result in the availability of only one conference system with only one design. If the automobile industry had been regulated in this manner, it might still be true that consumers could have any kind of car they wanted, as long as it were black, started with a crank, did not travel more than 30 miles an hour, and were not used for any international travel.

What about the Hybrid Systems?

There are many situations in which CC capabilities become buried within other applications. A computer-assisted instructional system is one example. Should such a system now be regulated or accredited by the postal system or FCC, or should those providing such systems be prevented from providing communications between and among students and teachers as an integral part of the CAI system? The latter situation would mean designers would be

ISBN 0-201-03140-X, 0-201-03141-8 (pbk.)

prohibited from designing what might be the most beneficial system for educational purposes, and the thought of the postal system or FCC controlling or regulating such systems is not in keeping with traditions of private enterprise, freedom, and encouragement of technological progress.

A more significant example in terms of current regulatory issues is electronic funds transfer. Today this area is largely viewed as a mere substitution for the transfer of checks, cash, sale transactions, bills, and receipts into an electronic equivalent. Thought of as a banking process by way of the substitution for checks, it is presumed that this process is a banking enterprise which should be regulated by those federal and state agencies that normally regulate banks. The evolution of this technology has been so rapid that it took the spread of remote banking terminals in a number of states to make people realize that there was no longer an understanding in a legal sense of what constituted a branch bank. For two years the Commission on Electronic Funds Transfer has supposedly reviewed the whole area. The reports of that commission prompted one critic to state that "the results of this commission's work prove that we need a commission to study EFT" (Coates, 1977b). The work of that commission is a good example of how commission efforts can ignore completely any future potential alternatives, and concentrate entirely on the issues that represent current popular impressions.

As we showed in Chapter 6 on public use, the process of transferring payments of one sort or another represents only one step in a chain of communications:

1. communications between those advertising and those seeking;
2. reaching agreement on a transaction;
3. delivery of goods or services;
4. transfer of payments;
5. analysis and accounting of what took place.

Given the alternative of an EFT service that provides only one of these steps and one that provides all of these steps, which would business and individuals utilize? Communications between individuals (bid and barter, negotiations) and institutions will become an integral part of EFT systems. We have the additional complication in the banking area of an existing set of federal and state regulatory mechanisms separate from those in the communications area, and liable to produce major policy conflicts between state and federal regulations.* This area also illustrates the potential for the

---

*This two-tier regulatory mechanism has allowed those regulated to defeat policy at one level by influencing the introduction of different policies at the other level—this is true in the communications and the education areas, in particular.

ISBN 0-201-03140-X, 0-201-03141-8 (pbk.)

emergence of a few large institutions that could create an oligopoly and squeeze out or assimilate small institutions offering the diversity of services just listed.

It is communications via computers and among institutions and individuals that will determine the course of these events. The branch bank will be at the person's home terminal and the individual will want to receive all financial services through that terminal.

In fact, it would now be technologically possible, for about a half-million dollars, to set up a single computer that merely allowed individuals to seek and transact exchanges of goods and services in which no actual money changed hands, and transactions were handled in some sort of credit unit other than dollars. The computer would run a settlement function, so that your plumber got credit for the job he did for you from the grocer for whom you performed legal services. It is also possible to introduce interest on credits on deposits and allow borrowing of credits. Is such a system a bank in the regulated sense, or is it a communication system in the FCC-regulated sense, or is it an information system? Or an employment agency? Or perhaps, illegal gambling?

What Is Information?

Perhaps at the heart of the policy problem is the root cause (suggested by David Snyder) that we do not understand the nature of information. We have accepted, to a large extent, the continuing trend toward an information-oriented society. However, we do not comprehend the dominant commodity of exchange in that society. In the industrial society we have been able to quantify and assign values via various mechanisms (e.g., marketplace, regulation) to material goods and services. We have no such existing assignment of value to or monetary scales for information, and the current mechanisms for information handling preclude the assignment, in the near future, of value by an open-market mechanism.

Estimates have been made that 50% of our GNP is involved in the production, storage, retrieval, mobilization, and transmission of information in the society. However, there is little careful analysis or available tracking data to examine issues of productivity, efficiency of use, resource expenditure, and resulting value as it concerns information. Until we pay more attention to these considerations it is doubtful that coherent policies can be formulated to deal with this area. From this standpoint, the process of "muddling through" may be better than setting the wrong policies.

Computerized conferencing holds the promise of a universal mechanism whereby information can be transferred among individuals and organizations. As an interface mechanism between information sources, it can be utilized to

ISBN 0-201-03140-X, 0-201-03141-8 (pbk.)

establish a marketplace and thereby clarify value and ownership. We foresee a potential for computerized conferencing as the vehicle to provide the economic marketplace for information that is ultimately needed to support an information society. It provides that opportunity to quantify on a compatible base the value of information now passed via a host of incompatible sources. In the long term this may be the cause rather than the effect of meaningful policies and regulation.

Who Regulates or Makes Policy for Information Systems?

The answer is everyone, and, in effect, no one. Internal federal policy is a troika of the Office of Management and Budget, the General Services Administration, and the National Bureau of Standards. So far this three-horse team often gives an impression of pulling in different directions. In particular problem areas that are popular, we find a host of agencies and special commissions involved. Responsibility is fragmented at the federal and state levels by specific problems and specific applications. Information systems represent an area today that is as important as communication systems, and in the long run there is going to have to be some centralization of policy formulation for this area. Whatever model results from this centralization has to be more workable than some of our earlier attempts in other areas.

Computerized conferencing is a merger of information and communications systems, but in its broadest perspective it is more of an information system incorporating communications. The policy problems associated with information systems are in themselves a major concern for the next few decades, and the movement of our societies concern from property rights to civil rights will probably be followed by the concern for information rights. The areas of interest here apply to information systems as well as conference systems; they include

Ownership of information and copyrights
Legality of liability for information content
Rights of access, sunshine laws, and freedom of information
Freedom of speech and censorship
Privacy and security

None of these issues are clear-cut for information and communications handled by computers. There is no understanding, for example, of how the copyright law that took effect in 1978 applies to written material entered into a computer terminal. One reflection is that if taken very literally the law would imply that material printed on a hard-copy terminal is copyrightable, whereas that displayed on a visual terminal is not. It is amazing but true that such vagueness can exist in a new law.

ISBN 0-201-03140-X, 0-201-03141-8 (pbk.)

Since material put in a terminal is not signed, what is the accountability or liability for what is said or transacted? Is an agreement a legally binding contract? Can a person be sued for "libelous" statements in a conference?

Possible interpretations of all the other issues above are a function of current specific applications of the particular information systems and what organization might or might not have the policy formulation responsibility. However, this does not appear to be the way society should allow this railroad to be run. Results of the current morass will almost certainly lead to many inconsistencies in policy.

The foregoing are the macroissues of policy. There are a host of micro-issues tied up with areas such as transportation/communication tradeoffs, conferencing/communication media tradeoffs, and work at home. If, for example, a person can trade off the use of communications for transportation, then she or he should be able to draw money from travel budgets to pay for communications. Currently, if you have a federal grant for some purpose, this substitution would require special justification. Work at home will require new personnel and work practice policies. Once we have the electronic equivalent of daily newspapers, we will have advertisements as well, and some of the regulatory issues associated with this.

It is time that areas of information systems that cut across all application areas began to receive some centralized attention in terms of an overall perspective.

In terms of computerized conferencing, the direction and extent of future application will depend critically on the resolution of the information rights issues. Authors, poets, free-lance journalists, and other writers will not contribute material to these systems unless there is some concept of ownership or copyright that allows them to obtain revenues for the use of their material in these systems. One major problem that has not occurred to the formulators of copyright legislation is the segmented and overlapping nature of material in these systems. An author has to be able to own one item, which may appear in many different places which may change dynamically over time, and the author might alter his item after it already is in the system. Delivering copies of the item to the copyright office whenever it is changed, or a copy of each and every "publication" of it, is going to lead to chaos. If the message is viewed as a letter, then ownership, according to legal precedent, resides with the receiver.

If individuals are going to conduct business through these systems, the legal responsibility to live up to the terms of a written communication or the responsibility for financial loss resulting from a lost piece of communication remain to be determined.

It is not clear that current wiretapping laws apply to this medium of communication, or what rights of privacy exist. If a computer is set to

ISBN 0-201-03140-X, 0-201-03141-8 (pbk.)

monitor all communications and automatically destroy any message with profanity or obscenity in it, has any privacy been violated, in the usual sense that another human or humans have had access to communications not intended for them? This automated censorship by a computer, without human intervention, could possibly be legal under the FCC laws, although probably not under Postal Service operations. The FCC does have the power to exercise censorship or denial of licenses for obscenity in the case of ham radio operators.

There has been at least one reported case of censorship of the content allowed in a CCS developed under a federal research effort, as reported in *Business Week* (March 16, 1974).

No Computer Talk on Impeachment

> The Pentagon is picking a fight with top universities by refusing to let its Advanced Research Projects Agency computer network be used for a study of impeachment. The network, located at civilian campuses but funded by the Pentagon, is being adapted to provide a nationwide "teleconferencing" system. By using a designated code number, scholars could tap in at any time to contribute data and ideas to an on-going conference on almost any subject.
>
> Some subjects apparently are taboo. Political science professor Stuart Umpleby of the University of Illinois says that the contractor for the network, Institute for the Future, turned him down when he proposed a study of impeachment and now is barring him from any access. "They were just scared they'd lose the contract and knuckled under to anything the Pentagon said," he charges.

This incident highlights the problem that censorship can sometimes be an indirect process based on such things as availability of funds and control of access to systems.

Is a computerized conference between some government and industrial officials subject to the sunshine laws, or is it considered a set of phone calls or private messages? Are the records of who communicated with whom public data when officials are involved?

As we have seen, the resolution of some of these issues may depend on the name applied to the technology. In most organizations different procedures and policies are applied to the treatment of correspondence, phone calls, and committee meetings. Does an organization fit computerized conferencing into one of these, or does it evolve a new set of policies to cover this medium?

Privacy and security are well-known issues. What is new, however, for CC and EFT systems is the indirect knowledge about individuals that can be gained by the records of an individual's activity: who the person communicates with; what types of discussion are enjoyed; with whom business is done or transactions are made; how he or she votes on issues or answers polls. In a

ISBN 0-201-03140-X, 0-20-03141-8 (pbk.)

recent court case the Supreme Court established that banks and not the individual owns the information in a person's history of his or her checking accounts. The new technology makes this a very uncomfortable situation. To date mailing lists have had a value of fractions of a cent per name. Now that check records are being computerized and the possibility emerges of computerizing communication records, it will be possible to compile very selective mailing lists on which a single name may be worth many dollars to the right merchant. The potential revenues from gathering these names become so great that it is not possible to rely for protection on the ethical feeling of some corporation, be it a bank or any other public or private enterprise. Even many of our state governments had no qualms about selling automobile registration data. This area will require new and specific legislation to head off potential major abuses.

The International Situation: Moves toward "Data Protectionism"

The only consoling aspect of the lack of U.S. policy is that the international situation is in a worse mess. Many countries have evolved their own separate laws on the use and transmission of data contained in computer systems. Since none of these laws agree, the introduction of data networks able to tie computers in one country with those in another has created a legal and diplomatic nightmare. For example, an employee of a multinational organization transferred to a different country creates the problem that the medical data in his personnel file may not be legally incorporated in a personnel file in the new country. The problem extends to the transmission of communications as well, and differences apply to both the nature of a communication and what might be included in the content.

The increasing availability of international data communication networks means it becomes very efficient for a computer installation in the United States to turn its nighttime slack hours to serving users in other countries. Also, the success of the U.S. computer industry in Europe and Asia has led to considerable fear by foreign governments that the United States will become the information capital of the world. As a result, foreign governments have begun adopting policies that have been labeled "data protectionism":

> In the name of privacy and the rights of the individual, the world may be erecting and tripping over barriers to the free flow of information across borders.
>
> Frightened that computer and communications technology is rendering borders meaningless, many nations are scurrying to reestablish those borders through data protection laws . . . . The major problem is that the definition of

ISBN 0-201-03140-X, 0-201-03141-8 (pbk.)

what data is to be protected has spilled out of the privacy realm to include any information affecting national sovereignty, cultural values, and technological advancement . . . . Eighteen nations have laws on the books or in the works [Pantages and Pipe, 1977, pp. 115–116].

In addition, foreign regulatory bodies have placed rather high and artificial rate structures on the use of international data networks. This is in part a protection against inroads these networks could make on local TWX and cable services. In many foreign countries, the governments run all the communications services and make the regulations as well.

The rates being set will inhibit the utilization of modern computer message systems as well as conferencing systems. In addition, regulations on data transfer will inhibit the development of integrated conferencing–information systems that incorporate pertinent data bases.

However, the actions of these countries may very well backfire, since they will delay the availability of computer and information technology to those sectors of their economies that need it to compete in the world market. If the United States follows this same path, we may see huge information utilities emerge under Liberian or Panamanian "flags." In other words, the technology is more portable than people often realize and those countries that offer the best regulatory environment could easily become world information centers. The stopping of information flow across borders has always been a government enterprise in many countries, but never a very successful one when it conflicts with international business needs. The information will travel to wherever there are people who will pay to collect it for dissemination to others who will pay more. The ones that lose in this atmosphere are the smaller concerns and the public that would benefit from local spinoffs of information technology if it were available on a competitive rather than a regulated basis. Foreign individuals and organizations may end up being unable to compete or participate in the world market for information and computer services, thus adversely affecting the economic development of the "data protectionist" nations.

In theory, regulation is in the public interest, but when we read what regulators, whether banking, communication, or whatever, listen to, most of the concern seems to be with which segment of the industry will get a bigger cut of the pie. If public interest *is* considered, it is often focused on the current headlined concerns. However, the major public concerns with areas illustrating rapid technological change are those that occur with future or potential applications. The current mechanisms of regulation are prime discounters of the future and somehow we are going to have to change from a reactive to an anticipatory process if the public interest is to get fair representation.

ISBN 0-201-03140-X, 0-201-03141-8 (pbk.)

What Are the Ethics of It All?

Computerized conferencing systems allow for the design of a complete human social system and the easy translation from the laboratory to the "real" world. Whereas the traditional view of changes in communication systems has been one of incremental improvements in current systems with a resultant minor impact on the behavior and values of those involved, CCSs offer the potential for promoting and influencing behavior and value changes on the part of their users through the design of the system. For those who practice what is called "social engineering" (the conscious effort to modify behavior and values in the hope of improving the human situation) this technology offers a new and powerful tool, transcending anything available to date. Even the computer person designing a CCS may unintentionally influence behavior and value changes.

There is likely to be a need for more explicit ethical commitments on the part of designers and other professionals who become involved in this area, and far more emphasis on careful experimentation with human subjects than has been typical in the past. In fact, it may be desirable to extend or develop regulations similar to the HEW regulations for the protection of human subjects to cover experiments with CCSs and other types of computerized information systems as well.

Conclusions

We are at a time in the evolution of CCSs when the widest possible range of experimentation with a diversity of different systems is called for. It would be unfortunate to see this area stifled by premature regulation and policies that limit the application potential of these systems.

Technically a computerized conference is a piece of software sitting in a single computer. It can be completely divorced from the issues associated with establishing common transmission networks for digital data. In fact, any information system can in principle be divorced from that issue, which is a valid one for an agency like the FCC. Such a transmission network can have hundreds of alternative computerized conferencing and information systems in many different computers plugged into it, and satisfy many different human groups with differing needs and problems.

Currently, major investment of capital in this area is being held up because potential investors are uncertain whether a stroke of a pen enacting a new regulation may wipe out their investment. This lack of announced policy or interpretation of existing laws is a hindrance to the experimentation that is needed. There is not the knowledge in existence today, nor will there be for a long time to come, to design a single or a few systems that could be

ISBN 0-201-03140-X, 0-201-03141-8 (pbk.)

considered the "best." It would be a sad mistake to establish a legal or de facto monopoly situation for this area. With current entrance fees of a half million dollars and prices continuing to fall, there is no rational justification for such monopolies or for closing the doors on the widest possible competition.

Our attention as a society must turn to considerations of public good and to changing our regulatory and policy mechanisms to fit today's technology. Trying to put today's technology into mechanisms evolved 30 and 40 years ago may result in the stillbirth of the baby. Today's generation of computerized conferencing, message systems, or whatever you call them, resembles future possibilities in this area about as much as the Wright brothers little biplane resembles the jumbo jet.

The most important impacts of this technology will result from new types of applications, rather than mere substitution for existing communications. The range of applications and designs to meet those applications is unforeseeable at this time.

Current names used for this technology may be inadvertently producing incorrect images of these systems for policy and decision makers as well as managers.

Who shall have access to CCSs?

Will they develop their full potential?

Will they evolve in a direction that supports individual choice and humanistic values, or one that curtails individual freedom and privacy?

The answers to these questions lies not with technology, but with policy and regulations, or lack of it.

What is needed is a more general approach to the whole area of information–communications systems and a period of innovation, proliferation, experimentation, development, evaluation, and assessment, during which maximum incentives operate to create a diversity of these systems.

For Discussion

1. Conceptualize the single CCS that would be offered by the following organizations if they respectively had a monopoly: The Postal Service; AT&T; IBM.

2. If the FCC were to assume regulatory powers for this area, what would be the likely consequences?

3. What sort of policy formulations or regulatory agency would you set up to handle information systems?

ISBN 0-201-03140-X, 0-201-03141-8 (pbk.)

4. What would you hypothesize as the most likely outcome of the current situation, based on current trends?

5. What would be some useful experiments in this area to provide data specifically for policy formulation?

6. What value changes would you predict for people using these systems?

7. Provide some examples other than education where the impact of CCSs would be accumulative over long time spans—three years or more.

8. What are some potential prescriptive impacts of CCSs? Consider cases where the system provides data on the communications behavior of users.

9. What types of professional expertise should be available to aid policy formulation or regulatory bodies for this technology?

10. What type of CC system *should* (normative view) the Postal Service provide for citizen use? Or, if you think that they should stay out of this area, why?

11. What would be the applications resulting from allowing the radio transmission or broadcasting of digital data by CB'ers, hams, and organizations?

12. How would you approach the determination of public interest in this area?

13. What name would you use for what we call computerized conferencing? If you were offering a service to the public, business, or banks, what respective names would you use for your particular service?

14. What would the CC equivalent of CB radio look like? What uses would people make of it?

15. What actions or policies would generate, in the current situation, a massive private–sector investment in this technology for public use?

16. What should be the ethical guidelines for professionals or others involved in developing, experimenting with, and offering these systems?

17. Determine some examples of commercially valuable selective mailing lists based on knowledge of the communication patterns of individuals?

18. What specific applications will be enhanced if authors can be assured ownership of information?

19. Develop a situation leading to major financial loss due to a lost piece of communication. Where would ultimate responsibility lie if it were a hardware error, undetected by the software? The possibilities are the organization providing the service, the company that produced the software, or the hardware manufacturer.

20. What types of new insurance policies will evolve to protect people and institutions in this area?

ISBN 0-201-03140-X, 0-201-03141-8 (pbk.)

21. Can you give a legal or regulatory definition that would separate communication systems from information systems? Where would computerized conferencing fall?

22. Give specific examples of situations where the free flow of information across national boundaries can give one country a significant advantage over another.

ISBN 0-201-03140-X, 0-201-03141-8 (pbk.)

THE BOSWASH TIMES

The Computerized News Summary Service of the Megalopolis

Nobels Swept by CC Groups

Stockholm, Sweden, July 14, 1995. A group of 57 social and information scientists today shared the Nobel Prize in economics, while 43 physicists and scholars in other disciplines captured the prize in physics. Many groups of scientists who worked cooperatively through a computer network have won the prize previously; this is the first time all prizes have been won by such electronically linked groups, or by such large and interdisciplinary groups.

The physics group, comprising a mixture of physicists, mathematicians, social scientists, philosophers, and computer scientists, won their prize for the now classic "Theory of Elementary Communication Structures." Through the development of an isomorphism that adapts the theory of elementary particles to human communication structures, this thesis allowed the development of a whole new branch of physics: the physics of human systems.

On a pragmatic basis the standardized measures that this theory has generated for computerized conferencing now allow an individual to be highly selective in terms of the specific conference he or she wishes to choose. A person can now indicate the charm, strange-ness, color, and parity, for example, of the group he or she wishes to join. In the business world this same theory allows fairly accurate predictions of mean lifetime, stability, fusion and fission behavior, etc. of conferences. This allows fairly accurate estimates of potential resource expenditures.

The economics prize was awarded for "The Value Theory of Information." This was the key to the inclusion in GNP of a measurement of increases in knowledge or information available to and utilized by the general population, thus producing a GNP that more accurately determined the relative changes in productivity and well-being among nations.

Earlier this month, prizes in literature, biology, and chemistry also went to CC groups. When the first such collective prize was announced eight years ago, the committee tried to convince the group involved to name the two or three of its members who were most responsible for the theory developed. However, the group insisted that this was impossible. Dr. Andrea Turoff, spokesperson for the collective, explained "We were engaged in what we call a 'synologue'—a process in which the synthesis of the dialogue stimulated by the group

ISBN 0-201-03140-X, 0-201-03141-8 (pbk.)

process creates something that would not be possible otherwise. Each of us made unique contributions to the final theory." Since that time, it has become commonplace for groups working on the frontiers of scientific knowledge to pool their findings and collectively create and present a synthesis to the scientific community as a whole.

There was no Peace Prize awarded this year. It is rumored that the Nobel Prize Committee will not award a Peace Prize until someone or some group achieves a solution to the "information war" that has escalated among nations in recent years.

Another debate that is said to have engaged the committee this year concerns the traditional disciplinary designations or distinctions for each of several prizes in an era when most significant scientific breakthroughs represent interdisciplinary efforts.

It is rumored that they have filed a court case asking for reinterpretation of Nobel's will. Beginning next year there could be simply a number of Nobel Prizes in science, without any attempt to categorize or classify the winning efforts according to the outmoded disciplinary distinctions in vogue at the time that Alfred Nobel initially established the prize.

AFL-CIO Disintegrating

Spokane, Washington, July 14, 1995. The American Association of Plumbers has recently been formed by plumbers who have separated from the AFL–CIO. The plumbers claim their trade is no longer well represented by unions dominated by white collar workers and it is now necessary to go back to the meaningful concept of a craft association to adequately represent the ideals of their trade.

This is one of a series of similar moves made in recent years by various blue collar occupational groups, widely interpreted as an attempt to regain the militancy and cohesiveness that characterized the "old style" unions of the mid-twentieth century. Analysts have attributed the decline of the labor movement to the growing phenomenon of individual workers employed at their own home terminals, rather than in a common location.

CC System Named in Divorce Suit

English Plains, NJ, July 13, 1995. In a bizarre divorce suit, a computerized conferencing system has been named as co-respondent. Mr. H. Ills contends that his wife's infatuation with a CC system has resulted in mental cruelty, alienation of affection, and loss of services. "She rushes right past me when she comes home from work", he complained to a reporter, "and turns on to that computer terminal. Every night, she indulges in hours of wanton interaction with that machine, right before my eyes. Meanwhile, the children and I are neglected."

Ms. Ills pleaded "no contest", and asked for custody of the terminal and maintenance payments.

ISBN 0-201-03140-X, 0-201-03141-8 (pbk.)

Electronic Hermits?

Cambridge, Mass., July 14, 1995. A Harvard professor of psychology, writing a graduation address for the mid-summer "class" completing its computer-based instruction, has sounded a warning note about the long-term effects of teleconferencing dependence on the American character. "We are seeing the emergence of a new kind of super-self-sufficient hermit," he warned. "Among those of you receiving this message today, several have told me that during the entire two and a half years that you have been engaged in completing the work of the accelerated curriculum of the electronic college, you have not left your home except to answer the door for delivery persons bringing goods you have ordered. What will happen to our society if its intellectual leaders completely isolate themselves from the segments of the society who do not use computer-based systems? What will happen to our children if they never see anyone but the members of their immediate household? I shudder to think that emotions and relationships as we have known them are likely to disappear."

ISBN 0-201-03140-X, 0-201-03141-8 (pbk.)

466

you constant advice; and help build your conscience by establishing guilt feelings.

2. The *Nagging Wife* or *Perfectionist Husband*: Have you been tempted to try matrimony? Experience the negative aspects of the relationship before making a permanent decision. Our pseudo-spouses are endlessly demanding of your time, attention, and affection. We guarantee that nothing you can do can satisfy them or make them happy for more than a short time.

3. *Borrow A Brat*: Before making an irrevocable decision, try the trial and tribulations of parenthood. We specialize in rebellious teenagers.

Lost Arts, Inc.

Classes now forming for in-person, face-to-face instruction in "The Lost Art of Handwriting" and "The Lost Art of Public Speaking." Reply to Box 1096.

Role Exchanges Incorporated

Tired of your role in our information-rich society? Seeking diversity of work and social experience? Looking for fresh new ideas and concepts? We broker role exchanges on the strictest of confidential bases. All parties are screened for the accuracy of documentation of their roles and we are fully insured to guarantee authenticity. The latest in computer-based writing stylizers are available to ensure that your communications cannot be distinguished from those of your exchangee. A complete line of roles available in business, industrial, governmental, family, ethnic, religious, and other sociological groupings. We cannot offer you the Pope or President, but let us say we have opportunities that can be as gratifying. Reply to Box 1099.

ISBN 0-201-03140-X, 0-201-03141-8 (pbk.)

CHAPTER 13

Societal Impacts of Computerized Conferencing

In Part II of this book we discussed the potential applications and impacts of computerized conferencing within specific institutional sectors of society—the management of complex organizations, social services, public participation, scientific and technical information exchange, and research. In Part III, we have reviewed the kinds of technological, intellectual, economic, and political developments that will determine the extent and areas of application of computer communications systems in the near future. In this concluding chapter, we ask the overall "what if" questions: What if technological and political policy developments proceed so as to produce the widespread application of the kinds of uses for computerized conferencing systems that are described in Part II? What kinds of effects, both good and bad, would this application have on the society as a whole?

Among the developments that we foresee implemented by 1990 are the following:

- Computer-based office automation systems that facilitate work at home or at decentralized neighborhood centers.
- The replacement of the current systems for the storage and retrieval of knowledge (consisting of periodicals, books, libraries, separate data bases, and abstract services on various computers, etc.) with an integrated, on-line system of abstract and bibliographic services, text, and data bases.
- Computer-assisted instruction routines incorporating discussion or conferencing capabilities between student and teacher, or among the members of a class, facilitating life-long learning from the home.
- Electronic funds transfer and shop-at-home services making it possible to search for, order, and pay for most goods and services from the home terminal.

Starr Roxanne Hiltz and Murray Turoff, The Network Nation: Human Communication via Computer

ISBN 0-201-03140-X, 0-201-03141-8 (pbk.)

- The implementation of new technologies, such as satellite communication and portable telephones, which will connect the entire world into one telecommunications-linked network, making a computer-based information and communications system accessible at low cost from any location.
- Inexpensive, high-quality home terminals and personal computers.
- Participatory democracy, with discussions of public issues and polls on public opinion allowing interested citizens constant and effective input into the political decision-making process on the local, state, national, and supra-national levels.
- Replacement of printed newspapers by constantly updated on-line news and classified advertising services, such as the *Boswash Times*.
- Emergence and sustenance of family, friendship, professional, and leisure-oriented networks through computerized conferences.

What Constitutes a Communications Revolution?

The implementation of such systems of human communication through computers will result in a "communications revolution" that will have profound effects on societies. Certainly, it would constitute a revolution in terms of the criteria set forth by futurist Gordon Thompson (1972).

- It will ease the access to stored human experience.
- It will increase the size of the *common communication space* shared by the communicants, and therefore increase the amount of shared information and interaction among them.
- It will increase the ease with which new ideas can be developed and spread throughout a society.

As proposed by Thompson, these predictions represent three highly aggregated measures, qualitative in nature, that in their current formulation allow only a rank ordering or subjective estimation. If, however, we follow Thompson's rationale and add computerized conferencing to his original considerations of books, TV, and telephone, the results are somewhat startling.

Ease of Access to Stored Human Experience

The first dimension is what is termed "the ease of access to stored human experience." Every significant communications revolution has increased the ease of access: examples are the phonetic alphabet, the printing press, and the telephone. The proposed test of significance for whether a change has occurred is whether or not the "advance," change, or alternative results in modifications in the way people index information for retrieval. According to

ISBN 0-201-03140-X, 0-201-03141-8 (pbk.)

Thompson, the book rates highest in this dimension (individual books have tables of contents and indexes; libraries use card catalogues and Library of Congress or other classification schemes), followed by the telephone, and much lower down, TV. Note that the telephone allows us to contact a person to obtain information, as opposed to going to and searching a library for that information, whereas TV or other broadcast mechanisms do not provide the selectivity of books or telephone. The computer, in terms of information retrieval systems, would seem to have the potential for much higher retrieval rating than even books. If we integrate conferencing with information systems, we would be able to combine retrieval systems and expert knowledge. Therefore an extrapolation on Thompson's rationale leads us to the following ordering of access from highest to lowest.

1. Combinations of computerized conferencing and computerized information retrieval systems;
2. Computerized information retrieval system only;
3. Computerized conferencing systems only (assuming they include keywords and search abilities for retrieval);
4. Books and physical libraries;
5. Telephone;
6. Face-to-face meetings;
7. Television.

Size of the Shared Information Space

The second dimension proposed is "the size of the common information space shared by the communicants." This includes the abilities to modify or contribute to the communications taking place and to interrupt one another.

The telephone permitted sharing of an acoustic space and face-to-face meetings permit sharing of both visual and acoustical space. The introduction of the book permitted a termendous increase in the sharing of ideas and information across space and time. On the other hand, it is hypothesized that the Picturephone failed because it did not allow a shared visual space; a person could not point to something on the screen and have the other person see it.

A computerized conference provides a completely shareable writing space for larger groups than either face-to-face or telephone exchanges. The ability to share that space in many different ways (e.g., new contributions, editing, footnotes, voting) dramatically increases the range of strategies available for the communicants. Before computerized conferencing, there never existed a truly shared writing space where all participants could be reading or writing

ISBN 0-201-03140-X, 0-201-03141-8 (pbk.)

simultaneously. Our ordering here would have to be:

1. Computerized conferencing;
2. Face to face;
3. Telephone;
4. Information retrieval systems;
5. Television and books.

Ease of Discovery and Development of Nascent Consensus

The third dimension, the "ease of consensus discovery and development," involves the size of the audience and/or how easily new ideas can be propagated throughout the society. Obviously TV would be very high on this dimension, followed by newspapers, large public meetings, books, and telephone.

According to Skinner, the evolution of a society relates to the increase of a culture's sensitivity to the remote consequences of its acts. TV, in this context, has had dramatic impact on American society. Were computerized conferencing to become a medium of mass communication, as we predict, then this dimension leads us to ask: Can systems be designed that will capitalize on the selectivity the computer provides to convey highly pertinent information to targeted subgroups, showing them how they are similar to or different from other groups or individuals?

One example relates back to our previous discussion on the ease of forming lobbying groups. The test of significance for this dimension proposed by Thompson is particularly apropos; the innovation "must increase the probability of transmitting or receiving an interesting but unexpected message." This is exactly what we expect future generations of CCSs to do by capitalizing on the system's inherent computation abilities. Jane Jacobs, in *The Death and Life of American Cities*, pointed out that wide sidewalks are a medium of communication that permit interesting and unexpected messages to be exchanged at a low level of commitment. The rural party line may be likened to an electronic sidewalk. Computerized conferencing may provide electronic central plazas, reminiscent of Italian cities on a Sunday afternoon. (Other analogies might be the Viennese cafe and the French salon.) On this dimension, and assuming our usual degree of optimism with respect to overcoming design and policy problems, we would predict a rating in terms of future potentials as:

1. Computerized conferencing;
2. Television;
3. Information retrieval systems;

ISBN 0-201-03140-X, 0-201-03141-8 (pbk.)

4. Books;
5. Face-to-face encounters;
6. Telephone.

Obviously Thompson's dimensions are not exhaustive. Another approach (Wish, 1974a, b; Wish and Kaplan, 1975) is to derive, via multidimensional scaling, the hidden dimensions that people are unconsciously using to rate communication situations.

The five dimensions that fall out are

1. Hostile versus friendly, or "cooperative versus competitive";
2. "Superficial versus intense" or uninterested and uninvolved versus completely engrossed;
3. "Very different roles versus very similar roles," "each treats other as an equal versus one totally dominates the other," and "autocratic versus democratic";
4. "Informal versus formal" and "reserved and cautious versus frank and open";
5. "Not at all task oriented versus entirely task oriented" and "unproductive versus productive."

If we actually mapped computerized conferencing into this five-dimensional space, the potential range over which it is applicable or useful would occupy a much larger five-dimensional volume in this space than any other mechanism, mode, or medium of communication.

Whatever morphology is chosen to evaluate the potential impact of computer-based communication–information systems, it is highly likely that they will appear at least as important and "revolutionary" as the book, the telephone, or television.

The telecommunications industry itself is generally concerned only with marketing services that can yield a profit, not with examining the social effects of services that it may provide. As Thompson points out (1977, p. 1):

> We, in the telecommunications industry, have been negligent in not assessing adequately the effect of our actions, or innovations . . . .
>
> Previous communications innovations, that have been recognized as being revolutionary, such as the phonetic alphabet, produced extensive impacts that were far reaching and wide ranging throughout the host societies. Havelock, 1963, reports on the profound effects of the adoption of the phonetic alphabet in Athens. These effects were very extensive, in that they penetrated far beyond merely increasing the efficiency with which records could be kept.

The fundamental effect of computerized conferencing, we believe, will be to produce new kinds of and more numerous social networks than ever before possible. Along with this will come massive shifts in the nature of the values and institutions that characterize the society. Hence we next consider some

ISBN 0-201-03140-X, 0-201-03141-8 (pbk.)

fundamental social impacts that we expect as a result of this innovation, including changes in values, social structure, and economic institutions; changes in sex roles, transportation, and ultimately the shape of urban areas themselves; and dangers, such as potential infringements on privacy, surveillance, and the use/abuse of "social engineering" through the new computer-telecommunications technologies.

In trying to make these predictions and projections we are leaving the realm of simple generalization of applications and impacts of existing systems and applications. As Joseph Coates notes:

> A major new technological invention generally passes through three stages. The first stage is substitution for a previous function or activity. In the second stage, one finds what I shall call the "back fill": The institution or the system in which the technology is embedded begins to evolve and change to utilize more efficiently the new technological capability. Finally the point is reached where the new technology has permeated the institution or society, as the case may be, to the extent that new undreamed-of uses for the technology can be effectively tried [Coates, 1977a, p. 202].

In Chapter 10, we pointed out that the philosophy behind EIES was adaptability of a computer-mediated communications system to the tasks and tastes of its users. It was pointed out that current CCer's almost immediately begin to imagine and innovate uses for the system that were not foreseen by its designers. A CCS as a computer-linked human network becomes an organic, growing thing, almost with a life and vitality of its own. Its current users feel that they cannot foresee its ultimate shape and capabilities; that the current systems are indeed just the embryo, the promise of growth and diversification to come.

These systems can be seen as social systems, formed and shaped by their members, with the resultant structure subsequently influencing the human relationships within it; or as an instance of people–computer symbiosis that grows and changes as its members find new ways to use it to enable them to work together across space and time.

Given the undreamed-of uses and the organic nature of the systems themselves, we cannot claim any precision for the nature and scope of the social impacts that will stem from such uses. What this chapter represents, therefore, is a projection of the general shape of things to come, based on the history and the current characteristics of a growing and changing technology.

Communicating Rather than Commuting to Work

Many organizations already have most of their records computerized, with data and programs entered remotely on terminals and the work actually executed at a central computer installation. There is thus no reason why

ISBN 0-201-03140-X, 0-201-03141-8 (pbk.)

employees using such systems could not do most of their work from terminals located in their homes or at neighborhood work centers.

Saving Time and Energy

Nilles et al. (1976b) calculate that as of 1970, 38 million white collar workers, or 48.3% of the work force, were "information industry" workers who would be candidates for "telecommuting to work," and that the "exponential growth" of this portion of the work force suggests that soon a majority of all workers will be in this category (pp. 9–10). As they put it:

> Given the capability of modern telecommunications and computer technologies to efficiently produce, transmit, and store information, it appears probable that many information industry workers could "telecommute." That is, they could perform their work, using communications and computer technologies, at locations much closer to their homes than is the case now. If telecommuting could be shown to be feasible, an alternative to commuting would be available to a significant portion of the central business district labor force [Nilles et al., 1976b, p. 4].

The possible advantage of this would be substantial savings in energy and time resources, and potentially an increase in the time that members of the family unit have to spend with one another. Possible problems are the attitudes of employers and employees toward making this kind of shift, and strains introduced into both the family and the work units as a result of such a massive relocation of work.

Even if only 15% or 20% of current urban commuting were replaced by telecommuting, time would be saved not only by the individuals who no longer have to spend one to three hours a day sitting in their automobiles, but also by the remaining commuters, as traffic thinned down during the rush hour. Nilles et al. (1976a) have calculated and presented the following set of observations:

1. Ninety-seven percent of urban passenger traffic uses the automobile; of this, over 40% of the trips are made by commuters, for an average round trip of 21.6 miles.
2. The ratio between energy consumption for the average automobile commutation and the delivered electrical energy for daily use of a telecommunication system such as computerized conferencing, which uses telephone bandwidths, is about 60 to one. For each 1% of the urban work force that replaced automobile commutation with telecommuting, there would be a net reduction in U.S. gasoline consumption of about 5.36 million barrels annually.

Note that unlike many energy-saving measures, this reduction would be achieved simultaneously with what most people would consider an improve-

ISBN 0-201-03140-X, 0-201-03141-8 (pbk.)

ment in the "quality of life" (namely, less time spent in traffic jams), rather than a decrease. In addition, the other fringe benefits of such a substitution also are very desirable: less air pollution from automobiles; less energy and land expended on an ever-increasing network of roads and parking lots; a probable slowdown in the rate of purchase of automobiles, as fewer families need two cars and as the decrease of car-punishing commutation increases the life of automobiles. The last factor is in itself quite significant, since on the average it takes about as much energy to manufacture an automobile as will be burned up in gas and oil during a whole year of driving.

The extent of such time and energy savings depends on the exact modes and forms in which telecommunication is substituted for travel. For example, Nilles et al. (1976b) have distinguished four types of organizational dispersion patterns.

The *centralized* organization, which is the current dominant mode, has a single large headquarters, either in the central business district or in an outlying suburb, and everyone must commute there. (It is interesting that in the latter case, upper-level executives can usually afford to live nearby, and thus have less commuting time than would be needed to travel to the central city, whereas clerical and lower-level employees probably have their driving distances greatly increased as compared to a central city commutation, and are forced to use energy-intensive automobiles rather than mass transport.)

In the *fragmentation* pattern, a few "coherent subunits" in the organization break off and are located elsewhere. (This seems to be very common for the computer departments. There is absolutely no need for them to be physically located in the central office, since most operations of data entry and retrieval are done over terminals, or in remote batch.*) This fragmentation generally results in an overall increase in the amount of commuting.

In *dispersion*, a number of work centers are located around the urban area, and an employee, regardless of function, reports to the nearest one, with telecommunications and computers tying the functional units together. This results in considerable commutation savings, on the average.

In *diffusion*, small multicompany neighborhood work centers and work at home are the norm, and most people can walk or bicycle to work (see Figure 13-1).

It is probable that only professionals with private studies in their homes can really choose the work-at-home option. The neighborhood work center would have audio and video conferencing facilities as well as terminals and computerized conferencing systems, and secretaries to help with routine entry and retrieval of information.

---

*In remote batch computer operations, a job may be fed into a card reader at a location remote from a time-shared computer, executed at some later time, and the results eventually printed out at the remote location.

ISBN 0-201-03140-X, 0-201-03141-8 (pbk.)

Figure 13-1 Patterns of decentralization (adapted from Nilles et al., 1976b, p. 12)

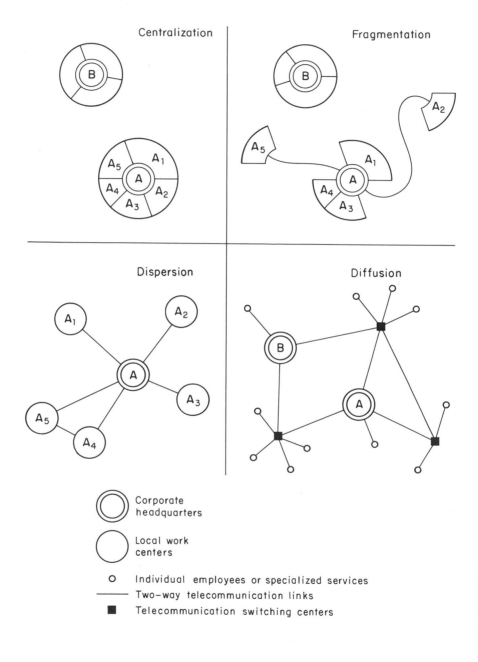

Centralization

Fragmentation

Dispersion

Diffusion

⊚ Corporate headquarters

○ Local work centers

o Individual employees or specialized services

—— Two-way telecommunication links

■ Telecommunication switching centers

ISBN 0-201-03140-X, 0-201-03141-8 (pbk.)

Either the dispersion or the diffusion model is compatible with the "new cities" or "new rural society" idea of planned communities of medium density, in which park and recreational facilities are interspersed with living, working, and commercial and cultural facilities (see, for instance, Goldmark, 1973). Although this arrangement seems attractive at first glance, there are possible accompaniments that could create problems. For instance, what is going to happen to the already decaying central cities if more and more corporations abandon central business district offices in favor of dispersed organizations tied together by computer communications networks? It is possible that they might be "renewed" or "rerurified," and that neighborhood work centers would be set up in ghetto and minority areas as well as in suburban and semirural areas, but not unless strong policy guidelines or financial support from the government encourages corporations to do so.

What if work at home does encourage people to move out to lovely wild, remote areas? They will not remain lovely and wild and remote very long.

In addition to "telecommuting" to work on some or all days, a significant impact on travel can be made by substituting for some of the intracity and intercity business trips that are made for short meetings. Studies by the Communications Studies Group and by Bell Canada indicate that approximately 2% of national energy consumption could be saved by substitution for a moderate amount of intercity travel to meetings (Day, 1977, p. 10). For this purpose, the technology is here now—the problem is to change attitudes. For instance, Day (1974a) reports that only one in five of Canadian travelers interviewed felt that they would have been willing to substitute telecommunications for their trip. A lot of this has to do with unfamiliarity, probably, but there are also many gratifying "fringe benefits" to traveling to a meeting, even within the same city—usually an expense account lunch, for instance, and a chance for a change of routine. Organizations would do well to serve festive banquets and drinks at meetings conducted by telecommunication, to try to make them just as much fun!

One problem standing in the way of dispersion or diffusion patterns is that organizations have little economic incentive to adopt such patterns. Currently, the costs of commutation to work are borne by the individual worker (who pays for the autos, gas, insurance, or mass transport tickets) and government (which builds the highways and heavily subsidizes mass transport). In order to switch to telecommutation, organizations would have to make considerable investments in computer terminals and other telecommunications equipment, at their expense. Though there are societal benefits that can potentially result (less energy consumption, less pollution, fewer deaths from automobile accidents), the individual firm does not necessarily benefit economically in the short run, unless the alternative is to build or expand new facilities, which are very expensive in the central business district.

ISBN 0-201-03140-X, 0-201-03141-8 (pbk.)

Table 13-1
Attitudes of Los Angeles Residents Toward Commuting [a, b]

Commuting is a necessary evil.			
Strongly agree	37	Mildly disagree	4
Agree	25	Disagree	16
Mildly agree	6	Strongly disagree	13

Commuting is a useful interlude between home and work.			
Strongly agree	19	Mildly disagree	5
Agree	24	Disagree	14
Mildly agree	16	Strongly disagree	21

I spend too much time commuting.			
Strongly agree	18	Mildly disagree	14
Agree	10	Disagree	26
Mildly agree	4	Strongly disagree	27

[a] From a systematic, random-sample survey of 200 adults. Results are percentages. Copy of questionnaires and detailed tabulations of results provided by Nilles et al. at the request of the authors.
[b] "Commuting" is defined as traveling back and forth to work.

Some sort of subsidy or incentive would seem to be necessary to encourage firms to make investments in telecommunication facilities that would support work at home.

Another problem is that not all employees dislike commuting as an activity. Nilles et al. conducted an areawide random probability sample of Los Angeles County adult residents by telephone, with 197 respondents. Some of their results are shown in Table 13-1. Over one third seem to like commuting.

Short-term economic incentives also stand in the way of substituting teleconferencing for intercity travel to meetings. The main source of "savings" is savings in time that people would otherwise expend in traveling. However, as Day (1977, p. 8) points out:

> The problem with these forms of calculations (of cost/benefit ratios) is that the institution has to spend more money on telecommunications systems in order to optimize the existing expenditures in salary charges. Illogical or not, many managers would rather have employees under-utilized than spend more money to optimize a "sunk" cost, namely salary.

ISBN 0-201-03140-X, 0-201-03141-8 (pbk.)

Effects on Productivity, Morale, and Family Relationships

One problem that mangers might foresee is that they will have less opportunity to supervise and physically oversee their workers who are at neighborhood work centers or at home. How will they know their employees are actually working, and not goofing off? Conceptually, it is easy enough to design measures of the quantity and quality of work performed over a remote terminal for routine clerical operations. The volume of transactions or operations completed can be monitored, and quality can be tested by, for instance, using a random sampling technique to generate test transactions for which there is a known "correct" series of operations.

How this will be viewed by the employee is questionable. On the one hand, the employees may feel spied upon, and the managers may feel that their function is being usurped by a computer. On the other hand, employees may feel that it gives them more autonomy, since they have objective evidence as to the amount and quality of their work, and are not subject to the perhaps unfair impressions of their managers. (See Nilles et al., 1977b, pp. 154–155 for a further discussion of these issues.)

Another problem is that employee morale and work-group cohesion may depend on the duration and number of face-to-face contacts, which decrease the further we go along the continuum from single centralized large offices to work at home. Whether "electronic socializing" or substitution of socialization with friends and neighbors rather than with co-workers will occur, and whether they will be adequate substitutes to maintain the interpersonal supports that people need, is an unanswered question.

Work at home or at neighborhood work centers is especially likely to be related to an increased number of part-time jobs. This situation is bound to have some negative impact on the labor movement. How about Social Security, health coverage, and other kinds of fringe benefits that are not usually paid to part-time workers? We would need to provide such coverage through the government rather than through private companies; require employers to cover part-time workers; or else see a much larger proportion of the population subject to disastrous decreases in their economic well-being due to conditions beyond their control. Such institutional impacts need to be foreseen and planned for.

What about family relationships, among husband, wife, children, and the extended family? Telecommuting would enable the parent responsible for child care to have a flexible schedule. Since this is usually the wife, it would mean that women could work without the constant crisis of what to do if the school closes for holidays or the child is sick or the baby-sitter does not come. Moreover, with the main wage earner working in or near the home, he or she can spend more time with other family members, and conceivably perform a

ISBN 0-201-03140-X, 0-201-03141-8 (pbk.)

greater share of the household maintenance tasks that are now mostly allocated to women. Spending more time together would increase interaction among family members and might strengthen their relationships. Certainly, working at home or at neighborhood centers would increase the amount of time the family could spend together. This change will require new arrangements for avoiding conflict and strain if it is to have positive rather than negative impact on family stability.

Most futurists say an emphatic "No!" to this idyllic image of a new cottage industry. Larry Day, for instance (1974a, p. 22) says, "Having the working husband and perhaps working wife, plus children, pets, etc. in the same environment continuously would likely be more of a strain than most people could survive." Dave Snyder (EIES conference 70) observes that "experiences with flextime in Europe and in this country have revealed that the increased presence of the household head in the home has led, in most instances to increased domestic problems, including alcoholism, physical conflict, and divorce."

Men's Work or Women's Work?

A tremendous proportion of the women employed in white collar work today perform essentially clerical tasks of moving and processing information —as secretaries, bank tellers, telephone operators, clerks, etc. Males, on the other hand, tend to occupy the higher-status white collar roles having to do with the creation and analysis of information (the staff functions) and the making of decisions on the basis of available information (the managerial roles). The new information and communication technologies will have the effect of greatly decreasing the amount of purely clerical, routine tasks involved in information handling and processing, and of removing the distinctions between the mere processing of information and communications and their use *as information* to aid in carrying out the decisions and routine operations of an organization. Old occupational roles will be changed, new ones created.

The most probable direction of development, unless training programs are devised to forestall the tendency, is that new jobs related to the new technologies will be seen either as "women's work" (because they involve sitting at things that look like typewriter keyboards) or as "men's work" (because they involve an ability to interact creatively using a very high technology, the computer, which has been seen as an essentially male machine). Either of these developments would be unfortunate, since they would tend to limit the use of the technology and to further delimit occupational opportunities for one sex or the other.

Our own hope is that the potential of this medium for work at home and for making the sex of participants totally inconsequential will mean that it

ISBN 0-201-03140-X, 0-201-03141-8 (pbk.)

will lead to a reorganization of occupational and household sex roles. However, it may also reinforce current distinctions—or paradoxically, reverse them:

> Sexism and CC—A Hunch:
> One possible impact—
>> premised on—
>> (a) People will continue to use CC increasingly at home.
>> (b) Men are increasingly over-represented in top managerial and professional jobs—those that will use CC (this over-representation is a statistical fact, before everyone starts telling about all the women managers they know—and it's increasing).
>> (c) Hence, women will continue to go out to work in low-level white collar and blue collar jobs; and the proportion of married women in such jobs is increasing.
>
> The payoff—
>
> In such families—man staying home, working via his terminal and woman out to work, we have some structural conditions for partial role inversion: will the man now be expected to do the wash, the dishes, repot the flowers, wait for the plumber, since he's at home anyway? [Barry Wellman, 4/19/77, EIES conference 72, comment 206.]

Larger, More Fluid Social Networks

The impact of the technological progress of transportation over the past century has been a significant increase in physical mobility and an erosion of traditional social networks such as families and ethnic and religious subcultures. The replacements for these affiliations have often been less emotionally satisfying, since they are functionally oriented (e.g., business groups, garden clubs, country clubs, fraternities) and are usually at a much more superficial level.

Because of the factors of geographic mobility and commutation to work, there has also been less overlapping of social ties among the various groups a person belongs to (less "cross-cutting of status sets," to use Robert Merton's phrase). Such social fragmentation leads to lack of social integration.

Now, even these less deep social ties are threatened with diminution, as travel becomes more difficult and expensive. We believe that computerized conferencing can serve to substitute electronic mobility for physical mobility, and permit a person to exchange communications on a fairly deep and meaningful level with many different interest-oriented groups. At the same time, there is likely to be considerable connectedness among the various groups, with many overlapping memberships.

ISBN 0-201-03140-X, 0-20 -03141-8 (pbk.)

Socially, a CC network can make it possible to maintain a larger number of ties to groups that correspond to a diversity of roles or occupational specialties or avocational interests. There would be no need to find local employment in your area of expertise, or enough people locally to form a special-interest group related to ecological activism or ornithological pursuits. The electronic network would form the common social interaction or work space for bringing persons together, wherever they are. At the same time, the cross-cutting nature of memberships in groups on a CC network would tend to prevent the isolation and segmentation that occur in modern societies when extreme specialization becomes the basis for the organization of work and play.

Such a scenario for the emergence of new forms of social and occupational and economic relationships depends on the assumption that the written word as it appears on one's terminal can indeed become the basis for forming and maintaining "electronic group life." That it is possible for people to "get to know one another" through their computer, to develop emotional and social ties, and to develop norms and rituals, has been illustrated in Chapter 3. Of course, groups formed and sustained by a CCS need not use only that means of communication; there could be telephone calls, and face-to-face meetings or video conferences or Picturephone calls to supplement the computer-mediated ties.

Whether or not computer-mediated human networks will have the effect of fragmenting the society into a more integrated middle and upper class that has more access to information and to the decision-making processes of the society, and a computer-powerless underclass that is cut off from this new resource, or the effect of decreasing inequality by bringing all segments of the society into information and communication networks will depend on the policies that are followed. We agree with Burt Cowlan's assessment at the World Future Society Conference, "Communications and Society: Policies for the Plannable Future" (New York, November 2, 1977):

> As the "44" was once the equalizer of the Old West, the telephone, the satellite and the computer are the equalizer not only of the New West, but of the Old East . . . . Adequate resources should be made available to all mankind for interactive communication.

The Electronic Tribe

Human organizations evolve to satisfy various physical, emotional, and social needs of individuals, groups, and society. These organizations are dependent on the technology available for supporting them. We can conjecture, for instance, that the automobile and telephone have allowed the formation of more intense peer grouping among teenagers, and have tended

ISBN 0-201-03140-X, 0-201-03141-8 (pbk.)

to reduce the prominence of the family as the basic unit in our society. A number of trends and circumstances are now emerging that at least allow us to hypothesize what the long-term impact of computerized conferencing might be on human groupings.

The first trend is to a society that must contend with physical resource limits and constraints. Energy, raw materials, and environmental impacts of pollution all represent ultimate constraints on the "wealth" that society can distribute among its members. The result of this trend over the long term will be either greater class distinctions or a pronounced change of values toward a greater sharing of the wealth available. For example, instead of everyone having her or his own car, a situation may result where cooperative ownership exists of such an item as a car or camper by a group of people. In other words, resource constraints will gradually force more efficient use of the resources we have and bring about new forms of ownership and sharing.

We are all familiar with the concept of the cooperative, which to date has been applied in such areas as farming and financing, where a large number of individuals having similar wealth can pool their resources. At present the principle of the cooperative needs a rather large scale to be economical when it deals with a single resource.

Computerized conferencing has two properties that could change this situation. The number of persons an individual can maintain active and close communications with on a CCS is between three and ten times that possible with current communications technology. The logical processing available in the computer makes it possible to account for and organize transactions involving very different goods and services on a very efficient basis without needing the scales that current cooperatives need. As a result, we can foresee the emergence of a new "tribal" form of economic organization involving groups of people by the hundreds who have banded together to share resources. Current homogeneous groupings, such as doctors or lawyers, who often waive fees for services to others in their local organization, will now be replaced by groups of a heterogeneous nature: a lawyer, a camper owner, a plumber, a gardener, a housekeeper, etc. As a result, if the housekeeper spends a day cleaning the house of the lawyer, who in turn writes a will for the camper owner, the housekeeper might then be entitled to utilize the camper for some equivalent time. The CC capability can easily be designed to allow for the finding and negotiation of such transactions and the processing of a generalized settlement function that provides appropriate credits and debits to the individuals involved.

The emergence of such public groupings on a local basis has many of the characteristics of the purposes of a tribal organization in primitive societies where each member supplies his or her talents and/or wealth to the collective. The computer eliminates the need for a bureaucracy or higher-level

ISBN 0-201-03140-X, 0-201-03141-8 (pbk.)

organization to carry out the process. Obviously any such practice on a wide scale will lead to dramatic changes in what we "mean" in the society by the exchange of goods and service. Are these exchanges taxable? Are the current laws on cooperatives, ownership, etc., compatible with the emergence of the "electronic tribe"?

The concept of the electronic tribe is similar to the idea of letting families incorporate in order that they may operate on a more efficient and equitable basis within a free-enterprise society. Whether we view this grouping as an instance of the "extended family," or free enterprise extended to citizen cooperatives, or a form of "local socialism" seems a moot point concerned more with semantics than substance. The pressures of growing resource constraints in the decades ahead are going to cause the emergence of considerable individual resource sharing and the growth of the communication–information systems to allow for this joint use. To what extent our values are modified and our laws and regulations adapt to the situation in either an anticipatory or reactive manner remains to be seen. The trends already exist in homogeneous organizations in legal service plans, and in terms of informal sharing of tractors and leaf mulchers, and well as car pooling, in the suburban environment. With all the emphasis placed on car pooling, for example, it is amusing that governments have yet to change the laws governing the ownership of automobiles to really encourage this form of resource sharing.

Future Fads, Foibles, Fallacies, Frauds, and Frolics

A rich diversity of subcultures will be fostered by computer-based communication systems. Social, political, and technical change produce the conditions likely to lead to the formation of groups with their own distinctive sets of values, activities, language, and dress. In recent times we have seen flappers, beatniks, hippies, jet setters, grey flannel suiters, hot rodders, swingers, and numerous other examples of groups that deviate from the societal norms. We suspect that any widespread use of computerized conferencing will cause the formation of distinctive subcultures, just as phenomena like the automobile and the Viet Nam War did. In this section, we are indulging in speculation about the sorts of subcultural forms and value shifts that might evolve as a result of computerized conferencing.

INVISIBLE WORKERS

Those who work through their terminals may emerge as a new class, sharing the condition of being physically invisible to others. Will some forms of work relegate these "information workers" to the role of cognitive cogs in

ISBN 0-201-03140-X, 0-201-03141-8 (pbk.)

the "information machine"? The computer can impose all sorts of "piece-work" and time and motion measures on performance, as is now done in some data-entry operations. Individuals in this type of environment can appear to others as some new type of computer element—the person in the machine. The usual human relationships would disappear, to be replaced by such feelings as that firing the cognitive cog is the equivalent of changing the tube in an old radio. Orders would be given to him/her/it with as much consideration as is involved in instructing a computer. We hope it will go the other way—that information workers will be freed from the constraints of the time clock and enabled to evolve as generalists, as opposed to assembly line specialists. However, it is more likely we will see a mixture of the two. Once the novelty of computer terminals wears off, as happened with typewriters, doing work through them may become a sign of lower status. It may become a high-status activity to go to an office and confront fellow employees on a face-to-face basis.

STYLISTS

The theme of the stylists is, it is not what you write, but the style in which you write it that counts. We can foresee groups that develop elements of snobbishness and "class" based on styles of writing. What is the writing style equivalent to the Bostonian or Brooklyn accent? No doubt some will be invented. Groups will evolve unique styles and there will be less homogeniza-tion of the language as newspapers and magazines exert less influence on the use of the written word. As with any human endeavor, a sense of pride and class will develop for such groups. There are already a host of familiar style types whose proponents have no group identity because of the difficulties of finding one another. Certainly for recreational purposes we can expect to see the following types of groups forming on CC networks: punsters; four-letter-word users; quoters; word inventors; esoteric or obscure word users; long winded types; concise writers; outliners; long sentence makers; users of slang; abbreviators; etc.

Some of these group word games may outsrip the crossword puzzle as a form of relaxation. The stylists could develop a miniature "Tower of Babel" that could set lexicography back decades.

EXTROVERTED ANARCHISTS

Computerized conferencing makes available to any participant a much larger population for potential communication. There may thus develop the feeling that out of all those potential contacts there must be some who agree with and will appreciate a particular individual's views and values, however unusual or iconoclastic. In addition, there are those who derive a certain amount of pleasure from expressing shocking, deviant, or unusual points of

ISBN 0-201-03140-X, 0-201-03141-8 (pbk.)

view to others. This behavior may be particularly tempting when the medium of communication protects the person from any physical expressions of displeasure on the part of the recipient, and from having to view the recipient's tears and other nonverbal indicators of emotional distress. As a result, computerized conferencing might foster the emergence of subgroups that pride themselves on having unusual views and on "letting it all hang out," regardless of whom the views or language might offend. Such tendencies could lead to intellectual anarchy, or competitions to see who can produce the wildest or most outrageous statements.

AMATEUR SOCIAL SCIENTISTS

The capturing of a social or human communication process does offer the temptation to analyze what is taking place in the interaction among people in any group. As a result we are likely to see a flood of amateur sociologists, psychologists, and anthropologists. Just as it is now possible to buy books on body language that are supposed to enable the layperson to tell when the boss likes you or when the other person agrees with your intentions, there will emerge guidance for the layperson on how to interpret the same sort of things in the computerized conferencing process. Those areas of the social sciences related to human and group communication are likely to become as popular as home electronics or amateur painting are today, once it becomes possible to take an electronic excursion into the world of another subculture, all within the comfort and security of a person's own home.

Privacy, Freedom, and Self-Determination: Some Potentially Negative Effects of Human Communication via Computers

There is no doubt that widespread adoption of this new form of technology would cause some additional unemployment, at least on a temporary and sectoral basis, among persons who are employed in current transportation, communications, and information-handling jobs. On hearing about "electronic mail," for instance, the head of the Canadian Union of Postal Workers predicted that the "labor movement would not take sitting down the introduction of electronic mail to the exclusion of the normal postal system" (Dotto, 1977). Unless some provision is made for retraining, there will be much bitterness and unnecessary suffering among some members of a society who are displaced by the new technology.

There is also no doubt that reliance on this form of technology could make a nation much more susceptible to disruption from natural disasters or sabotage, since the network can be shut down by failures in electricity or

ISBN 0-201-03140-X, 0-201-03141-8 (pbk.)

telephone service, or destruction or damage to key computers. If tasks relevant to national security in time of emergency are to be carried on such a computer communications network, then considerable (and expensive) redundancy must be built in (backup generators; radio-satellite communications; backup data bases on backup computers).

The new computer communications technology offers an opportunity to enable society to deal more effectively and more humanely with its deviants. Instead of putting them in jail (for nonviolent crimes) or on probation (a rather ineffectual means of social control), their whereabouts, communications, and transactions can simply be fully and constantly monitored electronically, so that any deviance or "crime" will be immediately detected and the person can be apprehended. Coates (1977a, pp. 198–199) explains the potential and the danger as follows:

> It is now technically feasible and probably economical to do continuous real-time electronic surveillance on individuals outfitted with appropriate receivers and transmitters. This may permit real-time monitoring on the movements, activities, and locations of people on parole, marginally dangerous mental cases, etc. How can society better manage the deviant who attacks nine year old girls? It may cost $20,000 per year to keep him incarcerated. But it may cost only $10,000 over a lifetime to wire him up and let him move freely and safely because he is monitored . . . . This technique may enrich with freedom lives that could never be enriched under present penitentiary, custodial or medical supervision. Without a thorough anticipatory analysis of the implications of utilizing this capability, one could expect it to be rejected as unfeasible or, if implemented, abused as every other physical or chemical approach to mental illness has been.

This final observation is the crux of the most severe of the potential negative effects that could follow from the application of computer telecommunications technology. Just as every other communications medium has been manipulated and abused for political or economic motives, so may computer-mediated communication be abused. The danger is that this new technology is potentially a much more powerful means of monitoring and controlling behavior.

Effects on Privacy and Freedom

Current information and communication systems are constantly abused, as a look at both the daily newspaper and scholarly studies will show. Unauthorized opening of mail and tapping of telephones by Hoover's FBI of "national security risks" whose crimes were belonging to a civil rights or women's liberation or other antiestablishment social movement have been detailed on the front pages of the *New York Times*. There is carelessness with very personal data stored in computers by the Social Security Administration

ISBN 0-201-03140-X, 0-201-03141-8 (pbk.)

and other government agencies. There is fairly blatant disregard for the principle that information about a person belongs to him or her, and should be kept confidential by credit companies and most organizations that keep computerized files about employees or customers or clients.

The unique element for abuse introduced by the coming computer information–communication technologies is that they make it so much cheaper, faster, and simpler to monitor activities and potentially to interfere with and attempt to direct them. It is very expensive to monitor telephone conversations by having a person sit around to record and listen to telephone conversations, for instance. But it would cost very little to intercept, screen, and automatically analyze all of the computer-stored communications that a person sends. It would also be much easier for an unscrupulous person or organization to electronically "frame" a dissident–certainly it would be easy enough to create a fake trail of electronic circumstantial evidence linking a person to a criminal act; all it would take is some clever programming and unauthorized access to systems and files, easily gained in the name of "national security."

Such intentional manipulation of computer-based communications and information systems does not stop at the national border, of course. Stuart Umpleby (1977) has painted the following scenario of a possible means of international propaganda:

> Three new technologies—computer networks, electronic newspaper typesetting machines, and covert propaganda operations—could converge in the future to produce a deleterious effect on the freedom and autonomy of the press . . . . With "paper" traveling at the speed of light, the writer or editor could be located half way around the globe without encountering significant delays in getting out a daily newspaper . . . . It offers the opportunity for interference in the press of other nations, particularly the developing countries [by making] it much easier for the intelligence agencies of large countries to coordinate propaganda themes and disinformation campaigns in other countries.

What these centralized communication–information systems make possible is manipulation on an unprecedented scale. This is just one example of a potential application of CCSs that could detrimentally affect the quality of life, if not prevented by policy decisions and legal safeguards. As Martin and Norman (1970, p. vi) put it:

> The impact of the computer and new telecommunication systems is going to be more sweeping than the impact of the automobile . . . . Will we anticipate and plan for the new machines, or will we let information technology race ahead undirected, leaving us to sort out the mess afterwards, as we are now doing with traffic in cities? If we permit the latter, then we have reason to be apprehensive.

ISBN 0-201-03140-X, 0-201-03141-8 (pbk.)

In *The Conquest of Will*, Mowshowitz (1976, pp. 161–169) summarizes the dangers to privacy posed by computer–information utilities, which are vast data banks accessible remotely through computer networks:

> Privacy has long been recognized as an important component of life in a democratic society. It is essential to the physical, psychological, and spiritual integrity of the individual . . . . The invasion of privacy is perhaps the most controversial issue associated with computer technology . . . . The use of computer-based information-retrieval systems by law-enforcement agencies is a dramatic illustration of the technology's potential for surveillance. Another area of intense public concern is the record-keeping activities of government and private organizations . . . .
>
> For example, no matter what arrangements are made for development, ownership, and regulation, the existance of a financial computer utility will result in the consolidation of private information about individuals. A computer-based funds transfer system would provide the means for recording a person's movements, the organizations he belongs to or supports, the material he reads, his amusements and patterns of consumption . . . .

The result, Mowshowitz points out, is that it will be possible for government easily and cheaply to monitor data on everyone:

> The primary concern here is with data surveillance. Like the other types, this one does not involve an entirely new concept; only the methods at our disposal are new. The lack of a real-time storage and retrieval medium forced governments in the past to be highly selective in their efforts to monitor individual activity. Nevertheless, the dossier system elaborated for purposes of controlling political dissidents was far from harmless. The Gestapo used it quite effectively, as did other secret police organizations. Now that the computer has replaced the filing cabinet, the need for selectivity is no longer so acute. Modern governments can avail themselves of vast stores of information documenting the activites of virtually everyone [p. 69].

Mowshowitz did not take into account the idea of actual communications being computer stored. Thus, the problems of potential abuse that he points out are magnified that much more, with the additional private communications in CCSs potentially subject to monitoring or surveillance activities.

Conclusion

We do believe that computer-based communications can be used to make human lives richer and freer, by enabling persons to have access to vast stores of information, other "human resources," and opportunities for work and socializing on a more flexible, cheaper, and convenient basis than ever before. This is the image behind the idea of a Network Nation; but unless proper

ISBN 0-201-03140-X, 0-201-03141-8 (pbk.)

policies and safeguards are introduced to select and guide the implementation of alternative forms of these services and alternative systems for guaranteeing their security from manipulation or abuse, they could also become the basis for a totalitarian network of control much more comprehensive and efficient than any that has ever been developed.

A Scenario
Some Crucial Problems for a New Century

Remarks by Dr. Harold Smith, U.S. Chief Planner

Presented before the Executive Committee of the United States Societal Planning Commission, January 15, 2001

Ladies and gentlemen, we are beginning a new administration and a new century. I want you to reflect with me for a few minutes on some of the fundamental changes that have taken place in the nature of communications and work in our society during the last 25 years, before turning to one of the key problems with which we are going to have to grapple during the next few years. Many of these developments have to do with advances in computer technology and decreases in cost, so that computers are now doing not only jobs that people used to do, but also many kinds of things that were never even dreamed of previously. The problem is, some of these advances seem to have gotten out of hand.

The dollar value of computer equipment sold to the home market first exceeded the amount spent on automobiles in 1988. Most homes now have the equivalent of the IBM 370, which was a popular commercial computer in the 1970s, for work and household management, plus a terminal to tie into large conferencing networks and data bases. Many homes have several terminals, to reduce disagreement about scheduling of times and activities.

Educators have begun to warn us that we are producing a generation of computer addicts or robots. The children of today do not seem to know how to do anything for themselves anymore. Instead of playing games with their friends, they play games with computers. Instead of learning how to spell or write a correct sentence or add or multiply, they let computers do this for them.

Travel has become very expensive since we ran out of petroleum in 1995. The average white collar employee spends only two days a week at a central office, and works at home the rest of the time. Actually, a better way to put it is that she or he oversees the computer's work from home. Almost all of the information-processing-related transactions are now automated and directed by computers. Beginning with electronic funds transfer systems, all consumer activities have become computerized, so that people no longer have to travel

ISBN 0-201-03140-X, 0-201-03141-8 (pbk.)

to stores; they receive pictures and descriptions of merchandise or services and their costs on their home TV screen or terminals, and order "direct."

We have had some problems with poorer people complaining about the Selective Advertising breaks to which they are exposed while using their terminals. People can use a terminal free of charge if they agree to receive these commercials for five minutes of each on-line hour. The better-off people can afford to buy their own terminals and avoid receiving the specially computer-selected advertisements, based on their past buying behavior and social characteristics. They can also afford to pay the $5 a month necessary to keep their charge account data from being released to advertising-planning companies. It seems that the only people who can afford privacy anymore are the rich.

However, I am wandering; let me get to the main problem with which we must deal. I understand that all of you in this room have stability clearance and that I can talk freely. You all recall that JAWS (Job Allocation and Welfare Service) was created in the mid-eighties to ensure the enforcement of the Right to a Job Law. JAWS's objective was to see to it that anybody who desired to work could find a job, and that those that could not find any employment would be provided with a subsistance allowance and a recreational computer.

By the early nineties, almost all manufacturing and information processes were supervised automatically by computer. We could not expand production because of limitations on energy and other resources. One solution would have been to employ the approximately 10% of the population needed to program these computers and perform the few remaining manual service tasks, and put the rest on welfare. Unfortunately, our technology changed faster than our values. Suicide and divorce rates were so high among those permanently retired in their early 40s that we decided to use JAWS to create a full-employment solution.

As far as the public knows, JAWS accomplishes its full employment task by imposing a tax on employers proportional to their productivity. The automation of the eighties made productivity an ugly word in the public's mind and this was the easiest approach. However, no one wanted to see government employment grow to more than the current third of the population as a result of utilizing this tax income to create additional government-sponsored jobs. Therefore JAWS undertook a secret method of creating jobs. In order to keep tabs on employment and commerce, JAWS imposed its own special Shark computers in all transactional information networks, supposedly to measure what was taking place. Unknown to most people outside this room, the Sharks generate errors in the data flowing in these networks, so that organizations are forced to hire people to correct these unexplained errors.

ISBN 0-201-03140-X, 0-201-03141-8 (pbk.)

We all realize that this has become a very delicate and difficult enterprise. JAWS employs the top computer talent in the country and buys over 30% of the large computers manufactured.

It is estimated that most of the top computer personnel whom we do not hire, plus 65% of the non-Shark large computers in the nation, are employed full time trying to solve the problems that we create. The estimated 50% of the total labor force that is employed in creating and solving these problems *must not* get the impression that they are employed in virtual jobs created by JAWS. Our computer analyses of the personality and values underlying their vocabulary and syntax have shown us that these people are likely to behave in an irrational manner if they discover that there is no need for their work. They might even go so far as to smash a computer. [The audience gasps at this sacrilege.]

Now, our problem is that the Russians have discovered our JAWS system. Beset by unemployment and dissatisfaction, they have demanded that we send them our Shark computers too, and all the secret plans for the JAWS operation. If we refuse, they have threatened to flood the nation with old-fashioned printed pamphlets, telling the whole story.

What are we to do?

For Discussion

1. The authors of this book seem to believe that computerized conferencing is potentially a panacea for human communications problems. How justifiable is their position?

2. Both authors prefer this method of human communication in most situations and self-perceptions ascribe this preference to a tendency toward introversion. Could widespread use of these systems produce a greater degree of introversion in society?

3. Try to hypothesize the makeup of an electronic tribe you would like to belong to. Would you consider a tribe based on intellectual and/or recreational needs?

4. How would your life-style be different in a future where computerized conferencing is all-pervasive?

5. What personal management functions could be replaced by the computer in a CC environment? Using analysis aids provided by the computer, could employees in some team-oriented functions evaluate one another more or less effectively than a manager?

6. At this point, what are your reflections on the degree of closeness and intensity of feeling that can be conveyed and exchanged by a group

ISBN 0-201-03140-X, 0-201-03141-8 (pbk.)

utilizing computerized conferencing? If this impression has been modified by the book, what specifically impacted that modification?

7. What do you see as the principal values and ethics that will be modified by widespread use of computerized conferencing?

8. With the growing number of information-handling and white collar occupations, will blue collar work and the jobs associated with physical labor take on a more prestigious position in the society?

9. Is "productivity" likely to remain a national economic goal or is it to become a somewhat undesirable goal? What might substitute for it?

10. What do you believe to be the desirability and/or feasibility of "social engineering" for, managing the changes likely to occur over the next few decades as a result of CC technology? How would things evolve if left to current societal mechanisms?

11. What is the feasibility of societal influence (propaganda) and control by widespread use of this technology? What forms might it take?

12. In what types of situations do you think a computer could successfully appear to members of a conference or message system as another human? What might be the impact of this occurrence?

13. What words are likely to take on new meanings as a result of the CC environment? What new concepts are likely to require new words?

14. How would you measure the significance of some of the dimensions proposed to measure communication systems?

15. Itemize all the different areas of interest for which you would like to join CC groups. Categorize them under work, recreation, political involvement, and family or friendship relations. How much per month do you now spend on communication, information, or transportation related to these activities?

16. Given a large mass of users of a CCS, individuals with very unusual interests will be able to find one another. What unique groups or "subcultures" can you imagine emerging?

17. Now that you have reached the final chapter, what applications of CC information systems have the authors omitted? What societal impacts, national or international, have they failed to discuss?

ISBN 0-201-03140-X, 0-201-03141-8 (pbk.)

References

References

Allen, Donald E., and Guy, Rebecca F. 1974. *Conversation Analysis*. Mouton, The Hague, The Netherlands.

———, 1977. "Ocular Breaks and Verbal Output," *Sociometry* **40**, 90–96.

Allen, V. L. 1965. "Situational Factors in Conformity," *Advan. Exptl. Social Psychol.* **2**, 133–176.

Allison, Paul D., and Stewart, John A. 1974. "Productivity Differences among Scientists: Evidence for Accumulated Advantage," *Amer. Sociol. Rev.* **39**, (4), 596–606.

Amarel, S., Brown, J. S., Buchanan, B., Hart, P., Kulikowski, C., Martin, W., and Pople, H. 1977. "Reports of Panel on Applications of Artificial Intelligence," *Proc. 5th Intern. Joint Conf. on Artificial Intelligence* (Cambridge, Mass., August).

Argyle, M., and Dean, J. 1965. "Eye Contact, Distance and Affiliation," *Sociometry* **28**, 298–305.

Asch, S. E. 1951. "Effects of Group Pressure upon the Modification and Distortion of Judgments," in *Groups, Leadership, and Men* (H. Guetzkow, ed.), pp. 177–190. Carnegie Press, Pittsburgh, Pa.

Aspen Systems Corp. 1974. *Editorial Processing Centers: A Study to Determine Economic and Technical Feasibility*. Nat. Tech. Information Service No. PB-234 = 959–964.

Baker, C. L. 1966. *JOSS: An Introduction to a Helpful Assistant*. Rand Memorandum RM-5058-PR.

Bales, Robert F. 1950. "A Set of Categories for the Analysis of Small Group Interaction," *Amer. Sociol. Rev.* **15**, 257–263.

Bales, R. F. 1955. "How People Interact in Conferences," *Scientific Amer.* **192**, (3), 31–35.

Bales, R. F., Strodtbeck, F. L., Mills, T. M., and Roseborough, M. E. 1951. "Channels of Communication in Small Groups," *Amer. Sociol. Rev.* **16**, 461–468.

Bamford, Harold 1972. "A Concept for Applying Computer Technology to the Publication of Scientific Journals," *J. Wash. Acad. Sci.* **62**, 306–314.

Barnes, J. A. 1972. "Social Networks," *Addison-Wesley Module* **26**, 1–29.

Barnes, S. B., and Dolby, R. G. A. 1970. "The Scientific Ethos: A Deviant Viewpoint," *European J. Sociology* **11**, 3–25.

Basu, Shankar, and Schroeder, Roger G. 1977. "Incorporating Judgments in Sales Forecasts: Application of the Delphi Method at American Hoist and Derrick," *Interfaces* **7**, (3), 18–27.

Bavelas, A., Hastorf, A. H., Gross, A. E., and Kitl, W. R. 1965. "Experiments on the Alteration of Group Structure," *J. Exptl. Social Psychol.* **1**, 55–70.

Bell, Daniel, 1973. *The Coming of Post-Industrial Society*. Basic Books, New York.

Bennett, John L. 1972. "The User Interface in Interactive Systems," *Ann. Rev. Information Sci. Technol.* **7**, ASIS Press, pp. 159–196.

Bennett, William, and Gurin, Joel 1977. "Science That Frightens Scientists: The Great Debate over DNA," *Atlantic Monthly* (Feb).

Bennis, Warren 1964. "Beyond Bureaucracy," *Trans-Action* (July).

Beres, Mary E., Koehler, E., Koehler, B. N., and Zaltman, G. 1975. "Communication Networks in a Developing Science: A Simulation of the Underlying Socio-Physical Structure," *Simulation & Games* **6**, (1), 3–38.

Bernard, Jessie 1973. *The Sociology of Community*. Scott-Foresman, Glenview, Ill.

Berul, L. H., and Krevitt, B. J. 1974. "Innovation in Editorial Procedure: The Editorial Processing Center Concept," Paper presented at 37*th Ann. Meeting Amer. Soc. Information Sci.* (Atlanta, Ga.).

Bezilla, R. 1977. Selected Aspects of Privacy, Confidentiality and Anonymity in Computerized Conferencing. (Unpublished.) Forthcoming Research Rept. New Jersey Inst. Technol. Computerized Conferencing and Commun. Center.

Blau, Eleanor 1978. "What Bleeps, Blinks, Rolls, Stops and Delivers Office Mail?" *New York Times* (Feb. 6, 1978), p. B1.

Blau, Judith 1976. "Scientific Recognition: Academic Context and Professional Role," *Social Studies of Sci.* **6**, (3-4), 533–545.

Blume, S. S., and Sinclair, R. 1973. "Chemists in British Universities: A Study of the Reward System in Science," *Amer. Sociol. Rev.* **38**, (Feb), 126–138.

494

Bogart, Dodd 1973. "Changing Views of Organization," *Technological Forecasting and Social Change*, 5, (3), 163–178.

Borden, G., Gregg, R., and Grove, T. 1969. *Speech Behavior and Human Interaction*, Prentice-Hall, Englewood Cliffs, N. J.

Bright, R. D. 1976. *Viewdata—An Interactive Information Retrieval System Developed by the UK Post Office*. Post Office Telecommunications Headquarters, London.

Brown, David 1976. "Teleconferencing and Electronic Mail," *Educom Bull.* 11, (4), 14–19.

Brush, S. 1974. "Should the History of Science be Rated X?," *Science* 183, 1164–1172.

Bureau of the Census, U. S. Dept. of Commerce 1975. "Characteristics of the Low-Income Population, 1973," *Current Population Reports* Ser. P-60, No. 98 (January).

Burnham, David 1977. "Nation Facing Crucial Decisions Over Policies on Communications," *New York Times* 126, (July 8, 1977), pp. 1 and 10.

Burns, J. C. 1977. "The Evolution of Office Information Systems," *Datamation*, 23, (4), 60–64.

Bush, Vannevar, 1945. "As We May Think," *Atlantic Monthly* (July), 101–108.

Carp, Frances M. 1974. "Position Effects on Interview Responses," *J. Gerontol.* 29, 581–587.

Carroll, J. D., and Wish, Myron 1975. "Multidimensional Scaling: Models, Methods and Relations to Delphi," in *The Delphi Method* (H. A. Linstone and M. Turoff, eds.), pp. 402–432.

Cerf, V., and Curran, Alex 1976. "The Future of Computer Communications," *Computers and Communications* (*Proc. Federal Commun. Commission Planning Conf.*, Nov. 8–9, 1977), pp. 141–176. AFIPS Press, Montvale, N. J.

Chapple, E. A., and Arensberg, C. M. 1940. "Measuring Human Relations: An Introduction to the Study of the Interaction of Individuals," *Genetic Psychol. Monogr.* 32 3–147.

Cherry, Colin 1966. *On Human Communication: A Review, a Survey and a Criticism*. MIT Press, Cambridge, Mass.

Chubin, Daryl E. 1975. On the Conceptualization of Scientific Specialties: The Interplay of Demography and Cognition. Paper presented at Ann. Meeting of Amer. Sociol. Association.

Churchman, C. W. 1971. *The Design of Inquiry Systems: Basic Concepts of Systems and Organizations*. Basic Books, New York.

Coates, Joseph 1977a. "Aspects of Innovation: Public Policy Issues in Telecommunications Development," *Telecommunications Policy* 1, (3), 196–206.

———, 1977b. The Need to Study the Social Impacts of EFT. Paper presented at Nat. Sci. Found. Conf. on EFT Research and Public Policy (Boston). Available from Public Systems Evaluation, Inc., Cambridge, Mass.

———, 1977c. What Is a Public Policy Issue? Paper presented at the annual meeting of the Amer. Association for the Advancement of Science (Denver, Colorado).

Cohen, Jean-Claude, Conrath, D., Dumas, P. and de Roure, G. 1977. Information and Communication: Is There a System? Paper presented at NATO Symposium on Telecommunications, Bergamo, Italy. To be published by Plenum, 1978.

Cole, Stephen and Cole, Jonathan 1973. *Social Stratification in Science*. Univ. of Chicago Press, Chicago, Ill.

Collins, B. E., and Guetzkow, H. 1964. *A Social Psychology of Group Processes for Decision Making*. Wiley, New York.

Communications Studies Group. 1975. "The Effectiveness of Person-to-Person Telecommunications Systems," *Long Range Research Rep.* 3. University College, London.

Crane, Diana, 1965. "Scientists at Major and Minor Universities: A Study of Productivity and Recognition," *Amer. Sociol. Rev.* 30, 699–714.

———, 1972. *Invisible Colleges: Diffusion of Knowledge in Scientific Communities*. Univ. of Chicago Press, Chicago, Ill.

Craven, Paul and Wellman, Barry 1973. "The Network City," Centre for Urban and Community Studies and Dept. of Sociology, Univ. of Toronto, Research Paper No. 59.

Dalkey, Norman C. 1975. "Toward a Theory of Group Estimation," in *The Delphi Method* (H. A. Linstone and M. Turoff, eds.), pp. 236–248. Addison-Wesley Advanced Book Program, Reading, Mass.

———, 1977. *Group Decision Theory*. UCLA School of Engineering and Applied Science Rep. No. 7749. Final Report for the Defense Research Projects Agency.

Davis, J. 1969. *Group Performance*. Addison-Wesley, Reading, Mass.

Davison, W. P. 1972. "Public Opinion Research as Communication," Public Opinion Quarterly **36**, (2), 311–322.

Day, Larry 1973. "An Assessment of Travel/Communications Substitutability," *Futures* **5**, (6), 559–572.

———, 1974a. Factors Affecting Future Substitution of Communications for Travel. Paper presented at Joint Meeting of Operations Research Soc. of America and Inst. of Management Sciences, San Juan, Puerto Rico.

———, 1974b. "Computerized Conferencing: An Overview," in *Views from ICCC* 1974 (N. Macon, ed.), pp. 53-70. Intern. Council for Computer Commun., Washington, D. C.

———, 1977. The Role of Telecommunications Policy Analysis in Service Planning. Paper presented at NATO Symposium on Telecommunications, Bergamo, Italy. To be published by Plenum, 1978.

Delbecq, Andre L., Van de Ven, Andrew H., and Gutafson, David H. 1975. *Group Techniques for Program Planning: A Guide to Nominal Group and Delphi Processes*. Scott-Foresman, Glenview, Ill.

DeSio, R. W., ed. 1966. *Proc. Man–Machine*. *IBM Scientific Computing Symposium, Commun. IBM, White Plains, N. Y.*

Deutsch, M., and Gerard, H. 1955. "A Study of Normative and Informational Social Influences upon Individual Judgment," *J. Abnormal and Social Psychol.* **51**, 629–636.

Dillman, Don A. 1972. "Increasing Mail Questionnaire Response in Large Samples of the General Public," *Public Opinion Quarterly* **36**, 254–257.

Dittman, Allen T. 1972. *Interpersonal Messages of Emotion*. Springer, New York.

Dotto, Lydia 1977. "Electronic Mail Foreseen as a Future Alternative to Letters, Phone Calls," *Globe and Mail* (Toronto, Canada), (Aug. 13), p. 4.

Dray, Sheldon 1966. Letter, *Science* (12 August), 694–695.

Drucker, Peter 1968. *The Age of Discontinuity, Guidelines to Our Changing Society*. Harper & Row, New York.

Duncan, Starkey, Jr. 1972. "Some Signals and Rules for Taking Speaking Turns in Conversations," *J. Personality and Social Psychol.* **23**, 283–292.

Elbing, Alvar O., Gadon, Herman, and Gordon, John J. 1975. "Flexible Working Hours: It's about Time," *Harvard Business Rev.* **52**, (1) 19–33 *and* 154–155.

Englebart, D. C., Watson, Richard W., and Norton, James C. 1976. "The Augmented Knowledge to Workshop," in *Computer Networking* (R. Blanc and I. Cotten, eds.), IEEE Press, pp. 228–240.

ERU/STU. (Expert Board for Regional Development/National Swedish Board for Technical Development). 1976. *Telecommunications and Regional Development in Sweden*. STU Rep. No. 48-1976.

ERU/STU 1977. *Telecommunications and Regional Development in Sweden: A Progress Report, April 1977*. STU Rep. No. 64-1977.

Etzioni, Amitai, Laudon, Kenneth, and Lipson, Sara, 1975. "Participating Technology: The Minerva Communications Tree." *J. Commun.* **25**, (*Spring*), 64–74.

Fast, J. 1971. *Body Language*. Pocket Books, New York.

Faust, W. L., 1959. "Group vs. Individual Problem-Solving," *J. Abnormal and Social Psychol.* **59**, 68–72.

Fields, Craig 1977. Puting the 'Tele' in Teleconferencing, Paper presented at the NATO Symposium on Telecommunications, Bergamo, Italy. [To be included in the Proceedings, *Evaluating New Telecommunications Services* (M. Elton, W. Lucas, and D. Conrath, eds.). Plenum, New York, 1978.

Forward, Robert L. 1974. "Far Out Physics." Seminar at Ann. Conf. of Science Fiction Writers of America, Los Angeles (27 April).

Freeman, Linton 1977a. Draft manuscript of a forthcoming book.

———, 1977b. "Electronic Information Exchange by a Social Networks Community," proposal submitted to Nat. Sci. Found.

French, R. L. 1950. "Verbal Output and Leadership Status in Initially Leaderless Group Discussions," *Amer. Psychologist* **5**, 310.

Fuller, C. 1974. "Effect of Anoymity on Return Rate and Response Bias in a Mail Survey," *J. Appl. Psychol.* **59**, 292–296.

Garvey, William D., and Griffith, Belver C. 1967. "Communication in a Science: The System and Its Modification," in *Ciba Found. Commun. in Science* 16–36.

Garvey, W. D., and Griffith, B. C. 1971. "Scientific Communication: Its Role in the Conduct of Research and Creation of Knowledge," *Amer. Psychologist* **26**, (4), 349–362.

Geertz, Clifford 1977. Toward an Ethnography of the Disciplines. (Unpublished.) Paper (dated November 1976) presented at the Seminar on the Sociology of Science, Institute for Advanced Study, Princeton, N. J., January 1977.

Gerbner, G., Gross, L. P., and Melody, W. H. eds. 1973. *Communications, Technology, and Social Policy.* Wiley, New York.

Glaser, Daniel 1964. *The Effectiveness of a Prison and Parole System.* Bobbs-Merrill, Indianapolis, Ind.

Goffman, Erving 1955. "On Face Work: An Analysis of Ritual Elements in Social Interaction," *Psychiatry* **18**, 213–231.

———, 1959. *The Presentation of Self in Everyday Life.* Doubleday (Anchor), Garden City, N. Y.

———, 1967. *Interaction Ritual: Essays on Face-to-Face Behavior.* Doubleday (Anchor), Garden City, New York.

———, 1969. *Strategic Interaction.* Univ. of Pennsylvania Press, Philadelphia.

———, 1971. *Relations in Public: Microstudies of the Public Order.* Basic Books, New York.

Goldstein, Joel W. 1975. "Computing and Our Relationship to Work," *ACM SIGSOC Bull.* **7**, 10–18.

Goldmark, Peter C. 1973. "A Rural Approach to Saving Energy: Technology Could Help Ease Urban Congestion," *New York Times*, (Nov. 12), p. 12.

Gordon, William J. 1961. *Synectics.* Harper & Row, New York.

Granovetter, Mark 1976. "Network Sampling: Some First Steps," *Amer. J. Sociology* **81**, (6), 1287–1303.

Gustafson, Thane 1975. "The Controversy over Peer Review," *Science* (December), 1060–1065.

Hagstrom, Warren O. 1965. *The Scientific Community.* Basic Books, New York.

———, 1970. "Factors Related to the Use of Different Modes of Publishing Research in Four Scientific Fields," in *Communication among Scientists and Engineers* (C. E. Nelson and D. K. Pollock, eds.) pp. 85–124. Lexington Books, Lexington, Mass.

———, 1976. "The Production of Culture in Science," *Amer. Behavioral Scientist* **19**, (6), 753–767.

Hamilton, Kenneth 1950. *Counseling the Handicapped in the Rehabilitation Process.* Ronald Press, New York.

Hammond, Kenneth R., and Adelman, Leonard 1976. "Science, Values, and Human Judgment," *Science* **194** (Oct. 22), 389–395.

Hardy, K. R. 1957. "Determinants of Conformity and Attitude Change," *J. Abnormal and Social Psychol.* **54**, 289–294.

Harrington, Michael 1963. *The Other America: Poverty in the United States.* Penguin Books, Baltimore, Md.

Heiser, John F., Colby, Kenneth, Faught, William S. and Parkinson, Roger C. 1977. Testing Turing's Test: Can Psychiatrists Distinguish a Computer Simulation of Paranoia from the Real Thing? Draft paper distributed at Workshop on Artificial Intelligence in Health, Rutgers University (July).

Heller, S., Milne, G. W. A., and Feldman, R. J. 1977. "A Computer Based Chemical Information System," *Science* **195**, (21 Jan.), 253–259.

Hiltz, Starr Roxanne 1976a. Computer Conferencing: Assessing the Social Impact of a New Communications Medium. Paper presented at Ann. Meeting of Amer. Sociol. Association, (New York); published in *Technological Forecasting and Social Change* **10**, (3), 225–238; (1977), and in *Ekistics*, **45** (268), 143–150.

———, 1976b. "A Social Scientist Looks at Computer Conferencing," *Proc. Intern. Computer Commun. Conf.* (Toronto,   ), *pp.* 203–207.

Hiltz, S. R., Johnson, K., and Agle, G. M. 1978. *Replicating Bales Problem Solving Experiments on a Computerized Conference: A Pilot Study.* New Jersey Inst. Technol. Computerized Conferencing and Commun. Center Research Rep. No. 8.

Homans, George 1956. "Social Behavior as Exchange," *Amer. J. Sociology* **62** (May), 597–606.
_____, 1961. *Social Behavior: Its Elementary Forms*. Harcourt, Brace, Jovanovich, New York.
House, J. S., Gerber, W., and McMichael, A. J. 1977. "Increasing Mail Questionnaire Response: A Controlled Replication and Extension," *Public Opinion Quarterly* **41**, (1), 95–99.
Hough, R. W. (with R. Panko) 1977. *Teleconferencing Systems: A State-of-the Art Survey and Preliminary Analysis*. Final report to Nat. Sci. Found., Stanford Research Inst., Menlo Park, Calif.
Huston, J. W. 1977. Constitutional Convention Communication Network, Inc. Two Way Communication Project Information. (Unpublished.) Honolulu, Hawaii.
Institute for the Comparative Study of History, Philosophy and the Sciences, Ltd. 1967. *Systematics* **5**, (3) (entire issue).
Jacobs, Jane. 1961. *The Death and Life of Great American Cities*. Random House, New York.
Johansen, Robert 1976. "Pitfalls in the Social Evaluation of Teleconferencing Media," in *The Status of the Telephone in Education* also, IFTF Report of Wisconsin Extension Press, Madison. (L. A. Parker and B. Ricommini, eds.), No. p–38.
Johansen, R., DeGrasse, R., Jr., and Wilson, T. 1977. *Group Communications through Computers, Vol. 5: Effects on Working Patterns*. Institute for the Future, Menlo Park, Calif.
Johansen, Robert, Vallee, Jacques and Collins, Kent 1977b. Learning the Limits of Teleconferencing: Design of a Teleconference Tutorial. Paper presented at NATO Symposium on Telecommunications, Bergamo, Italy. To be published by Plenum, 1978.
Johansen, R., Vallee, J., and Palmer, M. 1977c. *Computer Conferencing: Measurable Effects on Working Patterns*. Institute for the Future, Menlo Park, Calif.
Johansen, R., Vallee, J., Spangler, K., and Shirts, G. 1977d. *The Camelia Report: A Study of Technical Alternatives and Social Choices in Teleconferencing*. Institute for the Future, Menlo Park, Calif.
Kafafian, H. 1970, 1973. *Study of Man–Machine Communications Systems for the Handicapped*, Vols. I and II, 1970; Vol. III, 1973. Cybernetics Research Inst. Washington, D. C.
Kanter, Rosabeth Moss 1977. "Some Effects of Proportions on Group Life: Skewed Sex Ratios and Responses to Token Women," *Amer. J. Sociology* **82**, (5), 965–990.
Keller, Suzanne 1963. *Beyond the Ruling Class: Strategic Elites in Modern Society*. Random House, New York.
Kendon, A. 1967. "Some Functions of Gaze Direction on Social Interaction," *Acta Psychologica* **26**, 22–63.
Kerr, Elaine B. 1971. The Vocational Experiences of Physically Handicapped–Poorly Educated Workers after Job Placement. *Ph.D. dissertation, Columbia University*.
Kling, Rob 1977. Automated Welfare Client-Tracking and Service Intergration. The Case of Riverville. (Unpublished.) Information and Computer Sci. Dept., Univ. of California, Irvine.
Knapp, M. 1972. Non-verbal Communication in Human Interaction. Holt Rinehart and Winston, New York, N.Y.
Kochen, Manfred 1973. "Referential Consulting Networks," in *Towards a Theory of Librarianship* (C. Rawski, ed.), pp. 187–220. Scarecrow Press, Metuchen, N. J.
Kogan, N., and Wallach, M. A. 1967. "Effects of Physical Separation of Group Membership upon Group Risk Taking," *Human Relations* **20**, 41–49.
Kuhn, Thomas S. 1970. *The Structure of Scientific Disciplines*, Univ. of Chicago Press, rev. ed. Chicago, Ill.
Kupperman, Robert, Wilcox, Richard and Smith, Harvey 1975. "Crisis Management: Some Opportunities," *Science* **187**. (Feb. 7), 404–410.
Laudon, Kenneth C. 1976. Efficiency vs. Equity and Justice: Consequences of Public Sector Information Systems. Paper presented at meeting of Amer. Sociol. Association. (New York).
Lee, Alfred McClung 1973. *Toward Humanist Sociology*. Prentice-Hall, Englewood Cliffs, N. J.
Linstone, H. A., and Turoff, M. eds. 1975. *The Delphi Method: Techniques and Applications*. Addison-Wesley Advanced Book Program, Reading, Mass.
Linstone, H. A., Hays, J., Lendaris, S., Rogers, W., Wakeland, W. and Williams, M. 1978. *The Use of Structural Modeling for Technology Assessment*. Final report to Nat. Sci. Found., Portland State Univ., Portland, Oregon.
Luehrman, Arthur 1977. *PILOT User's Guide*. Dartmouth Time-Sharing TM031, January 1977.

McCurdy, H. G., and Lambert, W. E. 1952. "The Efficiency of Small Human Groups in the Solution of Problems Requiring Genuine Cooperation," *J. Personality* **20**, 478–494.

Maier, N. R. F., and Solem, A. L. 1952. "The Contribution of a Discussion Leader to the Quality of Group Thinking: The Effective Use of Minority Opinions," *Human Relations* **5**, 277–288.

Marquart, D. I. 1955. "Group Problem Solving," *J. Social Psychol.* **41**, 103–113.

Martin, James 1973. *Design of Man–Computer Dialogues*. Prentice-Hall, Englewood Cliffs, N. J.

Martin, James and Norman, Adrian R. 1970. *The Computerized Society*. Prentice-Hall, Englewood Cliffs, N. J.

Martin, Shirley M., Von Gehren, Edgar S., and Uhlig, Ronald P. 1976. Practical Experience in Computer Based Message Systems: Pros and Cons. (Unpublished.) U. S. Army Materiel Development and Readiness Command, Alexandria, Va.

McKendree, John D. 1975. Resource Interruption Monitoring System. (Unpublished.) Office of Preparedness, Washington, D. C.

———, 1977. "Decision Process in Crisis Management: Computers in a New Role," *Encyclopedia of Commun. Sci. Technol.* **6**, 115–156.

Mears, Peter 1974. "Structuring Communication in a Working Group," *J. Commun.* **24**, (1), 71–79.

Mehrabian, A. 1971. *Silent Messages*. Wadsworth, Belmont, Calif.

Merton, Robert K. 1968. "The Matthew Effect in Science," *Science* **159**, (Jan.) 56–63.

Milgram, Stanley 1965. "Some Conditions of Obedience and Disobedience to Authority," *Human Relations* **18**, (1), 57–75.

———, 1970. "The Experience of Living in Cities," *Science* **167**, 1461–1468.

Millard, Gord and Williamson, Hilary 1976. "How People React to Computer Conferencing," *Telesis* **1976**, (3), 214–219.

Miller, Judith 1977. "Self Paced Instruction, an Innovation That Stuck," *New York Times* (May 1), p. E9.

Mitchell, J. C. 1969. *Social Networks in Urban Situations*. University Press, Manchester, England.

Mitroff, Ian A. 1974a. *The Subjective Side of Science*. Elsevier, Amsterdam, The Netherlands.

———, 1974b. "Norms and Counter-Norms in a Select Group of the Apollo Moon Scientists: A Case Study of the Ambivalence of Scientists," *Amer. Sociol. Rev.* **34**, (4), 579–595.

Mitroff, I., and Turoff, M. 1975. "Philosophical and Methodological Foundations of Delphi," in *The Delphi Method* (H. A. Linstone and M. Turoff, eds.), pp. 17–36. Addison-Wesley, Reading, Mass.

Moore, Omar Khayyam 1957. "Divination—A New Perspective," *Amer. Anthropologist* **59** (1), 69–74.

———, 1977. *Discontinuity and Deontics*. Responsive Environments Foundation, Inc., Pittsburgh, Pa.

Mowshowitz, Abbe 1976. *The Conquest of Will: Information Processing in Human Affairs*. Addison-Wesley, Reading, Mass.

Mulkay, Michael J. 1972. "Conformity and Innovation in Science," *Sociol. Rev. Monogr.* **18**, 5–23.

Mulkay, M. J. 1976. "The Mediating Role of the Scientific Elite," *Social Studies of Sci.* **6**, 445–470.

———, 1977 "Sociology of the Scientific Research Community," in *Science, Technology and Society* (A. I. Spiegel-Rosing and D. de Solla Price, eds.), Sage Publications, pp. 93–148. Beverly Hills, Calif.

Mullins, Nicholas C. 1973. *Theories and Theory Groups in Contemporary American Sociology*. Harper & Row, New York.

National Science Foundation. 1977. Program Announcement: Operational Trials of Electronic Information Exchange for Small Research Communities. NSF 76-45, Division of Science Information.

Neisner, Ulric 1964. MAC and Its Users. Internal memorandum, Project MAC at MIT, MIT-MAC-M-185.

———, 1968. "Computers as Tools and Metaphors," in *Conversational Computers* (W. O. Orr, ed.), pp. 206–217. Wiley, New York.

Nelson, Theodore H. 1965. "A File Structure for the Complex, the Changing, and the Indeterminate" *ACM 20th Nat. Conf. Proc.* p. 440.

———, 1970. "No More Teachers' Dirty Looks," *Computer Decisions*; reprinted in Nelson, 1974, pp. DM16–DM19.

———, 1972. "As We Will Think," *Proc. On line 72 Conf.* pp. – Oxbridge, England.

———, 1973a. "A Conceptual Framework for Man–Machine Everything," *Proc. AFIPS* **42**, M21.

———, 1974. *Dream Machines & Computer Lib.* Hugo's Book Service, Chicago, Ill.

———, *Computopia and Cybercrud.* Rand Rep. R-718-NSF/CCOM/RC, pp. 185–199.

Nilles, Jack M., Carlson, F. R., Jr., Gray, P., and Hanneman, G. 1976a. "Telecommuting—An Alternative to Urban Transportation Congestion," *IEEE Trans. Systems, Man, and Cybernetics* **SMC-6**, 77–84.

———, Carlson, F. R., Jr., Gray, P., and Hanneman, G. 1976b. *The Telecommunications–Transportation Tradeoff: Options for Tomorrow.* Wiley (Interscience), New York.

Norfleet, B. 1948. "Interpersonal Relations and Group Productivity," *J. Social Issues* **2**, 66–69.

Orr, William D. 1968. *Conversational Computers.* Wiley, New York.

Osborn, A. 1963. *Applied Imagination*, 3d ed. Scribner, New York.

Panko, Raymond R. 1975. Computer Teleconferencing: History, Status, Trends and Questions. (Unpublished.) Stanford Research Inst., Menlo Park, Calif.

———, 1976. The Outlook for Computer Mail. (Unpublished.) Stanford Research Inst., Menlo Park, Calif.

Pantages, Angeline and Pipe, G. Russell 1977. "A New Headache for International DP," *Datamation* (June), 115–126.

Parnes, S. J. 1967. *Creative Behavior Guidebook.* Scribner, New York.

Parsons, Talcott 1951. *The Social System.* Free Press, New York.

Peters, Howard J. 1976. "The Electronic Aristotle," *Computer Decisions* **8**, 42.

Pfeiffer, J. W., and Jones, J. E. 1973. *Annual Handbook for Group Facilitators.* University Associates, Iowa City.

Powers, Edward A., Morrow, P. Willis and W. J., Goudy, Keith, P. 1977. "Serial Order Preference in Survey Research," *Public Opinion Quarterly* **41**, 80–85.

Price, Charlton 1975. "Conferencing via Computer," in *The Delphi Method* (H. A. Linstone and M. Turoff, eds.), pp. 497–509. Addison-Wesley Advanced Book Program, Reading, Mass.

Price, Derek 1963. *Little Science, Big Science.* Columbia Univ. Press, New York.

Price, Derek, J. De Solla, and Beaver, Donald De B. 1966. "Collaboration in an Invisible College," *Amer. Psychologist* **21**, 1011–1018.

Pye, R., and Williams, E. 1977. "Teleconferencing: Is Video Valuable or Is Audio Adequate?," *Telecommunications Policy* (June), 230–241.

Rahmstorf, Gerhard and Penniman, David 1977. "Scientific Communication and Knowledge Representation." Paper presented at NATO Symposium on Telecommunications, Bergamo, Italy. Proceedings to be published by Plenum, 1978.

Ravetz, Jerome 1971. *Scientific Knowledge and Its Social Problems.* Oxford (Clarendon Press), London and New York.

Reid, Alex A. L. 1970. "Electronic Person-to-Person Communications." Communications Studies Group, University College, London, Ref. No. B/70244/CSG.

Reif, F. 1961. "The Competitive World of the Pure Scientist," *Science* **134**, (3094), 1957–1962.

Renner, Rod L., Bechtold, R. M., Clark, C. W., Marbray, D. O., Wynn, R. L., and Goldstein, N. H. 1973. "EMISARI: A Management Information System to Aid and Involve People," Office of Emergency Preparedness TM-230.

Reskin, Barbara 1977. "Scientific Productivity and the Reward Structure of Science," *Amer. Sociol. Rev.* **42**, (June), 491–504.

Rhodes, Sarah N., and Bamford, Harold E., Jr. 1976. "Editorial Processing Centers: A Progress Report," *Amer. Sociologist* **115**, (August), 153–159.

Rubin, Sylvan 1973. "PILOT: A Simple Instructional Language," *Computer Decisions*, (Nov.), 17–18.

Sandals, L. H. 1974. "Computers Assist Handicapped Children with Learning Problems," *Educ. Manitoba* **1**, (2), 11–12.

Scheele, Sam 1975. "Reality Construction as a Product of Delphi Interaction," in *The Delphi Method* (H. A. Linstone and M. Turoff, eds.), pp. 37–71. Addison-Wesley, Reading, Mass.

———, 1977. Personal communication, EIES Conference 117, December 1977.

Scher, Julian M. 1976. Computerized Conferencing—A Getsalt Tool for Simulation and Game Designers. Paper presented at ORSA-TIMS Meeting, Philadelphia, Pa.

Shaw, J. C. 1964. "JOSS: A Designer's View of an Experimental On-Line Computing System," *Proc. AFIPS Fall Joint Computer Conf.* **26**, 455–464.

Shaw, Marvin E. 1960. "A Note Concerning Homogeneity of Membership and Group Problem Solving," *J. Abnormal and Social Psychol.* **60**, 448–450.

Sheridan, Thomas and Ferrell, William 1974. *Man–Machine Systems: Information, Control & Decision Models of Human Performance*. MIT Press, Cambridge, Mass.

Shinn, A. 1977. The Utility of Social Experimentation. Paper presented at NATO Symposium on Telecommunications, Bergamo, Italy. Proceedings to be published by Plenum.

Short, John 1975. "Residential Telecommunications Applications: A General Review," *Long Range Intelligence Bull.* **6**. Post Office Telecommunications Headquarters, Cambridge, England.

Short, John, Williams, Ederyn and Christie, Bruce 1976. *The Social Psychology of Telecommunications*. Wiley, New York.

Simmel, George 1950. "The Stranger," in *The Sociology of George Simmel* (Kurt H. Wolff, ed.), pp. 402–408. Free Press, New York.

Smith, Vallee, J., Johansen, R., and Turoff, M. 1976. "Computerized Conferencing: A New Medium," *Mosaic* **7**, (1), 16–22.

Stein, M. 1975. *Stimulating Creativity* Vol. 2. Academic Press, New York.

Steiner, Ivan D. 1972. *Group Process and Productivity*. Academic Press, New York.

Sterling, Theodore D. 1975. "Humanizing Computerized Information Systems," *Science* **190**, 1168–1172.

Sterling, Theodore D., and Laudon, Kenneth 1976. "Humanizing Information Systems," *Datamation* **22**, (12), 53–59.

Stoner, J. A. F. 1961. A Comparison of Individual and Group Decisions Involving Risk. (Unpublished.) Master's thesis, Massachusetts Institute of Technology.

Storer, Norman W. 1966. *The Social System of Science*. Holt, Rinehart and Winston, New York.

Teger, A. I., and Pruitt, D. G. 1967. "Components of Group Risk Taking," *J. Exptl. Social Psychol.* **3**, 189–205.

Thompson, D., and Johnson, J. D. 1971. "Touch Tutor at Hawkworth Hall," *Special Educ.* **60**, 11–12.

Thompson, Gordon B. 1970. "Moloch or Aquarius?," *THE* Number 4. Bell-Northern Research.

———, 1972. "Three Characterizations of Communications Revolutions," in *Computer Communication: Impacts and Implications* (S. Winkler, ed.), pp. 36–37. (Proceedings of the First Intern. Conf. on Computer Commun. Washington, D. C., October 1972).

Thompson, G. B. 1977. Information Technology and Society. Paper presented at NATO Symposium on Telecommunications, Bergamo, Italy. Proceedings to be published by Plenum, 1978.

Torrance, E. P. 1954. "Some Consequences of Power Differences on Decision Making in Permanent and Temporary Three-Man Groups," *State College of Washington Research Studies* **22**, 130–140.

Turoff, Murray 1966. "Immediate Access and the User," *Datamation* (August).

———, 1969. "Immediate Access and the User Revisited," *Datamation* **15**, 65–67.

———, 1970. "The Design of a Policy Delphi," *Technological Forecasting and Social Change* **2**, (2), 149–172.

———, 1971. "Delphi and Its Potential Impact on Information Systems," *Proc. AFIPS Nat. Computer Conf.* **39**, 317–326.

———, 1972a. "Delphi Conferencing: Computer-Based Conferencing with Anonymity," *Technological Forecasting and Social Change* **3**, 159–204.

Turoff M., 1972b. "Party-Line and Discussion: Computerized Conference Systems," *Proc. Intern. Conf. Computer Commun.* pp. 161–171.

———, 1974. "Potential Applications of Computerized Conferencing in Developing Countries," *Ekistics* **38**, 131–132.

———, 1975a. "The Future of Computer Conferencing," *Futurist* **9** 182–195.

———, 1975b. "Computerized Conferencing for the Deaf and Handicapped." Paper presented at Ann. Meeting of the American Association for the Advancement of Science. (A version of this paper is in *Urban Telecommunications Forum* **4**, 5–10.)

———, 1975c. "The Policy Delphi," in *The Delphi Method* (H. A. Linstone and M. Turoff, eds.), pp. 84–101. Addison-Wesley, Reading, Mass.

———, 1976. "The Costs and Revenues of Computerized Conferencing," *Proc. Intern. Computer Commun. Conf.* (Toronto, Canada), pp. 214–221.

———, and Spector, M. 1976. "Libraries and Implications of Computer Technology," *Proc. AFIPS-Nat. Computer Conf.* **45**, 701–707.

Turoff, M. 1977. "An On-Line Intellectual Community, or 'MEMEX' Revisited." Paper presented at Ann. Meeting of Amer. Association for the Advancement of Science, Denver, Colorado. (Published in *Technological Forecasting and Social Change*, July) **10**, (4), 401–412.

Turoff, M., Enslow, P., Hiltz, S. R., McKendree, J., Panko, R., Snyder D., and Wilcox, R. 1977. Human Communications via Computers: Research Options and Imperatives. Working draft, New Jersey Inst. Technol. Computerized Conferencing and Commun. Center.

Turoff, Murray and Scher, Julian 1975. Computerized Conferencing and Its Impact on Engineering Management. Paper Presented at 23rd Ann. Joint Engineering Management Conf. Washington, D. C.

Turoff, M., Whitescarver, J., and Hiltz, S. R. 1977. "The Human–Machine Interface in a Computerized Conferencing Environment," *Proc. IEEE Systems, Man and Cybernetics Conf.* pp. 145–157.

Turoff, M., and Hiltz, S. R. 1978. *Development and Field Testing of an Electronic Information Exchange System: Final Report on the EIES Development Project.* New Jersey Inst. Technol. Computerized Conferencing and Commun. Center Research Rep. No. 9.

Turoff, M., Enslow, P., Hiltz, S. R., McKendree, J. Panko, R. Snyder, D., and Wilcox, R. 1978. *Research Options and Imperatives in Computerized Conferencing.* New Jersey Inst. Technol. Computerized Conferencing and Commun. Center Research Rep. No. 10.

Tyson, Herbert L. and Kaplowitz, Stan A. 1977. "Attitudinal Conformity and Anonymity," *Public Opinion Quarterly* **41**, 226–234.

Uhlig, R. P. 1977. "Human Factors in Computer Message Systems," *Datamation* **23**, (5), 120–126.

Uhlig, Ronald P., Marten, Shirley M., and Von Gehren, Edgar S. 1975. The Role of Informal Communications in Computer Networks. Paper presented at the Pacific Area Computer Network System Symposium (Sendai, Japan).

Umpleby, Stuart A. 1977. A Potential Negative Secondary Effect of On-Line Intellectual Communities. Paper presented at a workshop on On-Line Intellectual Communities, Soc. for General Systems Research (Denver, Colorado).

Vallee, Jacques 1977. The Outlook for Computer Conferencing on ARPANET and PLATO. Proc. of the Ann. Northamer. Meeting of the Society for General Systems Research (Denver, Colorado), pp. 194–201.

Vallee, Jacques, Askevold G. and Wilson, T. 1977. Computer Conferencing in the Geosciences. Institute for the Future, Menlo Park, Cal.

Vallee, Jacques and Johansen, R. 1975. "An Evaluation of the Use of Computer Conferencing to Improve Group Productivity in Energy Research," proposal submitted to the Nat. Sci. Found.

Vallee, J., Askevold, G., and Wilson, T. 1978. *Computer Conferencing in the Geosciences*, a report prepared for the U. S. Geological Survey by the Institute for the Future, Menlo Park, Calif.

Vallee, J. Lipinski, H., and Miller, R. 1974a. *Group Communication through Computers* Vol. 1 (*Design and Use of the FORUM System*). Institute for the Future Rep. R-32.

Vallee, J. Johansen, R., Randolph, R., and Hastings, A. 1974b. *Group Communication through Computers Vol. 2* (*A Study of Social Effects*). Institute for the Future Rep. R-33.

Vallee, J. Johansen, R., Lipinski, H., Spangler, K., Wilson, T., and Hardy, A. et al. 1975a. *Group Communications through Computers* (Vol. 3: *Pragmatics and Dynamics*), Institute foɪ the Future Rept. R-35.

Vallee, J., Johansen, R., and Spangler, K. 1975b. "The Computer Conference: An Altered State of Communication?," *Futurist* **9**, (3), 116–121.

Vallee, J., Johansen, R., Lipinski, H. and Wilson, T. 1976. "Pragmatics and Dynamics of Computer Conferencing: A Summary of Indings from the Forum Project," *Proc. Intern. Conf. Computer Commun.* (Toronto, Canada), pp. 208–213.

Van de Ven, Andrew H. 1974. *Group Decision Making and Effectiveness*. Kent State Univ. Press, Kent, Ohio.

Van de Ven, Andrew H., and Delbecq, Andre L. 1974. "The Effectiveness of Nominal, Delphi and Interacting Group Decision Making Processes," *Academy of Management J.* **17**, (4), 605–621.

Voelker, A. H. 1977. "Power Plant Setting: An Appplication of the Nominal Group Process Technique," Oak Ridge Nat. Lab. ORNL/NUREG/TM-81.

Walker, Donald E. 1971. *Interactive Bibliographic Search: The User/Computer Interface*. AFIPS Press, Montvale, N. J.

Wallach, M. A., and Kogan, N. 1965. "The Roles of Information, Discussion and Consensus in Group Risk Taking," *J. Exptl. Social Psychol.* **1**, 1–19.

Walsh, John 1975. "NSF Peer Review Hearings: House Panel Starts with Critics," *Science* 435–437.

Warfield, J., Geschra, H., and Hamilton, R. 1975. *Approaches to Problem Solving Number 4: Methods Of Idea Management* Battelle Institute, Columbus, Ohio.

Webber, Melvin 1963. "Order in Diversity: Community without Propinquity." Reprinted in Environmental Psychology (*H. M. Proshansky, W. H. Illetson, and L. G. Rivlin, eds.*), Holt, Rinehart and Winston, New York, 1970.

Webber, Melvin 1973. "Urbanization and Communications," in *Communications, Technology and Social Policy* (G. Derber et al., eds.), pp. 293–303, Wiley, New York.

Whitley, Richard 1976. "Umbrella and Polytheistic Scientific Disciplines and Their Elites," *Social Studies of Sci.* **6**, 471–498.

Wicker, A. W. 1969."Size of Church Membership and Member's Support of Church Behavior Settings," *J. Personality and Social Psychol.* **13**, 278–288.

Wilcox, R., and Kupperman, R. 1972. "An On-Line Management System in a Dynamic Environment," *Proc. Intern. Conf. Computer Commun.* pp. 117–120.

Willard, Don and Strodtbeck, Fred L. 1972. "Latency of Verbal Response and Participation in Small Groups," *Sociometry* **35**, 161–175.

Williams, Ederyn 1975. "Medium or Message: Communications Medium as a Determinant of Interpersonal Evaluation," *Sociometry* **38**, 119–130.

Wiseman, Frederick 1972. "Methodological Bias in Public Opinion Survey," *Public Opinion Quarterly* **36** (1), 105–108.

Wish, M. 1974b. Toward a Conceptual Framework for Studying Interpersonal Communication. Paper presented at annual Meeting of Amer. Psychol. Association, New Orleans.

Wish, Myron 1974b. "Dimensions of Interpersonal Communication," *Proc. 18th Ann. Meeting of Human Factors Soc.* (Huntsville, Alabama), pp. 598–603.

Wish, Myron and Kaplan, Susan 1975. "Subjects Perceptions of Their Own Interpersonal Communication," Paper presented at 83rd Annual Meeting, Amer. Psych. Assn.

Zuckerman, Harriet 1970. "Social Stratification in American Science," *Sociol. Inquiry* **40**, 235–257.

Zuckerman, Harriet 1977. *Scientific Elite: Nobel Laureates in the United States*. Free Press, New York.

Zuckerman, Harriet, and Merton, R. K. 1971. "Patterns of Evaluation in Science: Institutionalization, Structure and Functions of the Referee System," *Minerva*, **9** (1), 66–100.

Bibliography of Computerized Conferencing

Bibliography of Computerized Conferencing

This bibliography is drawn from Murray Turoff, Philip Enslow, Starr Roxanne Hiltz, John McKendree, Raymond Panko, David Snyder, and Richard Wilcox, 1978, *Research Options and Imperatives in Computerized Conferencing,* New Jersey Inst. Technol. Computerized Conferencing and Commun. Center Research Rept. No. 10. It covers published material in the area of computerized conferencing, message systems, electronic mail, and other closely related subjects. The only exceptions are a few items in the area of electronic townhalls and citizen participation systems in which the authors refer to needs that seem to be met by this particular technology.

The major efforts in this area are represented by

(1) the early work at the Office of Emergency Preparedness (OEP)

(2) the efforts at the Institute for the Future (IFTF), and the New Jersey Institute of Technology (NJIT)

(3) the work at Bell Canada (BC) (not totally represented by publications because of the company-confidential nature of some of that effort)

(4) the emergence of concern in this area in recent years on the ARPANET and associated organizations such as the Massachusetts Institute of Technology (MIT), and the Stanford Research Institute (SRI).

Smaller efforts worthy of note are

(1) the early work from a Computer-Assisted Instruction (CAI) direction at Northwestern University (NU), and the University of Illinois (UI)

(2) the recent emergence of a program in this area at the University of Michigan (UM)

(3) a general increase in publications by individuals in an assortment of organizations and institutions.

An attempt has been made to list only papers published or accepted for publication, but in a few cases in which an unpublished manuscript represents the only source of information on the work, a citation has been included.

Initials following an entry indicate that the work was done in association with one of the above-cited organizations.

Agnew, Carson E., et al. 1974. "ARPANET Management Study: New Application Areas," Cabledata Associates Quarterly Tech. Rept. No. 2. (NTIS AD-787-039/7SL.) (ARPA)

Amara, Roy, and Vallee, Jacques. 1974. "FORUM: A Computer-Based System to Support Interaction among People," *Proc. 1974 IFIP Congr. 5 (Systems for Management and Administration);* IFTF Rept. No. P-39. (IFTF)

Anderson, Peter. 1975. "A Structured Approach to Computerized Conferencing," *Proc. IEEE Symposium on Computer Networks.* (NJIT)

———, 1976. "A Language for Describing Human Communication Structures," *Proc. 3rd Intern. Conf. Computer Commun.* (NJIT)

———, 1977. "An Investigation of Computer Language Structures Required for Describing Human Communication Processes," NJIT Computerized Conferencing and Commun. Center Research Rept. No. 5. (NJIT)

Arnold, George W., and Unger, Stephen H. 1977. "A Structured Data-Base Computer Conferencing System," *Proc. AFIPS Nat. Computer Conf.* **46**

Baran, Paul, et al. 1972. ARPA Policy-Formulation Interrogation Network: Semiannual Tech. Rept. (NTIS AD-749-800.) (IFTF)

Baran, Paul, Lipinski, Hubert M., Miller, Richard H., and Randolph, R. N. 1973. ARPA Policy-Formulation Interrogation Network: Semiannual Tech. Rept. to Advanced Research Projects Agency. (NTIS AD-758-716.) (IFTF)

Batchman, Ted E., and Wearing, A. J. 1976. "Application of Dynamic Modeling to the Social Consequences of Telecommunications," *IEEE Trans. Systems, Man, and Cybernetics* **SMC-6,** No. 9.

Bechtold, B., et al. 1972. "EMISARI; Management Information System Desigend to Aid and Involve People," *Proc. 4th Intern. Symposium on Computer and Information Sci.* (OEP)

Bedford, M. 1972. Future of Communication Services in the Home. Bell Canada. (BC)

———, 1975. "Trends in Teleconferencing and Computer-Augmented Management Systems," *Proc. IEEE Nat. Conf.* (BC)

Bewick, Joseph R. 1975. Telecommunications Substitutability for Travel: An Energy Conservation Potential, Appendix A. Office of Telecommun., and Office of Energy Conservation and Environment, Washington, D.C. (NTIS PB-249-511/7ST.)

Bhushan, A. K., et al. 1973. "Standardizing Network Mail Headers," ARPANET Request for Comments No. 561, Network Information Center No. 18516, Stanford Research Inst., Menlo Park, Calif. (ARPA)

Billingsley, Rat V. 1972. System Simulation as an Interdisciplinary Interface in Rural Development Research. (Unpublished.) Texas Agricultural Experiment Station Tech. Article 9915. College Station, Texas.

Bretz, Rudy, et al. 1976. "A Teleconference on Teleconferencing," Working Paper, USC Information Sci. Inst. ISI/WP-4. (ARPA)

Brown, David. 1976. "Teleconferencing and Electronic Mail," *Educom Bull.* **11**(4).

———, 1977. *MACC-TELEMAIL System: User Manual.* Academic Computing Center, Univ. of Wisconsin, Madison. (UM)

Canning, Richard. 1977. "Computer Message Systems," *EDP Analyzer* **15**(4).

Carlisle, James H. 1975. *A Selected Bibliography on Computer-Based Teleconferencing Communications.* Information Sci. Inst. USC and Annenberg School of Communications.

———, 1975. "A Tutorial for Use of the TENEX Electronic Notebook-Conference (TEN-C) System on the ARPANET," Information Sci. Inst. Rept. No. ISI/RR-75-38. Marina del Rey, Calif. ARPA Order No. 2930.

———, 1975. *Behavioral Problems in Computer-Based Conferencing.* Information Sci. Inst. USC, Marina del Rey, Calif.

———, 1976. "Evaluating the Impact of Office Automation on Top Management Communication," Information Sci. Inst. USC and Annenberg School.of Communications.

Carter, George E. 1973. *Second Generation Conferencing Systems.* Computer-Based Educational Research Laboratory, Univ. of Illinois, Urbana. (UI)

———, 1973. *Computer-Based Community Communications.* Computer-Based Educational Research Laboratory, Univ. of Illinois, Urbana. (UI)

———, 1974. "Confer — A Preliminary Design Concept," *Proc. IEEE Intern. Conf. Systems, Man, and Cybernetics,* (Dallas, Texas). (UI)

Cerf, Vinton G., and Curran, Alex. 1977. "The Future of Computer Communications," *Datamation* **23.**

Conrath, David W. 1972. "Teleconferencing: The Computer, Communication, and Organization," *Proc. 1st Intern. Conf. Computer Commun.* (BC)

Conrath, E. W., and Bair, J. H. 1974. "The Computer as an Interpersonal Communication Device: A Study of Augmentation Technology and Its Apparent Impact on Organizational Communication," *Proc. 2nd Intern. Computer Commun.* (BC)

Crickman, Robin. 1977. "The Value of Computer Conferencing in Values Discussion Held among Citizens," *Proc. Ann. North Amer. Meeting Soc. General Systems Research.* (UM)

Crocker, David. 1976. "User-Level Functions in MS: A Network-Oriented Message System for Personal Computing," Information Sci. Dept. R-2134-arpa. Rand Corporation. (ARPA)

Davies, M., and Shore, W. 1974. "Computer-Mediated Interaction – Computer Conferencing and Beyond," *Telesis* **3**(7). (BC)

Day, Lawrence H. 1972. "Electronic Mail Services in the Information Age," Bell Canada Business Planning Paper 1. (BC)

————, 1973. "Future of Computer and Communication Services," Bell Canada Business Planning Paper 6. (BC)

————, 1973. "An Assessment of Travel/Communications Substitutability," *Futures.* (BC)

————, 1974. "The Impact of Computer Conferencing on Less Developed Countries," Mexicon 74, Mexico City, Mexico (August). (BC)

————, 1974. "Factors Affecting Future Substitution of Communications for Travel," *Proc. Joint Meeting Operations Research Soc.* (BC)

————, 1975. "Computer Conferencing: An Overview," in *Views from ICCC 1974* (N. Macon, ed.). International Council for Computer Communications, Washington D.C., and Bell Canada, Montreal, Quebec, Canada. Also in *Proc. Conf. Telecommun. Policy Res.* (April, 1975). (BC)

————, 1975. "Future Opportunities in Telecommunications," *Proc. World Future Soc. Meeting.* (BC)

Dowson, Mark. 1975. *SCRAPBOOK Teleconferencing System: Interim Progress Rept. I.* Mark Dowson Associates, 23 Priory Terrace, London.

Edwards, Gwen C. 1974. *Group Communications through Computer Conferencing: An Analysis of Interaction.* Stanford Research Inst. Telecommun. Sci. Center, Menlo Park, Calif. (SRI)

————, 1975. "Computer Augmentation of Text Processing and Communication Functions," Bell Canada Business Planning Paper 43.

Englebart, Douglas C. 1970. "Intellectual Implications of Multi-Access Computer Networks," *Proc. Interdisciplinary Conf. Multi-Access Computer Networks* (Austin, Texas). (SRI)

————, 1971. "Network Information Center and Computer Augmented Team Interaction," Interim Tech. Rept. (June). (NTIS AD-737-131.) (SRI)

————, 1973. "Coordinated Information Services for a Discipline- or Mission-Oriented Community," *Proc. 2nd Ann. Computer Commun. Conf.* (San Jose). (SRI)

————, 1975. "NLS Teleconferencing Features: The Journal and Shared-Screen Telephoning," *Proc. IEEE COMPCON Fall.* (SRI)

Englebart, D. C., Watson, R. W., and Norton, J. C. 1973. "The Augmented Knowledge Workshop," *Proc. AFIPS Nat. Computer Conf.* **42.** Also in *Computer Networking,* IEEE Press, 1976. (SRI)

Etzioni, Amitai. 1972. "Minerva: An Electronic Town Hall," *Policy Sci.* **3**(4).

Featheringham, Tom. 1977. "Present and Potential Value of Computer Communications in Information Science," *Proc. Midyear Meeting Amer. Soc. Information Sci.* (NJIT)

————, 1977. "Computer Conferencing and Its Potential for Business Management," *Data Communications.* (NJIT)

————, 1977. "Teleconferences: The Message Is the Meeting," *Data Communications.* (NJIT)

————, 1977. "Computerized Conferencing and Human Communication," *IEEE Trans. Professional Commun.* **PC-** . (NJIT)

Ferguson, John. 1977. "Evaluation through a Case Study," *Behavior Research Methods and Instrumentation* **9** (2), pp. 92-95. (IFTF)

Ferguson, John, and Johansen, Robert. 1975. "Teleconference on Integrated Data Bases in Postsecondary Education. A transcript and summary," Institute for the Future, Menlo Park, Calif. (IFTF)

Fields, Craig. 1977. "Terminal Access Technology of the 1990's," *IEEE Trans. Professional Commun.* **PC-20.** No. 1.

Flood, Merrill M. 1974. Working Paper FRPUB/mmfl, 8 July 1974, 6 pp. (This untitled paper suggested that "computer-aided teleconferencing may enable a university community to restructure itself organically for the governance function [p. 9]".) (UM)

———, 1974. "Let's Redesign Democracy," Working paper FRPUG/mmf26, 20 July 1976, 49 pp. (UM)

———, 1975. "Can our universities govern themselves better?" *Univ. of Michigan School of Education Innovator,* January, pp. 6-7. (UM)

———, 1976. "Dynamic Voting Computer Programs." Working Paper FRPUB/mmf28, 1 April 1977, 6 pp. (UM)

Freeman, Linton. 1977. "Computer Conferencing and Productivity in Science," *Transnat. Associations, (J. Union of Intern. Associations)* **10**.

Gage, Howard, and Turoff, Murray. 1975. "Computerized Conferencing and the Physically Handicapped," NJIT Computerized Conferencing and Commun. Center Research Rept. No. 6. (NJIT)

Gusdorf, et al. 1971. "Puerto Rico's Citizen Feedback System," *Proc. AFIPS Spring Joint Computer Conf.* **38**.

Hall, Thomas. 1971. "Implementation of an Interactive Conference System," *Proc. AFIPS Spring Joint Computer Conf.* **38**. (OEP)

Hall, Thomas, and Grover, James. 1974. "CONFAB: A Computerized Conference System," Language and Systems Development, Inc., Silver Spring, Maryland. (OEP)

Harkness, R. C. 1975. "Telecommunications Substitutes for Travel," U.S. Dept. of Commerce Office of Telecommun. Rept. No. COM-74-10075.

Haverty, J. 1976. "Message Services Data Transmission Protocol," RFC 713, NIC 34739. (ARPA)

Helmer, Olaf. 1970. "Toward the Automation of Delphi," Internal Tech. Memo., Institute for the Future, Menlo Park, Calif. (IFTF)

Hiltz, Starr Roxanne. 1975. "Communications and Group Decision-Making: Experimental Evidence on the Potential Impact of Computer Conferencing," NJIT Computerized Conferencing and Commun. Center Research Rept. No. 2. (NJIT)

———, 1975. "The Potential Social Impacts of Some Near Future Developments in Computer Conferencing."

———, 1976. "A Social Scientist Looks at Computer Conferencing," *Proc. 3rd Intern. Conf. Computer Commun.* (NJIT)

———, 1977. "Computer Conferencing: Assessing the Social Impact of a New Communications Medium," *Technological Forecasting and Social Change* **10**(3). (NJIT)

———, 1977. "The Impact of a Computerized Conferencing System upon Scientific Research Specialities," Paper presented at AAAS, Denver, Feb., 1977. Forthcoming in *J. Research-Commun. Studies,* 1978. (NJIT)

———, 1977. The Human Element in Computerized Conferencing Systems. Paper presented at Amer. Soc. Information Sci. Meeting, Chicago. Forthcoming in *Computer Networks,* 1978. (NJIT)

Hiltz, Starr Roxanne, and Turoff, Murray. 1977. "Effective Communications Structures for Technology Assessment," *Technology Assessment/Textile Adjuvants* (C. Moore, G. Inglett, J. Craver, and R. Carlson, eds.), pp. 67-85. Chemical Marketing and Economics Reprints, Amer. Chem. Soc., New York. (NJIT)

———, 1977. "Computerized Conferencing: Meeting through Your Computer," *IEEE Spectrum,* May, pp. 58-64. (NJIT)

———, 1977. "Potential Impacts of Computer Conferencing upon Managerial and Organizational Styles," short version published in *Transnat. Associations (J. Union of Intern. Associations)* **10**, pp. 437-439. (NJIT)

Hollander, Stephen. 1977. "Computer Conferencing and the Transfer of Information," *Proc. 15th Canadian Conf. Information Sci.*

Hopewell, Lynn, Cerf, Vinton G., Curran, A., and Dunn, D. A., eds. 1976. *Computers and Communications (Proc. FCC Planning Conf.)* AFIPS Press, Montvale, N.J.

Horman, Aiko M. 1971. "A Man-Machine Synergistic Approach to Planning and Creative Problem-Solving," *Intern. J. Man-Machine Studies* **3**(2), 167-184.

Hough, R. W., and Panko, R. R. 1977. "Teleconferencing Systems: A State-of-Art Survey and Preliminary Analysis," SRI Project 3735.

Hunt, L. G. 1973. *Characteristics of Computer Information Systems: A Theoretical Construct and Case Studies.* U.S. Govt. Office of Telecommun.

Institute for the Future. 1973. "Development of a Computer-Based System to Improve Interaction among Experts," IFF Special Rept. SR-25. *Proc. 1st Ann. Rept. Nat. Sci. Found.*

———, 1974. *Social Assessment of Mediated Group Communication: A Workshop Summary.*

———, 1976. *Planet User Guide*

Irving, R. H. 1976. "Usage of Computer Assisted Communication in an Organizational Environment," Research Rept., Dept. of Management Sci., Univ. of Waterloo, for Canadian Dept. of Commun.

Jillson, Irene N. 1975. *Final Evaluation Report, Nonmedical Use of Drugs Computer Conferencing System Pilot Phase.* Nonmedical Use of Drugs Directorate, Ottawa, Ontario, Canada.

Johansen, Robert. 1972. "Sociological Applications of Interactional Computer Systems: Toward a Dialogical Research Model," in *Sociology in Action* (Arthur B. Shostak, ed.) 2nd ed., pp. 185-191. New York: David McKay Co.

———, 1975. Developing New Communications Media: Can We Account for the Human Dimensions? Paper presented at 2nd General Assembly, World Future Soc., Washington, D.C. (NW)

———, 1976. "Pitfalls in the Social Evaluation of Teleconferencing Media," in *The Status of the Telephone in Education.* Parker, L.A., and Riccomini, B., eds., Univ. of Wisconsin Extension Press, Madison. *Proc. 2nd Ann. Intern. Commun. Conf.* (IFTF Rept. No. P-38.)

Johansen, Robert, and Miller, Richard H. 1973. *Commentary on One Use of FORUM in a Research Environment.* Institute for the Future, Menlo Park, Calif.

———, 1974. "The Design of Social Research to Evaluate a New Medium of Communication." Paper presented at *Proc. Ann. Meeting Amer. Sociol. Assoc.* (IFTF)

Johansen, Robert, and Schuyler, J. A. 1975. "Computerized Conferencing in an Educational System: A Short-Range Scenario," in *The Delphi Method: Techniques and Applications* (H. A. Linstone and M. Turoff, eds.), pp. 550-562. Addison-Wesley, Reading, Mass. (NW)

Johansen, Robert, and Vallee, Jacques. 1976. "Impact of a Computer-Based Communications Network on the Working Patterns of Researchers," IFTF Rept. No. P-46.

Johansen, Robert, DeGrasse, Robert, and Wilson, T. 1978. *Group Communication through Computers* Vol. 5 *(Effects on Working Patterns).* Institute for the Future, Menlo Park, Calif.

Johansen, Robert, Miller, Richard, and Vallee, Jacques. 1974. "Group Communication through Electronic Media: Fundamental Choices and Social Effects," *Educ. Technol.,* August, pp. 7-20. (IFTF Rept. No. P-27.)

Johansen, Robert, Vallee, Jacques, and Palmer, Michael. 1976. "Computer Conferencing: Measurable Effects on Working Patterns,"IFTF Paper P-44.

Johansen, Robert, Vallee, Jacques, and Collins, Kent. 1977. "Learning the Limits of Teleconferencing: Design of a Teleconference Tutorial," *Proc. NATO Symposium on Evaluation and Planning of Telecommun. Systems* Univ. of Bergamo, Italy, September. (To be published by Plenum, 1978.) (IFTF)

Johansen, Robert, Vallee, Jacques, Spangler, K., and Shirts, G. 1977. "The Camelia Report: A Study of Technical Alternatives and Social Choices in Teleconferencing," IFTF Intermedia Project Rept. No. R-36.

Johnson-Lenz, Peter, and Johnson-Lenz, Trudy. 1977. "Conference Facilitation by Computer-Aided Sharing," *Transnat. Associations (J. Union of Intern. Associations* **10,** pp. 440-445. (NJIT)

Kochen, Manfred. 1977. "On-Line Intellectual Communities," *Ann. North Amer. Meeting Soc. General Systems Research,* pp. 165-167. (Also published in *Transnat. Associations (J. Union of Intern. Associations* **10,** pp. 425-427.) (UM)

Kollen, James H. 1973. Teleconferencing: Present Research and Future Expectations. Bell Canada Business Planning Group.

Kollen, James H., and Edwards, Gwen C. 1974. *Group Communication through Computer Conferencing: An Analysis of the First International Computer-Based Conference on Travel/Communication Relationships.* Institute for the Future, Menlo Park, Calif., and Bell Canada Business Planning Group.

Kollen, James H., and Garwood, John. 1975. Travel/Communication Tradeoffs: The Potential for Substitution among Business Travellers. Report published by Bell Canada Business Planning Group.

Kollen, James H., and Vallee, Jacques. 1974. *Proc. 1st Intern. Computer-Based Conf. on Travel/Commun. Relationships.* Bell Canada Business Planning Group and Institute for the Future, Menlo Park, Calif.

Krutar, R., et al. 1975. *Proposal for a Question-Based Information Retrieval System (A Controlled Confusion Reactor).* Univ. of Utah, Salt Lake City.

Kupperman, Robert H., and Goldman, Steven C. 1975. "Toward a Viable International System: Crisis Management and Computer Conferencing," in *Views from ICCC 1974* (N. Macon, ed.), pp. 71-80. Intern. Council for Computer Commun., Washington, D.C. (OEP)

Kupperman, Robert H., and Wilcox, Richard H. 1972. "EMISARI – An On-Line Management System in a Dynamic Environment," *Proc. 1st Intern. Conf. Computer Commun.* pp. 117-120. (Also available as OEP Rept. ISP-108 and NTIS AD 744-348.)

Kupperman, Robert H., and Wilcox, Richard H. 1974. "Interactive Computer Communications Systems – New Tools for International Conflict Deterrence and Resolution," in *Computer Communication Today and Up to 1985 (Proc. 2nd Intern. Conf. Computer Commun.),* pp. 469-471. (OEP)

Kupperman, Robert H., Wilcox, Richard H., and Smith, Harvey A. 1975. "Crisis Management: Some Opportunities," *Science* **187,** pp. 404-410. (OEP)

Lamont, Valerie C. 1972. "New Directions for the Teaching Computer: Citizen Participation in Community Planning," Univ. of Illinois Computer-Based Educational Research Lab. Rept. X-34.

Lathey, Charles E., and Bawick, Joseph E. 1975. "Bibliography of Selected Abstracts of Documents Related to Energy Conservation through Telecommunications," Office of Telecommun. Rept. No. OT-SP-75-5. (NTIS COM-75-11376/OST.)

Leonard, Eugene, et al. 1971. "Minerva: A Participatory Technology System," *Bull. Atomic Scientists* (November).

Licklider, J. C. R., Taylor, R. W., and Herbert, W. 1968. "The Computer as a Communication Device," *Intern. Sci. Technol.* **76,** pp. 21-23.

Liffman, B. 1974. "Generalized Conferencing System: Participant's Documentation," Memo from Turner Corp., Toronto, Canada.

Linstone, Harold A., and Turoff, Murray. 1975. *The Delphi Method: Techniques and Applications.* Addison–Wesley, Reading, Mass.

Lipinski, Hubert, and Miller, R. H. 1974. "FORUM: A Computer-Assisted Communications Medium," *Proc. 2nd Intern. Conf. Computer Commun.* pp. 143-147. (IFTF) (IFTF Rept. No. P-26.)

Lipinski, Hubert M., and Randolph, Robert H. 1973. "Conferencing in an On-Line Environment: Some Problems and Possible Solutions," *Proc. 2nd Ann. Computer Commun.*

Lipinski, Andrew, Lipinski, Hubert, and Randolph, Robert H. 1972. "Computer-Assisted Expert Interrogation. A Report on Current Methods Development," *Proc. 1st Intern. Conf. Computer Commun.* pp. 147-154. (Also available as Stanford Research Inst. Augmentation Research Center Catalog Item 11980; *Technological Forecasting and Social Change* 5, 1974, pp. 3-18; and as IFTF Rept. No. P-24.

Lukasik, Stephen J. 1974. "Organizational and Social Impact of a Personal Message Service," *Proc. IEEE Commun. Soc. Nat. Telecommun. Conf.*

McKendree, John. 1973. *RIMS Systems Expanations.* Office of Emergency Preparedness, Washington, D.C.

McKendree, John D. 1977. "Decision Process in Crisis Management: Computers in a New Role," in *Encyclopedia of Computer Science and Technology* (J. Belzer, et al., eds.), 7, pp. 115-156. Marcel Dekker, New York.

———, 1977. "The Future of Computerized Conferencing," *Proc. Data Processing Management Assoc. INFO/EXPO* (October). (Also published in *Data Management,* Jan. 1978.) (OEP)

McKendree, John, Wynne, R., and Macon, Nathaniel. 1975. "Computer Conferencing in Emergencies: Some Reliability Considerations," *Proc. IEEE 1975 Symposium on Computer Networks: Trends and Applications,* pp. 69-73.

Macon, Nathaniel, ed. 1975. *Views from ICCC 1974.* Intern. Council for Computer Commun., Washington, D.C.

Macon, Nathaniel, and McKendree, John D. 1974. "EMISARI Revisited: The Resource Interruption Monitoring System," *Proc. 2nd Intern. Conf. Computer Commun.* pp. 89-97, Office of Emergency Preparedness Rept. No. GSA/OP ISG-116.

Meals, Donald W. 1969. "Games as Communications Devices," in *Proc. 8th Symposium, Nat. Gaming Council.* Boston: Booz Allen Applied Research, Inc.

Millard, Gord C., and Williamson, Hilary. 1975. *Computer Mediated Interaction (CMI): User Guide.* Bell Canada Computer Commun. Group, Ottawa.

———, 1976. "How People React to Computer Conferencing," *Telesis* 4(7), pp. 214-219. (BC)

Miller, Richard H. 1973. "Trends in Teleconferencing by Computer," *Proc. 2nd Ann. Computer Commun. Conf.* (IFTF)

Myer, Theodore H. 1976. *Hermes Users Guide.* Bolt, Beranek and Newman Inc., Cambridge, Mass. (ARPA)

Myer, Theodore H., and Dodds, D. W. 1976. Notes on the Development of Message Technology. Paper presented at Berkeley Workshop on Distributed Data and Computer Networks. Write authors at Bolt, Beranek and Newman Inc., Cambridge, Mass. (ARPA)

Myer, Theodore H., and Henderson, D. Austin. 1975. *Message Transmission Protocol.* Network Working Group RFC 680, NIC 32216. (ARPA)

Ness, David N. 1976. "Office Automation Project: Automated Conference Support," Working Paper 76-0302, Dept. of Decision Sci., Wharton School, Univ. of Pennsylvania.

———, 1977. "Office Automation Today and Tomorrow," *Educom Bull.* 12(3), pp. 2-12.

———, 1977. "Office Automation Project: Responsive Mail," Working Paper 77-01-07, Dept. of Decision Sci., Wharton School, Univ. of Pennsylvania.

Nilles, Jack M., Carlson, F., Gray, P., and Hanneman, G. 1976. *The Telecommunications-Transportation Tradeoff: Options for Tomorrow.* Wiley, New York. (NTIS PB-241-87.)

Nixon, Helen R. 1974. "The Resource Interruption Monitoring System: Basic Reporting Procedures and Sample Reports," Office of Preparedness, Rept. No. GSA/OP ISG-117.

Noel, Robert. 1972. *POLIS: A Computer-Based System for Simulation and Gaming in the Social Sciences.* Dept. of Political Sci., Univ. of California at Santa Barbara.

———, 1973. "POLIS: A Resource Sharing Network for Instructional Simulation and Gaming," Paper presented at the *Proc. Ann. Meeting Amer. Political Sci.,* New Orleans.

———, 1974. "The 'POLIS Network' for Resource-Sharing in Simulation and Gamin," Political Institutions Simulation Lab. 1974. Dept. of Political Sci., Univ. of California at Santa Barbara.

Noel, Robert, and Jackson, T. 1972. "An Information Management System for Scientific Gaming in the Social Sciences," *Proc. AFIPS Spring Joint Computer Conf.* 40, pp. 897-905.

North, Steve. 1977. "Computer Conferencing: A Personal View," *Creative Computing* 3, 5, p. 60.

Panko, R. R. 1975. Computer Teleconferencing: History, Status, Trends and Questions. Stanford Research Inst. Draft.

———, 1976. The Outlook for Computer Message Services: A Preliminary Assessment. Stanford Research Inst. Draft.

Parnes, Robert, Hench, Chris H., and Zinn, Karl. 1977. "Organizing a Computer-Based Conference," *Transmat. Associations (J. Union of Intern. Associations* 10, pp. 418-422.) (UM)

Pogran, K., Vittal, J., Crocker, D., and Henderson, A. 1977. *Proposed Official Standard for the Format of ARPA Network Messages.* RFC 724, NIC 37435.

Prager, D. A. 1972. "A Proposal for a Computer-Based Interactive Scientific Community," *Commun. ACM* (Feb.) 15, pp. 2, 71-75.

Price, Charlton. 1973. "Computer Conferencing: Innovative Networks for Innovation," Program for Policy Studies in Sci. and Technol., George Washington Univ., Washington, D.C.

———, 1975. "Conferencing via Computer: Cost Effective Communication for the Era of Forced Choice," in *The Delphi Method: Techniques and Applications* (H. A. Linstone and M. Turoff, eds.), pp. 497-516. Addison-Wesley, Reading, Mass. (George Washington Univ., Washington, D.C., January 1974.) Special Rept. No. IIAP/PPSST.

Remp, R. 1974. "The Efficacy of Electronic Group Meetings," *Policy Sci.* (May).

Renner, R. L. 1972. *Conference: System User's Guide.* Office of Emergency Preparedness Tech. Manual No. 225. (NTIS AS-756-813.)

Renner, Rodney, and Turoff, Murray. 1973. Conference System Tape for 1100 Series Univacs. NTIS Conference-MT-1/2. (OEP)

Renner, R. L., Bechtold, R. M., Clark, C. W., Marbray, D. O., Wynn, R. L., and Goldstein, N. H. 1972. *Proc. 4th Intern. Symposium Computers and Information Sci.* [Office of Emergency Preparedness Rept. TM-230, February 1973; *Information Systems* (COINS IV) Plenum Press, New York, 1974; NTIS PB-224 852.] (OEP)

Rothenberg, Jeff. 1975. "An Editor to Support Military Message Processing Personnel," USC Information Sci. Inst. ISI/RR-74-27. (ARPA)

Rouse, William B. 1973. "SOLVER, A Group Computer Interactive Package for Problem-Solving," MIT Community Dialog Project Rept. No. 4.

Scheele, D. Sam, De Sante, Vincent, and Glasser, Edward. 1971. *GENIE: Government Executives Normative Information Expediter.* State of Wisconsin, and Information Services Division, U.S. Office of Economic Opportunity.

Scher, Julian M. 1975. *Simulation, Simulation-Games and Computerized Conferencing: A Laboratory Synthesis.* Paper presented at Co Ed (Computers in Education), ASEE meeting. (NJIT)

——, 1976. "Computerized Conferencing and the Simulation Study of Human Communication Processes," *Proc. Summer Computer Simulation Conf.*, pp. 861-867.

——, 1976. "The Constrained Computer Conference — A Methodological Tool for the Implementation Phase of Simulation-Games," *Proc. 3rd Intern. Conf. Computer Commun.*, pp. 230-235.

——, 1977. "Communication Processes in the Designed Implementation of Models, Simulations and Games: A Selective Review from the Vantage Point of Computerized Conferencing," NJIT Computerized Conferencing and Commun. Center Research Rept. No. 4.

Scher, Julian, and Turoff, Murray. 1975. "Computerized Conferencing and Its Impact on Engineering Management," *Proc. 23rd Ann. Joint Engineering Management Conf.*, (Washington, D.C., October); also in *Professional Engineer* 45(12), pp. 16-19. (NJIT)

Schuyler, J., and Johansen, Robert. 1972. "ORACLE: Computerized Conferencing in a Computer-Assisted Instruction System," *Proc. 1st Intern. Conf. Computer Commun.*, pp. 155-160. (NW)

Scientific Time Sharing Corp. 1975. *APLplus Message Processing System: Mailbox User's Guide.* Bethesda, Md.

Sheridan, Thomas B. 1971. "Technology for Group Dialogue and Social Choice," *Proc. AFIPS Fall Joint Computer Conf.* 39, pp. 327-335. [Reprinted in *The Delphi Method: Techniques and Applications* (H. A. Linstone and M. Turoff, eds.), pp. 535-549. Addison-Wesley, Reading, Mass., 1975.] (MIT)

——, 1971. "Citizen Feedback: New Technology for Social Choice," *Technol. Rev.* (January). (MIT)

Shore, J. W., and Davies, M. A. 1973. *An Instruction Manual for Conferencing on BNR's Experimental C.M.I. System.* Bell-Northern Tech. Rept. TR-3c30-1-73.

Smith, Vallee J., Johansen, R., and Turoff, M. 1976. "Computerized Conferencing — A New Medium," *Mosaic* 7(1), pp. 16-22.

Treu, Siegfried. 1975. "On-Line Student Debate: An Experiment in Communication Using Computer Networks," *Intern. J. Computer and Information Sci.* 4(1).

Treu, Siegfried, and Pyke, Thomas N., Jr. 1973. "Project-Oriented Collaboration via a Computer Network," *Proc. Computer Sci. Conf. (Columbus, Ohio).* (Nat. Bur. of Standards Project) NBS-6502121.

Tugender, Ronald, and Oestreicher, Donald. 1975. "Basic Functional Capabilities for a Military Message Processing Service," USC Information Sci. Inst. ISI/RR-74-23. (ARPA)

Turoff, Murray. 1969. "User and Program Management Considerations in the Computer Analysis of Strategic War," *Proc. 8th Symposium, Nat. Gaming Council,* Booz Allen Applied Research, Inc. (OEP)

——, 1970. "The Design of a Policy Delphi," *Technological Forecasting and Social Change* 2(2), pp. 149-172. (Reprinted as "The Policy Delphi" in *The Delphi Method: Techniques and Applications* [H.A. Linstone and M. Turoff, eds.], pp. 84-101. Addison-Wesley, 1975.) (OEP)

——, 1971. "Delphi + Computers + Communications = ?," in *Industrial Application of Technological Forecasting* (M. Cetron and C. Ralph, eds.), pp. 243-258. Wiley-Interscience, New York. (OEP)

——, 1971. "The Delphi Conference," *Futurist* 5(2), pp. 55-57. (OEP)

———, 1971. "Delphi and Its Potential Impact on Information Systems," *Proc. AFIPS Fall Joint Computer Conf.* **39**, pp. 317-326. (OEP)

———, 1972. "Delphi Conferencing: Computer Based Conferencing with Anonymity," *Technological Forecasting and Social Change* **3**(2), pp. 159-204. (OEP)

———, 1972. "An Alternative Approach to Cross Impact Analyses," *Technological Forecasting and Social Change* **3**(3), pp. 309-339. (Reprinted in *The Delphi Method: Techniques and Applications* [H.A. Linstone and M. Turoff, eds.], pp. 338-368. Addison-Wesley, Reading, Mass., 1975.) (OEP)

———, 1972. " 'Partyline' and 'Discussion' Computerized Conference Systems," *Proc. 1st Intern. Conf. Computer Commun.*, pp. 161-171. (OEP Rept. ISP-109; NTIS AD-744-349.) (OEP)

———, 1972. "Meeting of the Council on Cybernetic Stability: A Scenario," *Technological Forecasting and Social Change* **4**, pp. 121-128. *Data Exchange,* pp. 5-6, and (Reprinted in *The Delphi Method: Techniques and Applications* [H.A. Linstone and M. Turoff, eds.], pp. 563-569. Addison-Wesley, Reading, Mass., 1975.) (OEP)

———, 1972. "Conferencing via Computer," *Proc. IEEE Information Networks Conf.*, pp. 194-197. (OEP)

———, 1973. "Human Communication via Data Networks," *Computer Decisions* **5**(1), pp. 25-29. Also published in Blanc, R., and Cotton I., eds., *Computer Networking,* IEEE Press, 1976, pp. 241-245; and in *Ekistics* **35**(211), pp. 337-341. (OEP)

———, 1973. "Communication Procedures in Technological Forecasting," *Intercom Papers* **7**, paper 28/4. (OEP)

———, 1973. "Opposing Views of the Future," *Proc. AFIPS; Nat. Computer Conf.* **42**, pp. 717-722. (NJIT)

———, 1973. "Little Things That Count," ACM SIG Information Retrieval, *Forum* **8**(2), (Summer). (OEP)

———, 1974. "Potential Applications of Computerized Conferencing in Developing Countries," *Ekistics* **38**(225), pp. 131-132. [Also published in *Proc. Special Rome Conf. Futures Research; Herald of Library Sci.* **12**(2-3), pp. 157-160; and *Fields within Fields within Fields* **13**, pp. 62-66.] (OEP)

———, 1974. "Introduction to Delphi," in *Information Science: Search for Identity* (A. Debons, ed.), pp. 225-237. Marcel Dekker, New York.

———, 1974. "Interview on Computer Communications," *Data Exchange* (Oct.), pp. 3-9. (NJIT)

———, 1974. "Computerized Conferencing and Real Time Delphis; Unique Communication Forms," *Proc. 2nd Intern. Conf. Computer Commun.*, pp. 135-142. (NJIT)

———, 1975. "The State of the Art: Computerized Conferencing," in *Views from ICCC 1974* (N. Macon, ed.), pp. 81-86. Intern. Council for Computer Commun., Washington, D.C. (NJIT)

———, 1975. "The Future of Computer Conferencing," *Futurist* **9**(4), pp. 182-195. [Also published in *Law Office Economics and Management* **18**(2), pp. 235-248; and in *Transnational Associations* **10** (1977) pp. 404-408. (NJIT)

———, 1975. "Innovations: New Uses for the Public Library," *Educ. Tomorrow* **1**(1), pp. 1-2. (NJIT)

———, 1975. "Computerized Conferencing: Present and Future," *Intellect* (September/October), pp. 130-133. (NJIT)

———, 1975. "All at Once," *nce alumnus* **28**(4), pp. 19-20. NJIT: NCE stands for Newark College of Engineering, the institution's previous name.

———, 1975. "Initial Specifications, Electronic Information Exchange System (EIE)," NJIT Computerized Conferencing and Commun. Center Research Rept. No. 1.

———, 1975. Computerized Conferencing for the Deaf and Disabled. Paper presented at 141st Meeting of the AAAS. [Published in *Urban Telecommun. Forum* **4**(33), pp. 5-8; and in *ACM SIGGAPH Newsletter* **16** pp. 4-11.] (NJIT)

———, 1976. "Assessing the Future Impact of Computers and Communication," in *Discoveries (Proc. Intern. Discoveries Symposium),* pp. 113-129. Honda-Fujusawa Memorial Found., Tokyo. (NJIT)

———, 1976. "The Costs and Revenues of Computerized Conferencing," *Proc. 3rd Intern. Conf. Computer Commun.*, pp. 214-221. (NJIT)

——, 1977. "Computerized Conferencing: Present and Future," *Creative Computing,* pp. 3, 5, 54-57. (NJIT)

——, 1977. An On-Line Intellectual Community or MEMEX Revisited. Paper presented at the Ann. Meeting of the AAAS. [Published in *Technological Forecasting and Social Change* 10(4), pp. 401-412.] (NJIT)

Turoff, Murray, and Hiltz, Starr Roxanne. 1977. "The Threat of Premature Regulation of Computer Conferencing," *Transnat. Associations (J. Union of Intern. Associations)* **10,** pp. 436-437. (NJIT)

——, 1977. "Impacts of Computer Conferencing upon Organizational Styles," Abstracted in *Transnat. Associations (J. Union of Intern. Associations)* **10,** pp. 437-439. (NJIT)

——, 1977. "EIES Development and Evaluation Project Final Report," Computerized Conferencing and Commun. Center Research Rept. No. 9. (NJIT)

Turoff, Murray, and Spector, Marion. 1976. "Libraries and the Implications of Computer Technology," *Proc. AFIPS Nat. Computer Conf.* **45,** pp. 701-707. (NJIT)

Turoff, Murray, Whitescarver, James, and Hiltz, Starr Roxanne. 1977. "The Human-Machine Interface in a Computerized Conferencing Environment," *Proc. IEEE Conf. Intern. Systems, Man, and Cybernetics,* pp. 145-157. Abstracted in *Transnat. Associations* **10,** pp. 431-433. (NJIT)

Uhlig, Ronald P. 1977. "Human Factors in Computer Message Systems," *Datamation* **23**(5), pp. 120-126. (ARPA)

Uhlig, Ronald P., Martin, Shirley M., and von Gehren, Edgar S. 1975. The Role of Informal Communications in Computer Networks. *Proc. of Pacific Area Computer Commun. Network System Symposium,* Japan, pp. 117-119. (ARPA)

Uhr, Leonard. 1977. "Toward Computer-Based Communication Systems That Allow Users to Choose Alternatives Best Suited to Their Needs," *Proc. Ann. North Amer. Meeting Soc. General Systems Research,* pp. 186-193.

Umpleby, Stuart. 1969. "The Delphi Exploration: A Computer-Based System for Obtaining Subjective Judgments on Alternative Futures," Univ. of Illinois Computer-Based Educational Research Lab. Rept. F-1.

——, 1971. "Structuring Information for a Computer-Based Communication Medium," **39,** pp. 337-350. (UI)

Umpleby, Stuart, and Carter, G. 1973. *Handbook on Computer-Based Communications Media.* Computer-Based Education Research Lab., Univ. of Illinois, Urbana.

Umpleby, Stuart, and Lamont, Valarie C. 1973. "Social Cybernetics and Computer-Based Communications Media," Univ. of Illinois Computer-Based Educational Research Lab., Urbana.

USC Information Sci. Inst., Marina del Rey. 1977. *Reference Manual for the SIGMA Message System.*

Vallee, Jacques. 1972. "The FORUM Project: Network Conferencing and Its Future Applications," *Proc. 1st Intern. Conf. Computer Commun.* (IFTF Rept. No. P-41.) (Also published in *Computer Networks.)* **1,** pp. 39-52.

——, 1974. "Network Conferencing," *Datamation* **20,** 5, pp. 85-86. (IFTF Rept. No. P-28.)

——, 1976. "There Ain't No User Science," *Proc. Ann. Meeting Amer. Soc. Information Sci.* published in *People's Computers* Nov.-Dec. issue, 1977. (IFTF)

——, 1977. "The Outlook for Computer Conferencing on Arpanet and Plato," *Proc. Ann. North Amer. Meeting Soc. General Systems Research,* pp. 194-201. (IFTF)

——, 1977. "Functional Characteristics of Computer Communication," paper presented at *Proc. Ann. North Amer. Meeting Soc. General Systems Research.* (IFTF)

——, 1977. "Les Conferences par Ordinateur: Planet and Forum," *Transnat. Associations (J. Union of Intern. Associations)* **10,** pp. 409-411. (IFTF)

——, 1977. "Modelling as a Communication Process: Conferencing Offers New Perspectives." *Technological Forecasting and Social Change* **10,** pp. 391-400.

Vallee, Jacques, and Askevold, Gerald. 1975. "Geologic Applications of Network Conferencing: Current Experiments with the FORUM System," in *Computer Networking and Chemistry,* Amer. Chem. Soc., New York.

Vallee, Jacques, and Gibbs, Bradford. 1976. "Distributed Management of Scientific Projects," *Telecommun. Policy* **1**(1), pp. 75-85. (IFTF)

Vallee, Jacques, and Wilson, Thaddeus. 1975. "Computer Networks and the Interactive Use of Geologic Data: Recent Experiments in Teleconferencing," *Proc. COGEODATA Symposium (Paris)* (IFTF Rept. No. P-45.) (IFTF)

_____, 1976. "Computer-Based Communication in Support of Scientific and Technical Work," IFTF Rept. No. NASA CR 137879.

Vallee, Jacques, Askevold, Gerald, and Wilson, Thaddeus. 1977. "Computer Conferencing in the Geosciences," IFTF Special Rept. SR-78.

Vallee, Jacques, Johansen, Robert, and Spangler, Kathleen. 1975. "The Computer Conference: An Altered State of Communication?," *Futurist* (June) pp. 116-121. [IFTF Rept. No. P-33] ; also published in *Creative Computing* **3,** pp 58-59 (1977). (IFTF)

Vallee, Jacques, Lipinski, Hubert M., and Miller, Richard H. 1974. *Group Communication through Computers* Vol. 1 *(Design and Use of the FORUM System)*. Institute for the Future, Menlo Park, Calif. (IFTF Rept. No. R-32.)

Vallee, Jacques, Johansen, R., Randolph, R., and Hastings, A. 1974. *Group Communication through Computers* Vol. 2 *(A Study of Social Effects)*. Institute for the Future, Menlo Park, Calif. (IFTF Rept. No. R-33.)

Vallee, Jacques, Johansen, R., Lipinski, H., Spangler, K., Wilson, T., and Hardy, A. 1975. *Group Communication through Computers* Vol. 3 *(Pragmatics and Dynamics)*. Institute for the Future, Menlo Park, Calif. (IFTF Rept. No. R-35.)

Vallee, Jacques, Johansen, R., Lipinski, H., and Wilson, T. 1976. "Pragmatics and Dynamics of Computer Conferencing: A Summary of Findings from the FORUM Project," *Proc. 3rd Intern. Conf. Computer Commun.* (P. Verma, ed.) pp. 203-213. (IFTF Rept. No. P-42.)

_____, 1977. *Group Communication through Computers* Vol. 4 *(Social, Managerial, and Economic Issues)*. Institute for the Future, Menlo Park, Calif.

Vallee, Jacques, et al. 1973. ARPA Policy-Formulation Interrogation Network: Semiannual Tech. Rept. NTIS AA-767-438/5. (IFTF)

Vezza, A., and Broos, M. S. 1976. An Electronic Message System: Where Does It Fit? Paper presented at NCC. (Write authors at MIT Lab. for Computer Sci. document sys. 16.03, Cambridge, Mass.)

Wallenstein, Gerd. 1974. An Experiment in Conducting a Computer-Based International Conference. Cybernetic Systems Monograph No. 3. San Jose State Univ., Calif. (IFTF)

_____, 1975. "Inter-Human Aspects of Telecommunications," *Proc. Intern. Computer Conf.* (IFTF)

White, J. E. 1976. "Recorded Dialog: the NLS Journal, Identification, and Number Systems," in *Knowledge Workshop Development,* Stanford Research Inst., Menlo Park, Calif.

Wilcox, Richard. 1972. "Computerized Communications, Directives, and Reporting Systems," *Proc. Information and Records Admin. Conf. (IRAC)* (May). (OEP)

Wilcox, Richard H., and Kupperman, R. A. 1972. "An On-Line Management System in a Dynamic Environment," in S. Winkler, ed. *Proc 1st Intern. Conf. Computer Commun.,* pp. 117-120. (OEP)

Williamson, H., and Bocking, D. 1976. *CMI Reference Guide: User Facilities and Commands.* Bell-Northern Research.

Wilson, Stan. 1973. A Test of the Technical Feasibility of On-Line Consulting Using APL. APL Lab., Texas Agricultural Experiment Station, Texas A&M Univ., College Station, Texas.

Winkler, Stanley, ed. 1972. Computer Communication: Impacts and Implications. *Proc. 1st Intern. Conf. Computer Communication,* Washington, D.C.

Wolff, Edward A. 1977. "Public Service Communications Satellite System Review and Experiment Definition Workshop," Rept. Commun. and Navigation Division, Applications Directorate, NASA Goddard Space Flight Center.

Wynn, R. L. 1974. "Survival of Data for On-Line Data Acquisition Systems," Office of Preparedness Rept. No. GSA/OP TR-80. (Also among Univac Scientific Exchange Tech. Papers of USE Conf., San Francisco, Calif.)

Yoshpe, Harry B., Allums, J. F., Russell, J. E., and Atkin, B. A. 1972. *Stemming Inflation: The Office of Emergency Preparedness and the 90-day Freeze.* Office of Emergency Preparedness, U.S. Govt. Printing Office, Stock No. 4102-00008.

Zinn, Karl L. 1977. "Computer Facilitation of Communication within Professional Com-
    munities," *Behavioral Research Methods and Instrumentation*, Spring. (UM)
———, 1977. "Confer at the ISTA Congress," *Transnat. Associations, (J. Union of Intern.
    Associations* **10,** pp. 412-417. (UM)
Zinn, Karl L., Parnes, Robert, and Hench, Helen. 1976. "Computer-Based Educational Com-
    munications at the University of Michigan," *Proc. 31st ACM Nat. Conf.* (UM)

Indexes

Author Index

Numbers set in *italics* indicate pages on which complete literature citations are given. Names cited in the Bibliography are not included in this Index.

Subject Index

abstracting, automation, 272-273
academia, 73
accountability, 149. 431
action tracking, 158-159
adaptive systems, 445
adaptive text, 156, 337, 357, 369
addiction, 101-104
advanced features. EIES, 368-374
advertising, 274
aged, 176-177, 180
agreement, 105, 109
    see also disagreement, consensus
amateur social scientists, 486
anarchists (extroverted), 485-486
anonymity, 62, 94-95, 128-129, 269-270, 288-289, 360
APL (*A Programming Language*), 389
appearance, 91-92
applications and impacts conference, 90-92
ARPANET (*Advanced Research Projects Agency Network*), 138
artificial expert, 202
artificial intelligence, 272-274
assembly lining, 162
AT&T, 447, 449
attitudes toward
    commuting and travel, 477-478
    computers, 162, 184, 340, 433-434
audio conferences, 34, 89, 117-121
authority, 289

bargaining, 118
barriers, 123, 249-250
bid and barter, 183, 205, 207
bilingual interface, 65
board meetings, 149-150
brainstorming, 300
brainwriting, 301
bridge, 308
budgeting process, 146-147
bulletproofing, 349
bully effect, 343
bureaucracy, 141-142

cable TV, 449
candor, 100, 127
cashless society, 169
causality, 305
CCS (*Computerized Conferencing Systems*)
    advantages, 89, 139-140, 217, 298, 338
    applications, 278
    capabilities, 8-9, 121
    conventions, 82

costs, *see* cost-benefits
description, 7-8
disadvantages, 10, 140, 151-152, 223, 250, 487-489
functions, 392
impacts
    see also cognitive impacts, economic impacts, psychological impacts, social impacts
interface, 355-360
learning patterns, 81-83, 102-104, 111, 120, 123-124
motivations, 24, 104, 123-124, 222, 249
usage patterns, 103, 123-124
censorship, 457
centralization, 11, 475
    see also decentralization
chilling effect, 114
circulation ratio, 425
citations, 219
citizen band radio (CB), 35, 190, 345, 393, 448
citizen participation, 200
civil service, 40
clustering, 302-303
clutter effect, 342
codes, 407
cognitive inputs, 28, 82-83, 90, 277, 304, 353, 407
collective intelligence, 38, 43, 273, 290
collectivities, 121
college education, 195-197
command and control, 347
command interface, 351, 368
commitment, 344, 431
commodities market, 188
communications(,)
    act of 1934, 447
    mediated, 87
    morphology, 32-39
    networks, 403-406
    non-verbal, 78-80
    revolution, 469-473
    technology, 13
Communications Studies Group (CSG), 117-121
community
    groups, 200
community, user, 360
commuting, 473-478
comprehension
    of design, 350
    phase, users, 307

521

DATE DUE